Milestones in Drug Therapy
MDT

COX-2 Inhibitors

Edited by M. Pairet and J. van Ryn

Springer Basel AG

Editors

Michel Pairet
Boehringer Ingelheim Pharma GmbH & Co. KG
Department of Pulmonary Research
Birkendorfer Str. 65
D-88397 Biberach a. d. Riss
Germany

Joanne van Ryn
Boehringer Ingelheim Pharma GmbH & Co. KG
Department Drug Discovery Support
Birkendorfer Str. 65
D-88397 Biberach a. d. Riss
Germany

Library of Congress Cataloging-in-Publication Data
COX-2 inhibitors / edited by M. Pairet and J. van Ryn.
p. ; cm. -- (Milestones in drug therapy)
Includes bibliographical references and index.
ISBN 3-7643-6901-9 (alk. paper)
1. Cyclooxygenase 2--Inhibitors--Therapeutic use. I. Pairet, M. (Michel), 1959-
II. Van Ryn, J. (Joanne), 1958- III. Series.
[DNLM: 1. Anti-Inflammatory Agents, Non-Steroidal--therapeutic use. 2.
Cyclooxygenase Inhibitors--pharmacokinetics. QV 95 C877 2004]
RM666.C977C69 2004
616'.0473--dc22 2003063560

Bibliographic information published by Die Deutsche Bibliothek
Die Deutsche Bibliothek lists this publication in the Deutsche Nationalbibliografie;
detailed bibliographic data is available in the Internet at <http://dnb.ddb.de>.

ISBN 978-3-7643-6901-9 **ISBN 978-3-0348-7879-1 (eBook)**
DOI 10.1007/978-3-0348-7879-1

Part of Springer Science+Business Media
Printed on acid-free paper produced from chlorine-free pulp. TCF ∞

Cover illustration: This view of the COX-2 channel is from the bottom looking up from the membrane binding domain. COX-2 is shown as a ribbon diagram and the amino acid residues of COX-2 that define the binding site are represented as sticks. The COX-2 inhibitor, colored red, is bound in the channel and represented as sticks. The editors wish to thank Günter Trummlitz for the picture and Gerald Birk for the stylized rendition.

ISBN 3-7643-6901-9
9 8 7 6 5 4 3 2 1 www.birkhäuser-science.com

Contents

List of contributors

Phillip R. Bennett, Imperial College Parturition Research Group, Institute of Reproductive and Developmental Biology, Hammersmith Hospital Campus, Du Cane Road, London W12 ONN, UK; e-mail: pbennett@ic.ac.uk

Jack H. Botting, The William Harvey Research Institute, Charterhouse Square, London EC1M 6BQ, UK; e-mail: r.m.botting@qmul.ac.uk

Regina M. Botting, The William Harvey Research Institute, Charterhouse Square, London EC1M 6BQ, UK; e-mail: r.m.botting@qmul.ac.uk

Christianne J. Buskens, Department of Surgery, Academic Medical Center, University of Amsterdam, 1105 AZ Amsterdam, The Netherlands; e-mail: c.j.buskens@amc.uva.nl

Frank Degner, CDept. Medical Affairs, Boehringer Ingelheim Pharma GmbH, Binger Str. 173, 55216 Ingelheim, Germany; e-mail: degner@ing.boehringer-ingelheim.com

Tiina-Liisa Erkinheimo, Department of Obstetrics and Gynecology and Helsinki University Central Hospital and Molecular and Cancer Biology Research Program, Biomedicum Helsinki, University of Helsinki, P.O. Box 63, FIN-00014 Helsinki, Finland; e-mail: erkinhei@mappi.helsinki.fi

Caj Haglund, Department of Surgery, Biomedicum Helsinki, University of Helsinki, P.O. Box 63, FIN-00014 Helsinki, Finland; e-mail: caj.haglund@hus.fi

Koichi Hashimoto, Department of Microbiology, School of Medicine, Fukushima Medical University, 1st Hikariga-oka Fukushima, Fukushima, Japan 960-1295; e-mail: don@fmu.ac.jp

Amit S. Kalgutkar, Pharmacokinetics, Dynamics and Metabolism Department, Pfizer Global Research and Development, Groton, CT 06340, USA

Lawrence J. Marnett, Department of Biochemistry, Vanderbilt Institute of Chemical Biology, Vanderbilt-Ingram Cancer Center, and Center in Molecular Toxicology, Vanderbilt University School of Medicine, Nashville, Tennessee 37232-0146, USA; e-mail: larry.marnett@vanderbilt.edu

Brendan F. McAdam, Division of Cardiovascular Medicine, Vanderbilt University Medical Center, Nashville, TN 37232-8802, USA; e-mail: brendan.mcadam@vanderbilt.edu

Aarthi R. Mohan, Imperial College Parturition Research Group, Institute of Reproductive and Developmental Biology, Hammersmith Hospital Campus, Du Cane Road, London W12 ONN, UK; e-mail: a.mohan@imperial.ac.uk

Rahul Nayak, Division of Cardiovascular Medicine, Vanderbilt University Medical Center, Nashville, TN 37232-8802, USA

G. Johan A. Offerhaus, Department of Pathology, Academic Medical Center, University of Amsterdam, 1105 AZ Amsterdam, The Netherlands; e-mail: g.j.offerhaus@amc.uva.nl

Michel Pairet, Department of Pulmonary Research, Boehringer Ingelheim Pharma GmbH & Co. KG, Birkendorfer Str 65, D - 88397 Biberach, Germany; e-mail: Michel.Pairet@bc.boehringer-ingelheim.com

R. Stokes Peebles, Jr., Center for Lung Research, T-1217 MCN, Vanderbilt University Medical Center, Nashville, TN 37232-2650, USA; e-mail: stokes.peebles@vanderbilt.edu

Leena Pohjola, Department of Pathology and Helsinki University Central Hospital and Molecular and Cancer Biology Research Program, Biomedicum Helsinki, University of Helsinki, P.O. Box 63, FIN-00014 Helsinki, Finland; e-mail: lkpohjol@mappi.helsinki.fi

Kim D. Rainsford, Biomedical Research Centre, Sheffield Hallam University, Howard Street, Sheffield S1 1WB, UK; e-mail: k_d.rainsford@shu.ac.uk

Johanna Rintahaka, Department of Pathology and Helsinki University Central Hospital and Molecular and Cancer Biology Research Program, Biomedicum Helsinki, University of Helsinki, P.O. Box 63, FIN-00014 Helsinki, Finland; e-mail: rintahak@mappi.helsinki.fi

Ari Ristimäki, Department of Pathology and Helsinki University Central Hospital and Molecular and Cancer Biology Research Program, Biomedicum Helsinki, University of Helsinki, P.O. Box 63, FIN-00014 Helsinki, Finland; e-mail: ari.ristimaki@hus.fi

Kirsi Saukkonen, Department of Pathology and Helsinki University Central Hospital and Molecular and Cancer Biology Research Program, Biomedicum Helsinki, University of Helsinki, P.O. Box 63, FIN-00014 Helsinki, Finland; e-mail: kisaukko@mappi.helsinki.fi

Anna Sivula, Department of Pathology and Helsinki University Central Hospital and Molecular and Cancer Biology Research Program, Biomedicum Helsinki, University of Helsinki, P.O. Box 63, FIN-00014 Helsinki, Finland; e-mail: anna.sivula@helsinki.fi

Dirk O. Stichtenoth, Institute of Clinical Pharmacology, Medizinische Hochschule Hannover, 30623 Hannover, Germany; e-mail: Stichtenoth.Dirk@mh-hannover.de

Guenter Trummlitz, Boehringer Ingelheim GmbH, Birkendorferstr. 65, 88397 Biberach, Germany; e-mail: trummlitz@t-online.de

J. Jan B. van Lanschot, Department of Surgery, Academic Medical Center, University of Amsterdam, 1105 AZ Amsterdam, The Netherlands; e-mail: j.j.vanlanschot@amc.uva.nl

Bastiaan P. van Rees, Department of Pathology, Academic Medical Center, University of Amsterdam, 1105 AZ Amsterdam, The Netherlands; e-mail: b.van.rees@slingeland.nl

Joanne van Ryn, Department of Drug Discovery Support, Boehringer Ingelheim Pharma GmbH & Co. KG, Birkendorferstr. 65, 88397 Biberach, Germany; e-mail: Joanne.vanRyn@bc.boehringer-ingelheim.com

Timothy D. Warner, TheWilliam Harvey Research Institute, Bart´s and the London Queen Mary's School of Medicine and Dentistry, London EC1M 6BQ, UK; e-mail: t.d.warner@qmul.ac.uk

Preface

The pioneer work by Sir John Vane and collaborators demonstrating that aspirin-like drugs work by blocking cyclooxygenase (COX) had a huge impact on our understanding of the mechanism of action of the so-called non steroidal anti-inflammatory drugs (NSAIDs) and on their rational use. The subsequent discovery that two forms of COX exist, COX-1 and COX-2, had further massive implications. It helped us understand the mechanism of the wanted *versus* unwanted effects of NSAIDs. It was also the basis for a generation of new drugs, the COX-2 inhibitors. In addition, the discovery reactivated scientific work on the physiological and pathological roles of prostanoids and on the potential use of COX-2 inhibitors in indications other than pain and arthritis, such as cancer, pre-term delivery and Alzheimer's disease.

Much has been written in this field over the past 10 years; the danger of writing again on this topic is to come out with "just another book on COX-2". We are convinced, however, that more than 10 years after the discovery of COX-2 and with the first selective COX-2 inhibitors now on the market for several years, we may just now have the needed objectivity to answer a few key questions, based on both scientific and clinical data and far from the loud marketing communications which prevailed during the launch phases of the first selective COX-2 inhibitors.

We know for example that the theory of a bad COX-2 and a good COX-1 needs to be qualified, that COX-2 is not solely an inducible enzyme during inflammation, but can also be constitutively expressed in various tissues and is essential for physiological functions such as reproduction. In addition, a large clinical experience on the gastrointestinal tolerability of COX-2 inhibitors and new experimental data on the respective roles of COX-1 and COX-2 in mucosal protection and ulcer healing, allow us to better assess the level of improvement achieved by COX-2 inhibitors.

We also know that COX-2 inhibitors lack efficacy in patients with Alzehimer's disease but have a clear effect in Familial Adenomatous Polyposis and have therapeutic potential in various forms of cancer. All this could not have been predicted a few years ago based only on scientific data on the expression and function of COX-2 in tissues of interest. We also now realise that COX-2 inhibitors may have side-effects standard NSAIDs do not have, such as an increase in thrombotic risk. This is still a matter of intensive debate, but if confirmed may significantly change our risk/benefit analysis between COX-2 inhibitors and NSAIDs. Finally, the recent discovery of a COX-3 isoenzyme will further widen and possibly complicate our understanding of the physiopathology of COX.

We realise that this book will not answer all the questions the reader may have on the true therapeutic benefit of COX-2 inhibitors, on their potential in cancer prevention and/or therapy, or on the respective roles of the various COXs in pain. We simply hope that this book will help the reader to better understand the facts, limitations and hypotheses on COX and its inhibitors.

Michel Pairet and Joanne van Ryn

Biberach a.d. Riss,
October 2003

COX-2 Inhibitors
Edited by M. Pairet and J. van Ryn

The discovery of COX-2

Regina M. Botting and Jack H. Botting

The William Harvey Research Institute, Charterhouse Square, London EC1M 6BQ, UK

History of cyclooxygenase

For over 3000 years various herbal medicines, such as extracts of willow bark, or dried leaves of myrtle or meadowsweet have been used with apparent anecdotal success for the treatment of the pain and swelling of inflammatory disease. By the middle of the nineteenth century the active constituent of these various herbal remedies was found to be salicylate, and the better tolerated acetylated derivative, acetylsalicylate was introduced to medicine as aspirin over a hundred years ago. The manifest efficacy of aspirin resulted in its wide use as an antipyretic and as an analgesic for the pain associated with inflammatory conditions [1]. However, it soon became apparent that aspirin could cause gastrotoxicity, a fact demonstrated unequivocally in 1938 by the endoscopic studies of Douthwaite who observed erosions and ulcers in the gastric mucosa of patients on long-term aspirin therapy [2].

The search for an aspirin-like drug without gastrotoxicity resulted in the synthesis of many compounds of diverse chemical nature with anti-inflammatory actions. About 30 have entered the market over the last 30–40 years, but none of these so-designated nonsteroidal anti-inflammatory drugs (NSAIDs) was totally free of a gastrotoxic effect. Gastrotoxicity due to NSAID therapy is a significant problem. It has been estimated that in the USA alone between 10,000 and 20,000 patients die each year from NSAID-associated toxicity such as perforations and bleeding from the stomach [3].

The interesting phenomenon that chemicals of diverse structure should share the common therapeutic effects of antipyresis, analgesia and anti-inflammatory actions (these themselves of a diverse nature), and also share a common toxic action, suggested that a single biochemical effect could explain all the pharmacological actions of NSAIDs. Over the years, many theories for this common mechanism were proposed and subsequently rejected. Eventually, the enigma was resolved in 1971 by the elegantly simple experiments of Vane, using guinea pig lung homogenates who showed that aspirin and other NSAIDs inhibited the synthesis of the lipid mediators, prostaglandins (PGs) [4]. PGs were known to be pyrogenic, and were subsequently shown to be hyperalgesic and present at inflammatory foci, thus explaining the therapeutic effects of NSAIDs. That the common side-effects of NSAIDs were also due to inhibition of PG synthesis was apparent when PGs were shown to be cytoprotective on the gastric mucosa, and could maintain renal blood flow when renal circulation was compromised [1].

PGs are formed from arachidonic acid mobilised from membrane phospholipids by a phospholipase enzyme. Arachidonic acid is acted upon by the microsomal enzyme cyclo-oxygenase (COX) to form the cyclic endoperoxides PGG_2 and PGH_2. These unstable endoperoxides are then isomerised, enzymatically or non-enzymatically, into various prostanoids such as thromboxane A_2 (TxA_2), prostacyclin (PGI_2) and PGD_2, E_2 and $F_{2\alpha}$ (Fig. 1).

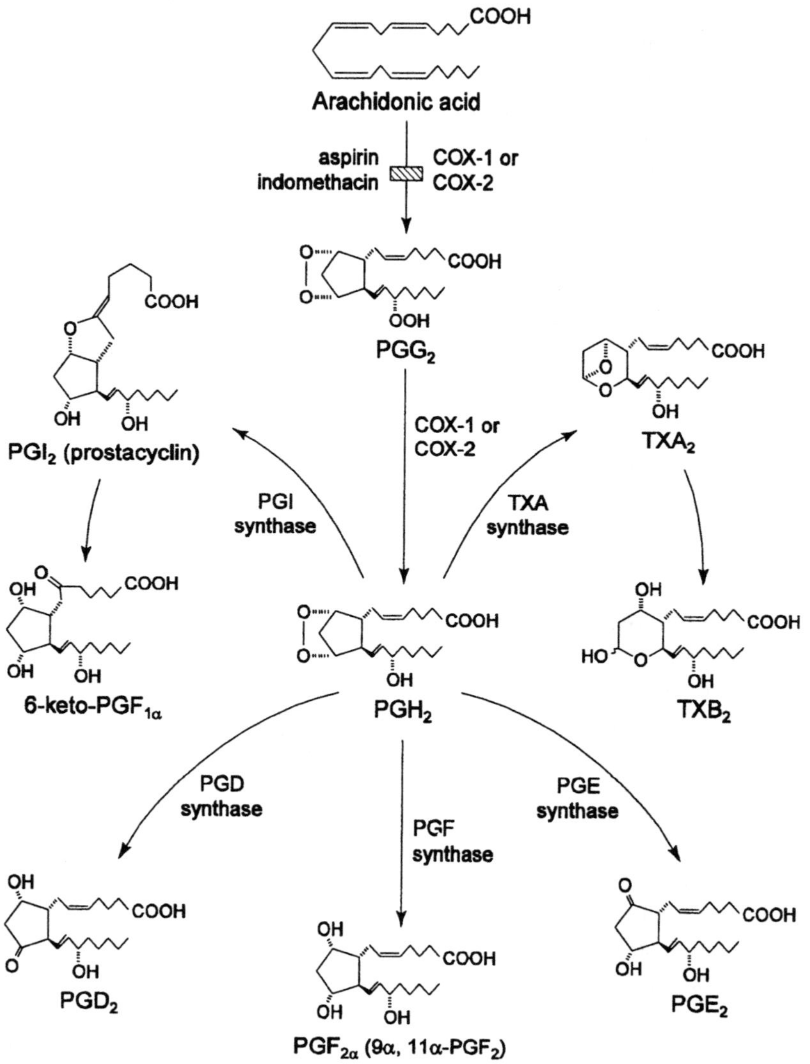

Figure 1. The arachidonic acid cascade.

Implicit in the establishment of the mechanism of action of NSAIDs was the assumption that since all the pharmacological actions depended on the inhibition of COX, it would be impossible to divorce the therapeutic effects from the toxic actions of these agents. However, there were some minor inconsistencies in this theory. Epidemiological studies showed that the severity of gastric toxic effects of different NSAIDs varied when the drugs were used at comparable anti-inflammatory doses. Ibuprofen, for example, causes less damage to the stomach than ketoprofen [5]. Similarly, the NSAID nimesulide was effective in models of inflammatory disease such as the carrageenan injected rat paw, yet was poorly active in inhibiting PG synthesis in conventional systems [6].

The existence of COX isozymes was a likely explanation for these inconsistencies, and over 30 years ago Flower and Vane suggested that paracetamol, which lacks anti-inflammatory actions, exerted its antipyretic effect by inhibition of a distinct isoform of COX in brain tissue [7]. More recently a number of workers showed that PG synthesis could be upregulated in inflammatory conditions by induction of synthesis of more COX enzyme. Needleman and colleagues, for example, showed that in a model of the inflammatory condition of hydronephrosis, produced in perfused rabbit kidney by prior ureteric obstruction, there was a marked increase in PGE_2 release following injection of bradykinin. This increased release of PG could be prevented by prior treatment with the protein synthesis inhibitors cycloheximide or actinomycin D. Of significance, aspirin inhibited the release of PG during the early stages of perfusion but became less effective as the experiment progressed. Needleman concluded that the increase in PG production in this model of inflammation was secondary to increased synthesis of COX enzyme [8].

Similar results were obtained from experiments using isolated cells stimulated with various inflammatory cytokines or growth factors. Human umbilical vein endothelial cells treated with interleukin-1 (IL-1) or tumour necrosis factor resulted in increased production of PGI_2 together with an increase in COX protein and mRNA [9], effects reduced by treatment with actinomycin D. Induction of COX protein accompanied by an increase in PG production was also demonstrated in murine smooth muscle cells and osteoblasts treated with epidermal growth factor [10, 11], 3T3 cells treated with PDGF [12], and in human dermal fibroblasts treated with IL-1 [13].

Needleman and his group [14–16] extended their earlier observations by demonstrating increased PG synthesis in human monocytes *in vitro* and in mouse macrophages *in vivo* due to activation with lipopolysaccharide (LPS). The increase in PG release was inhibited by prior treatment with dexamethasone, yet the basal level of COX activity was not affected. Needleman concluded that there may be various "pools" of COX within cells, one constitutive, another inducible, being upregulated by inflammatory stimuli [17]. The existence of such an inducible COX enzyme was firmly established by the discovery by Simmons and his colleagues at Brigham Young University of a novel PG synthase, encoded by an entirely different gene, induced in chick embryo fibroblasts by v-src, serum or phorbol esters [18, 19]. Subsequently, Herschman and his colleagues [20] found a similar gene in the mouse, as did Simmons et al. who, by cross hybridisation, cloned and characterised the murine homologue of the v-src-inducible COX [21] and showed the mRNA to be induced in an immediate early fashion by various mutagens. The deduced protein structure of the inducible enzyme was found to be 60% homologous to constitutive COX, and the two enzymes were designated COX-1 (constitutive) and COX-2 (inducible).

Although COX-1 is constitutive in most cells, very little COX-2 is present in resting cells, however, it is powerfully induced in fibroblasts, endothelial cells and vascular smooth muscle by mitogens or some cytokines, and in monocytic cells by LPS. These observations have led to the general hypothesis that the constitutive COX-1 is involved in normal physiological processes. This includes synthesising thromboxane A_2 involved in platelet aggregation, PGE_2 and PGI_2 to protect the gastric mucosa and PGI_2 in the vascular endothelium to maintain dilatation of blood vessels and to inhibit formation of inappropriate platelet thrombi. Although recent work suggests that formation of PGI_2 from endothelial cells may be due to COX-2 constantly induced in the cells by shear stress [22].

The inducible COX-2, on the other hand, is believed to be primarily responsible for the production of PGs in inflamed tissues, which cause the swelling and hyperalgesia associated with inflammatory disease (see Fig. 2). Certainly, increased COX-2 expression, which parallels increased PGE_2 synthesis, is demonstrable in models of inflammatory disease such as the murine air pouch, whilst COX-1 expression is unchanged [23]. Increased expression of COX-2 has also been demonstrated in synovial tissue taken from patients with rheumatoid arthritis, compared to tissues from patients with osteoarthritis or no arthritic pathology [24].

The acceptance that the COX-2 isoform was responsible for the synthesis of PGs at pathological foci and COX-1 for the synthesis of the beneficial, "housekeeping" PGs, implied that selective inhibitors of COX-2 would exert the therapeutic actions of NSAIDs without the previously regarded inevitable side-effect of gastrotoxicity. This led many workers to undertake studies to establish the relative activity of established NSAIDs on COX-1 and COX-2 by comparison of the concentration of NSAID necessary to inhibit the activity of COX by 50% (IC_{50}). Although simple in concept these measurements of selectivity were problematic in practice. Differences in source of enzyme, time of incubation, method of induction of COX-2 and protein concentration resulted in variation in ratios and even differences in IC_{50} values. However, the human whole blood assay described by Patrignani et al. [25] and subsequently modified by Warner [5], provided a reliable method with similar incubation times, appropriate human target cells and performed in the presence of plasma proteins. This method was reproducible and the selectivity of established NSAIDs such as nimesulide, meloxicam and etodolac towards inhibition of COX-2 rather than COX-1 was shown and this correlated well with data from large scale clinical trials and post-marketing surveillance studies, showing the substances to be accompanied by relatively low gastrotoxicity.

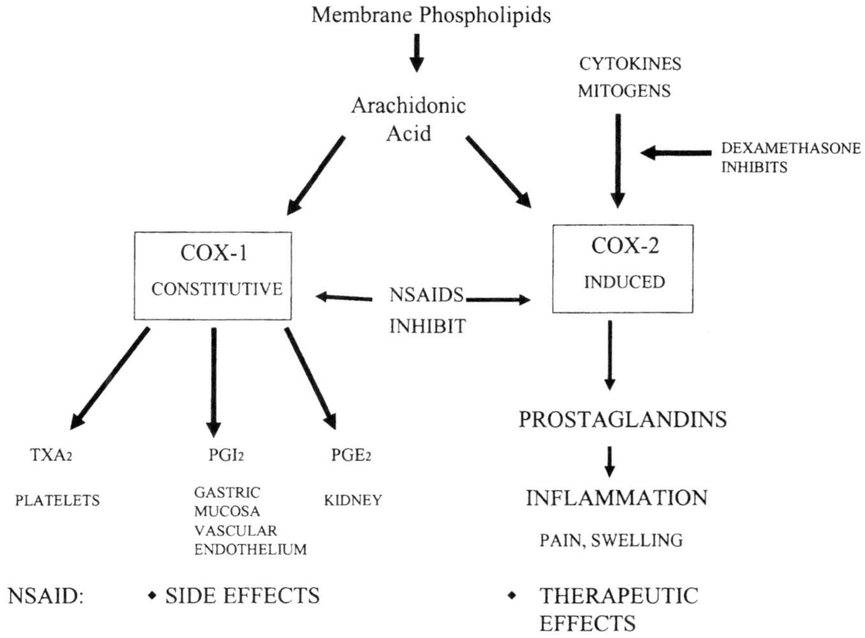

Figure 2. The side-effects of NSAIDs are caused by inhibition of the constitutive enzyme COX-1, which synthesises prostaglandins that serve essential physiological functions such as causing appropriate platelet aggregation, protection of the gastric mucosa, inhibition of thrombogenesis and maintenance of renal function. The therapeutic effects of NSAIDs are due to inhibition of COX-2, an enzyme induced by various factors released by bacteria, the vascular endothelium or other cells involved in the inflammatory response.

Structural basis for COX-2 selectivity

The three-dimensional structures of both COX-1 [26] and COX-2 [27] have been determined and are remarkably similar. Each enzyme is composed of three distinct folding units: an epidermal growth factor-like domain, a membrane binding section and a C-terminal enzymatic domain containing both the COX and peroxidase active sites. Through the membrane-binding domain the enzymes integrate into a single leaflet of the membrane lipid bilayer, thus positioning the COX site to allow access of the arachidonic acid substrate that is mobilised from membrane phospholipids.

Whereas NSAIDs generally act competitively to prevent access of arachidonic acid to the active site, aspirin acts in a unique manner to acetylate a serine residue at position 530 in COX-1, or the serine at an analogous position (516) in COX-2. While acetylation with aspirin will completely prevent the metabolism of arachidonic acid by COX-1, with COX-2 the substrate can reach the peroxidase site and be metabolised to 15-hydroperoxyeicosatetraneoic acid (15-HPETE) [28].

Although the active sites of COX-1 and COX-2 are similar, there are differences which have been utilised by medicinal chemists to synthesise molecules which have a selective action on COX-2. Of crucial significance is position 523, which in COX-2 is valine and in COX-1 isoleucine. This difference of a single methyl group is sufficient to allow access of a potential inhibitor to a side pocket in COX-2. In addition, proximity to an arginine residue at position 513 (histidine in COX-1), which provides hydrogen bonding for an inhibitor of a structure that enables it to extend into the side pocket [29].

A second significant difference between COX-2 and COX-1 is at position 503, which is the aromatic amino acid phenylalanine in COX-1 but the relatively small, nonaromatic leucine in COX-2; this allows leucine at position 384 to reorient its methyl side chain away from the enzymatic site and thus extend the space available for a larger inhibitor molecule at the active site of the COX-2 enzyme [30]. The differences in the structure of the COX channels of the COX isozymes are illustrated diagrammatically in Figure 3, in which the same numbering of the amino acids in the two enzymes is adopted for simplicity.

Drugs specifically designed to inhibit COX-2 have entered the market and two such have undergone comprehensive clinical investigation. Celecoxib is the prototype of a number of highly selective COX-2 inhibitors produced by Pharmacia (Pfizer). Celecoxib was shown to be an effective analgesic as assessed in patients after tooth extraction, and endoscopic studies after 3

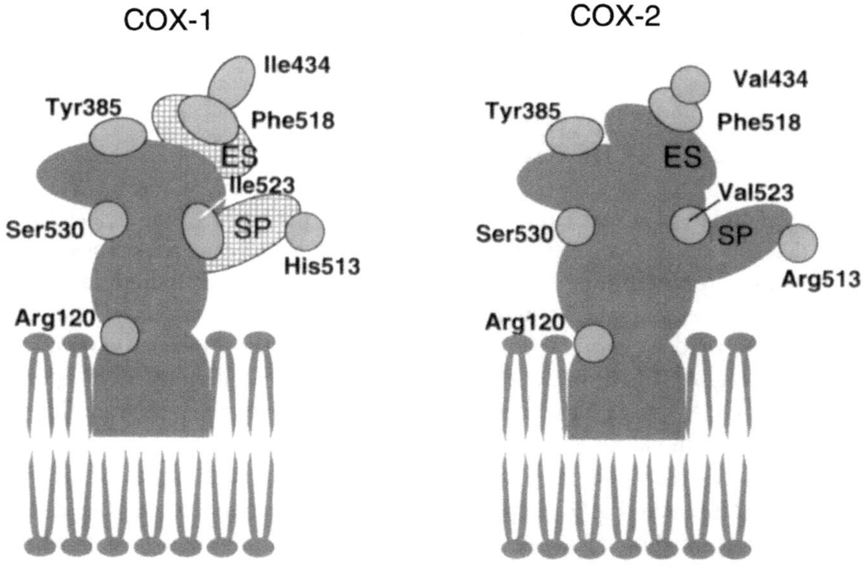

Figure 3. Amino acid substitutions between COX-1 and COX-2. The hatched areas in COX-1 are those that are more accessible in COX-2 due to the amino acid substitutions. SP: side pocket, ES: extra space.

months chronic usage showed no more gastric damage than with placebo. In a large scale clinical trial (the Celecoxib Long-term Arthritis Safety Study or CLASS, in 8000 patients) it was shown that, at 6 months, celecoxib produced less gastrotoxicity than the comparator drugs ibuprofen and diclofenac (though this failed to reach statistical significance at 12 months) [31]. Celecoxib is approved both for osteo- and rheumatoid arthritis in many countries.

Rofecoxib, produced by Merck, Sharp and Dohme, has been approved by many regulatory authorities for arthritis and acute pain. Rofecoxib in doses of 25–50 mg was an effective analgesic for dental pain, dysmenorrhoea and osteoarthritis. In a 6-month endoscopic study, with a dose of rofecoxib 10-fold greater than that required for an effective anti-inflammatory effect, gastric damage was no greater than that seen with placebo. In a large-scale, 9-month trial involving 8000 patients (the Vioxx Gastrointestinal Outcome Research, VIGOR) rofecoxib was compared with naproxen [32]. Rofecoxib was shown to produce substantially less gastrotoxicity than naproxen. Of possible significance, in the VIGOR study the incidence of myocardial infarction in the rofecoxib group was 0.4%, compared with 0.1% in the naproxen treated patients. This observation requires further study and is addressed in Chapter 7.

Newer even more selective COX-2 inhibitors such as valdecoxib [33], etoricoxib [34] and lumiracoxib [35] are currently either in Phase III development or have been introduced on the market. While there is less clinical experience with these agents to date, they all seem to be effective anti-inflammatory agents with reduced gastrointestinal toxicity.

Additional uses for selective COX-2 inhibitors

The availability of NSAIDs with clearly a vastly reduced propensity for causing gastrointestinal toxicity raised the possibility of the chronic use of these drugs for additional indications. Thus far, selective COX-2 inhibitors have been postulated to have potential in some cancers, premature labour and Alzheimer's disease.

COX-2 inhibitors and cancers

There is persuasive epidemiological evidence for an association of chronic use of NSAIDs with reduced incidence of gastrointestinal carcinomas [36–42]. Regular use of aspirin and other NSAIDs can halve the risk of development of colon cancer. Clinical and experimental studies have provided further evidence to justify this beneficial effect of NSAIDs; colonic tumour tissue has been shown to contain high levels of COX-2 protein but contiguous, normal tissue contains only COX-1 [43]. Patients with the inherited disorder of familial adenomatous polyposis (FAP), a condition characterised by a plethora of colonic polyps with eventual progression to tumour, manifest a marked reduction in polyp number on treatment with the NSAID sulindac [44, 45]. Similarly, COX inhibitors reduce tumours in azoxymethane-induced colorectal carcinogenesis in rat [46] and selective COX-2 inhibitors reduce tumour or polyp number in animal models of FAP such as minimal intestinal neoplasias (min) mice or APC (adenomatous polyposis coli gene mutant) knockout mice.

COX-2 is expressed in tumour tissue in many cancers other than colorectal carcinomas. Increased levels of COX-2 have been shown in human lung squamous cell and adenocarcinomas, squamous skin carcinomas and in tumours of the breast, liver, pancreas, prostate and cervical lymph nodes (reviewed in [47] and further discussed in Chapter 11).

Premature labour

The content of COX-2 mRNA in the uterus and the production of PGs increases just prior to and during labour, indicating an involvement of PGs (particularly $PGF_{2\alpha}$) in parturition. Premature labour, perhaps induced by inappropriate upregulation of COX-2 due to intrauterine infection, can be suppressed by COX inhibitors such as indomethacin, but inhibition of COX-1 will result in early closure of the ductus arteriosus which is maintained patent until birth by PGs synthesised by constitutive COX-1 [48]. Selective COX-2 inhibitors were thus thought to have a use in premature labour, producing a tocolytic effect without the problem of early closure of the ductus. The COX-2-selective inhibitor nimesulide has been used with some success in premature labour [49] but has been shown subsequently to produce oligohydramnios in 50% of exposed fetuses [50]. The side-effects of nimesulide are reversible, however, it is accepted that the use of this drug for premature labour should be reserved for patients at high risk of early preterm delivery and where adequate patient surveillance can be maintained. Since some of the renal side-effects of nimesulide may be due to actions other than COX inhibition, such as calcium channel blockade, thus other selective COX-2 inhibitors are at present under trial for preterm labour ([51], see Chapter 10).

Alzheimer's disease

Retrospective epidemiological studies suggest that long-term use of standard NSAIDs reduces the incidence of Alzheimer's disease (AD) [52], a benefit not seen with paracetamol or low-dose aspirin. Due to lack of suitable animal models there is little direct experimental confirmation, but it has been assumed that there is an inflammatory component in the pathogenesis of AD and *post mortem* studies have shown an increase of COX-2 expression in the cortex of AD patients [53]. Prolonged treatment with ibuprofen reduced the number of amyloid plaques in the brain of APP transgenic mice secondary to a lowered brain content of soluble Aβ peptide, which on aggregation forms the amyloid plaques [54]. Large placebo controlled trials of the highly selective COX-2 inhibitors celecoxib and rofecoxib have apparently shown no beneficial effect on the progress of AD [55, 56], which suggests that COX-2 may not be an appropriate target. It has been suggested that some NSAIDs may lower plaque formation through a mechanism independent of COX inhibition, possibly by inhibition of γ-secretase, which is the enzyme that cleaves the amyloid precursor protein into the plaque-forming Aβ42 peptide [57]. Such a prophylactic effect of NSAIDs would be in accord with the Cache County epidemiological study, which showed that general NSAID therapy does reduce incidence of AD if the drugs are used well before the onset of symptoms of dementia, whereas more recent exposure offered little protection [58].

Other cyclooxygenases

Although the discovery of COX-2 resolved the puzzle of why some NSAIDs manifested marked gastrotoxicity yet others, more selective for COX-2 were relatively less toxic, some enigmas remained. As mentioned above, acetaminophen, like conventional NSAIDs, has marked antipyretic and analgesic activity yet lacks the typical anti-inflammatory and antiplatelet activities of this group of drugs. Twenty years ago it was suggested that acetaminophen might inhibit a cyclooxygenase in the brain, and that various, perhaps tissue-specific, cyclooxygenases may exist that are selectively inhibited by different NSAIDs [7].

Some light has recently been shed upon this field by Simmons and his colleagues who have identified a new COX-1 variant in dog and human brain which is sensitive to acetaminophen and which has been designated COX-3 [59]. COX-3 has the same structure as COX-1 except that the mRNA has retained intron-1, which is translated into a 30 amino acid extension to the enzyme protein. COX-3 was ectopically expressed in insect cells and found to be more sensitive to inhibition by acetaminophen than either COX-1 or COX-2. At least two other partial COX-1 mRNAs were characterised but only one, PCOX-1a, had retained intron-1 in transcription (see Fig. 4). The partial COX-1 proteins did not form prostanoids and their functions remain unknown.

Human COX-3 was found in highest concentrations in the cerebral cortex followed by the heart and skeletal muscle. The possible function of COX-3 in organs other than the brain is as yet unknown. Simmons had previously found another COX isoform, sensitive to inhibition by acetaminophen, formed by a murine macrophage cell line (J774.2) treated with high concentrations of diclofenac [60]. However, unlike COX-3, this enzyme was insensitive to aspirin, and was considered to be a variant of COX-2, since a protein immunoreactive to COX-2 was co-induced together with the cyclooxygenase activity. Thus there may be yet other acetaminophen-sensitive cyclooxygenases derived from COX-2.

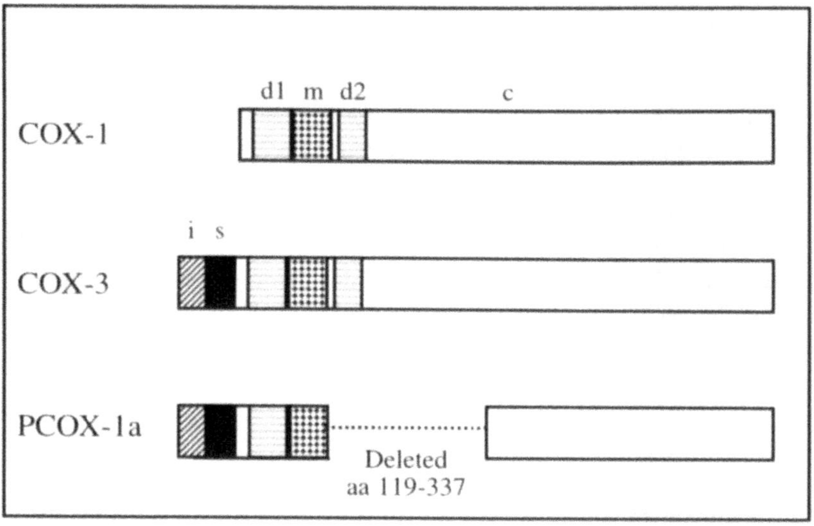

Figure 4. Structure of COX-3 and PCOX-1a. A schematic representation of the domains of COX-3 and PCOX-1a in comparison to COX-1. s: signal peptide; d1: dimerization domain/EGF-like domain; d2: dimerization domain 2; m: membrane binding domain; c: catalytic domain; i: 90-bp sequence encoded by intron 1. PCOX-1b is identical to PCOX-1a except that it lacks intron 1. Amino acid numbering is according to residues in sheep seminal vesicle COX-1. (Reproduced from [59] with permission.)

Conclusions

Clearly, the establishment of the existence of the isoform COX-2 has provided a means of reaping the long-recognised benefits of aspirin-like drugs without the sometimes lethal side-effects. Further research will no doubt provide refined treatments for inflammatory conditions and pain, possibly by establishing the optimal selectivity for COX-2, or by targeting novel isoforms of cyclooxygenase.

References

1 Vane JR, Botting RM (1996) The history of anti-inflammatory drugs and their mode of action. In: N Bazan, J Botting, J Vane (eds): *New Targets in Inflammation Inhibitors of COX-2 or Adhesion Molecules*. Kluwer Academic Publishers and William Harvey Press, London, 1–12
2 Douthwaite A, Lintott CA (1938) Gastroscopic observations of the effect of aspirin and certain other substances on the stomach. *Lancet* 2: 1222
3 Fries J (1996) Towards an understanding of NSAID-related adverse events: the contribution of longitudinal data. *Scand J Rheumatol* 25: 3–8
4 Vane JR (1971) Inhibition of prostaglandin synthesis as a mechanism of action for the aspirin-like drugs. *Nature* 231: 232–235
5 Warner TD, Guiliano F, Vojnovic I, Bukasa A, Mitchell JA, Vane JR (1999) Nonsteroid drug selectivities for cyclo-oxygenase-1 rather than cyclo-oxygenase-2 are associated with human gastrointestinal toxicity: a full *in vitro* analysis. *Proc Natl Acad Sci USA* 96: 7563–7568

6 Carr DP, Henn R, Green JR, Böttcher I (1986) Comparison of the systemic inhibition of thromboxane synthesis, anti-inflammatory activity and gastro-intestinal toxicity of non-steroidal anti-inflammatory drugs in the rat. *Agents and Actions* 19: 374–375
7 Flower RJ, Vane JR (1972) Inhibition of prostaglandin synthetase in brain explains the antipyretic activity of paracetamol (4-acetamidophenol). *Nature* 240: 410–411
8 Morrison AR, Moritz H, Needleman P (1978) Mechanism of enhanced renal biosynthesis of prostaglandins in ureter obstruction. Role of *de novo* protein biosynthesis. *J Biol Chem* 253: 8210–8212
9 Kawakami M, Ishibashi S, Ogawa H, Murase T, Takaku F, Shibata S (1986) Cachectin/TNF as well as interleukin-1 induces prostacyclin synthesis in cultured vascular cells. *Biochem Biophys Res Commun* 141: 482–487
10 Bailey JM, Muza B, Hla T, Salata K (1985) Restoration of prostacyclin synthase in vascular smooth muscle cells after aspirin treatment: regulation by epidermal growth factor. *J Lipid Res* 26: 54–61
11 Yokota K, Kusaka M, Ohshima T, Yamamoto S, Kurihara N, Yoshino T, Kumegawa M (1986) Stimulation of prostaglandin E_2 synthesis in cloned osteoblastic cells of mouse (MC3T3-E1) by epidermal growth factor. *J Biol Chem* 261: 15410–15415
12 Habenicht AJR, Goerig M, Grulich J, Rothe D, Gronwald R, Loth U, Schettler G, Kommerell B, Ross R (1985) Human platelet-derived growth factor stimulates prostaglandin synthesis by activation and by rapid *de novo* synthesis of cyclooxygenase. *J Clin Invest* 75: 1381–1387
13 Raz A, Wyche A, Siegel N, Needleman P (1988) Regulation of fibroblast cyclooxygenase synthesis by interleukin-1. *J Biol Chem* 262: 3022–3028
14 Honda A, Raz A, Needleman P (1990) Induction of cyclo-oxygenase synthesis in human promyelocytic leukemia (HL-60) cells during monocytic or granulocytic differentiation. *Biochem J* 272: 259–262
15 Fu J-Y, Masferrer JL, Seibert K, Needleman P (1990) The induction and suppression of prostaglandin H_2 synthase (cyclooxygenase) in human monocytes. *J Biol Chem* 265: 16737–16740
16 Masferrer JL, Zweifel BS, Seibert K, Needleman P (1990) Selective regulation of cellular cyclooxygenase by dexamethasone and endotoxin in mice. *J Clin Invest* 86: 1375–1379
17 Seibert K, Masferrer JL, Jiyi F, Hond A, Raz A, Needleman P (1990) The biochemical and pharmacological manipulation of cellular cyclooxygenase (COX) activity. *Adv Prostaglandin Thromboxane Leukotriene Res* 21: 45–51
18 Simmons DL, Levy DB, Yannoni Y, Erikson RL (1989) Identification of a phorbol ester-repressible v-src-inducible gene. *Proc Natl Acad Sci USA* 86: 1178–1182
19 Xie W, Chipman JG, Robertson DL, Erikson RL, Simmons DL (1991) Expression of a mitogen-responsive gene encoding prostaglandin synthase is regulated by mRNA splicing. *Proc Natl Acad Sci USA* 88: 2692–2696
20 Kujubu DA, Fletcher BS, Varnum BC, Lim RW, Herschman HR (1991) TIS10, a phorbol ester tumor promoter-inducible mRNA from Swiss 3 T3 cells, encodes a novel prostaglandin synthase/cyclooxygenase homologue. *J Biol Chem* 266: 12866–12872
21 Simmons DL, Xie W, Chipman J, Evett G (1991) Multiple cyclooxygenases: cloning of a mitogen-inducible form. In: M Bailey (ed.): *Prostaglandins, Leukotrienes, Lipoxins and PAF*. Plenum Press, London, 67–68
22 McAdam BF, Catella-Lawson F, Mardini IA, Kapoor S, Fitzgerald GA (1999) Systemic biosynthesis of prostacyclin by cyclooxygenase (COX-2): The human pharmacology of a selective inhibitor of COX-2. *Proc Natl Acad Sci USA* 96: 272–277
23 Willoughby DA, Tomlinson A, Gilroy D, Willis D (1996) Inducible enzymes with special reference to the inflammatory response. In: JR Vane, JH Botting, RM Botting (eds): *Improved Non-steroid Anti-inflammatory Drugs. COX-2 Enzyme Inhibitors*. Kluwer and William Harvey Press, London, 67–83
24 Sano H, Hla T, Maier JAM, Crofford LJ, Case JP, Maciag T, Wilder RL (1992) *In vivo* cylooxygenase expression in synovial tissues of patients with rheumatoid arthritis and osteoarthritis and rats with adjuvant and streptococcal cell wall arthritis. *J Clin Invest* 89: 97–108
25 Patrignani P, Panara NR, Greco A, Fusco O, Natoli C, Iacobelli S, Cipollone F, Ganci A, Creminon C, Maclouf J et al (1994) Biochemical and pharmacological characterisation of the cyclooxygenase activity of human blood prostaglandin endoperoxide synthases. *J Pharmacol Exp Ther* 272: 1705–1710

26 Picot D, Loll PJ, Garavito RM (1994) The x-ray crystal structure of the membrane protein prostaglandin H_2 synthase-1. *Nature* 367: 243–249
27 Luong C, Miller A, Barnett J, Chow J, Ramesha C, Browner MF (1996) Flexibility of the NSAID binding site in the structure of human cyclooxygenase-2. *Nat Struct Biol* 3: 927–933
28 Mancini JA, O'Neill GP, Bayly C, Vickers PJ (1994) Mutation of serine 516 in human prostaglandin G/H synthase-2 to methionine or aspirin or aspirin acetylation of this residue stimulates 15-R-HETE synthesis. *FEBS Lett* 342: 33–37
29 Wong E, Bayly C, Waterman HL, Reindeau D, Mancini JA (1997) Conversion of prostaglandin G/H synthase-1 into an enzyme sensitive to PGHS-2-selective inhibitors by a double His^{513} to Arg and Ile^{523} to Val mutation. *J Biol Chem* 272: 9280–9286
30 Browner MF (1998) The structure of human COX-2 and selective inhibitors. In: J Vane, J Botting (eds): *Selective COX-2 Inhibitors. Pharmacology, Clinical Effects and Therapeutic Potential.* Kluwer and William Harvey Press, London, 19–26
31 Silverstein FE, Faich G, Goldstein JL, Simon LS, Pincus T, Whelton A, Makuch R, Eisen G, Agrawal NM, Stenson WF et al (2000) Gastrointestinal toxicity with celecoxib vs nonsteroidal anti-inflammatory drugs for osteoarthritis and rheumatoid arthritis: the CLASS study: a randomised controlled trial. Celecoxib Long-term Arthritis Safety Study. *J Am Med Assoc* 284: 1247–1255
32 Bombardier C, Laine L, Reicin A, Shapiro D, Burgos-Vargas R, Davis B, Day R, Ferraz MB, Hawkey CJ, Hochberg MC et al, VIGOR Study Group (2000) Comparison upper gastrointestinal toxicity of rofecoxib and naproxen in patients with rheumatoid arthritis. *N Engl J Med* 343: 1520–1528
33 Ormrod D, Wellington K, Wagstaff AJ (2002) Valdecoxib. *Drugs* 62: 2059–2071
34 Gottesdiener K, Schnitzer T, Fisher C, Bockow B, Markenson J, Ko A, DeTora L, Curtis S, Geissler L, Gertz BJ (2002) Results of a randomized, dose-ranging trial of etoricoxib in patients with osteoarthritis. *Rheumatology (Oxford)* 41: 1052–1061
35 Sorbera LA, Castaner J, Bayes M, Silvestre JS (2002) Lumiracoxib. *Drugs of the Future* 27: 740–747
36 Giovannucci E, Egan KM, Hunter DJ (1995) Aspirin and the risk of colorectal cancer in women. *N Engl J Med* 333: 609–614
37 Greenberg ER, Baron JA, Freeman DH, Mandel JS, Haile R (1993) Reduced risk of large bowel adenomas among aspirin users. The Polyp Prevention Study Group. *J Natl Cancer Inst* 85: 912–916
38 Thun MJ, Namboodiri MM, Heath CWJ (1991) Aspirin use and reduced risk of colon cancer. *N Engl J Med* 325: 1593–1596
39 Thun MJ, Namboodiri MM, Calle EE, Flandes WD, Heath CW (1993) Aspirin use and risk of fatal cancer. *Cancer Res* 53: 1322–1327
40 Peleg II, Maibach HT, Brown SH, Wilcox CM (1994) Aspirin and nonsteroidal antiinflammatory drug use and the risk of subsequent colorectal cancer. *Arch Int Med* 154: 394–399
41 Giavannucci E, Rimm EB, Stampfer MJ, Colditz GA, Ascherio A, Willett WC (1994) Aspirin use and the risk of colorectal cancer and adenoma in male health professionals *Ann Intern Med* 121: 241–246
42 Eberhart CE, Coffey RJ, Radhika A (1994) Up-regulation of cyclooxygenase 2 gene expression in human colorectal adenomas and adenocarcinomas. *Gastroenterology* 107: 1183–1188
43 Giardello FM, Hamilton SR, Krush AJ (1993) Treatment of colonic and rectal carcinomas with sulindac in familial adenomatous polyposis. *N Engl J Med* 328: 1313–1316
44 Waddell WR, Loughry RW (1983) Sulindac for polyposis of the colon. *J Surg Oncol* 24: 83–87
45 Waddell WR, Gasner GF, Cerise EJ, Loughry RW (1989) Sulindac for polyposis of the colon. *Am J Surg* 157: 175–178
46 Reddy BS, Rao CV, Rivenson A, Kellof G (1993) Inhibitor effect of aspirin on azoxymethane-induced colon carcinogenesis in F344 rats. *Carcinogenesis* 14: 1493–1497
47 Ristimaki A, Narko K, Nieminen, Saukkonen K (2001) Role of cyclooxygenase-2 in carcinogenesis other than colorectal cancer. In: JR Vane, RM Botting (eds): *Therapeutic Roles of Selective COX-2 Inhibitors*. William Harvey Press, London, 418–441
48 Bennett PR, Elder MG (1992) The mechanisms of preterm labour: common genital tract pathogens do not metabolise arachidonic acid to prostaglandins or other eicosanoids. *Am J Obstet Gynecol* 166: 1541–1545
49 Sawdy R, Slater D, Fisk N, Edmonds DK, Bennett P (1997) Use of a cyclo-oxygenase type-2-

selective non-steroidal anti-inflammatory agent to prevent preterm delivery. *Lancet* 350: 265–266
50 Peruzzi L, Gianoglio B, Porcellini MG, Coppo R (1999) Neonatal end-stage renal failure associated with maternal ingestion of cyclo-oxygenase-type-1 selective inhibitor nimesulide as tocolytic. *Lancet* 354: 1615
51 Allport V, Bennett P (2001) Cyclooxygenase enzymes and human labour. In: JR Vane, RM Botting (eds): *Therapeutic Roles of Selective COX-2 Inhibitors*. William Harvey Press, London, 252–273
52 Stewart WF, Kawas C, Corrada M, Metter EJ (1997) Risk of Alzheimer's disease and duration of NSAID use. *Neurology* 48: 626–632
53 Pasinetti GM, Aisen PS (1998) Cyclooxygenase-2 expression is increased in frontal cortex of Alzheimer's disease brain. *Neuroscience* 87: 319–324
54 Lim GP, Yang F, Chu T, Chen P, Beech W, Teter B, Tran T, Ubeda O, Ashe KH, Frautschy SA et al (2000) Ibuprofen suppresses plaque pathology and inflammation in a mouse model for Alzheimer's disease. *J Neurosci* 20: 5709–5714
55 McGeer PL (2000) Cyclo-oxygenase-2 inhibitors: rationale and therapeutic potential for Alzheimer's disease. *Drugs Aging* 17: 1–11
56 Sainati SM, Ingram DM, Talwalker S, Geis GS (2000) Results of a double-blind, randomized, placebo-controlled study of celecoxib in the treatment of progression of Alzheimer's disease. Abstracts of the Sixth International Stockholm/Springfield Symposium on Advances in Alzheimer Therapy. April 5–8 (abstract)
57 Weggen S, Eriksen JL, Das P, Sagi SA, Wang R, Pietrzik CU, Findlay KA, Smith TE, Murphy MP, Bulter T et al (2001) A subset of NSAIDs lower amyloidogenic Aβ42 independently of cyclooxygenase activity. *Nature* 414: 212–216
58 Zandi PP, Anthony JC, Hayden KM, Mehta K, Mayer L, Breitner JC (2002) Reduced incidence of AD with NSAID but not H2 receptor antagonists. The Cache County study. *Neurology* 59: 880–886
59 Chandrasekharan NV, Dai H, Roos KL, Evanson NK, Tomsik J, Elton TS, Simmons DL (2002) COX-3, a cyclooxygenase-1 variant inhibited by acetaminophen and other analgesic/antipyretic drugs: cloning, structure, and expression. *Proc Natl Acad Sci USA* 99: 13926–13931
60 Simmons DL, Botting RM, Robertson PM, Madsen ML, Vane JR (1999) Induction of an acetaminophen-sensitive cyclooxygenase with reduced sensitivity to nonsteroid antiinflammatory drugs. *Proc Natl Acad Sci USA* 96: 3275–3280

COX-2 Inhibitors
Edited by M. Pairet and J. van Ryn

Structural diversity of selective COX-2 inhibitors

Lawrence J. Marnett[1] and Amit S. Kalgutkar[2]

[1] *Department of Biochemistry, Vanderbilt Institute of Chemical Biology, Vanderbilt-Ingram Cancer Center, and Center in Molecular Toxicology, Vanderbilt University School of Medicine, Nashville, Tennessee 37232-0146, USA*

[2] *Pharmacokinetics, Dynamics and Metabolism Department, Pfizer Global Research and Development, Groton, CT 06340, USA*

Introduction

Non-steroidal anti-inflammatory drugs (NSAIDs) are among the most widely used agents in the treatment of pain and inflammation, especially rheumatoid arthritis [1]. The first NSAID aspirin was introduced over a century ago, and continues to be a popular over-the-counter medication in the treatment of the symptoms of pain and inflammation. The global production of aspirin is approximately 50,000 tons annually [2]. The pharmacological effects of NSAIDs arise from their inhibition of the prostaglandin synthase or cyclooxygenase (COX) isozymes [3]. COX isozymes catalyze the rate-limiting step in the arachidonic acid cascade involving the conversion of arachidonic acid to prostaglandin H_2 (PGH_2), the common biosynthetic precursor to prostaglandins and thromboxane (Fig. 1). These bioactive lipids mediate numerous physiological and pathophysiological effects, including pain, fever, inflammation, hemostasis, and regulation of renal function and maintenance of mucosal integrity in the stomach.

In vitro evaluation of COX-2 selectivity

Because of the numerous and widely differing experimental protocols available for the *in vitro* evaluation of COX-1 and COX-2 activity, comparison of

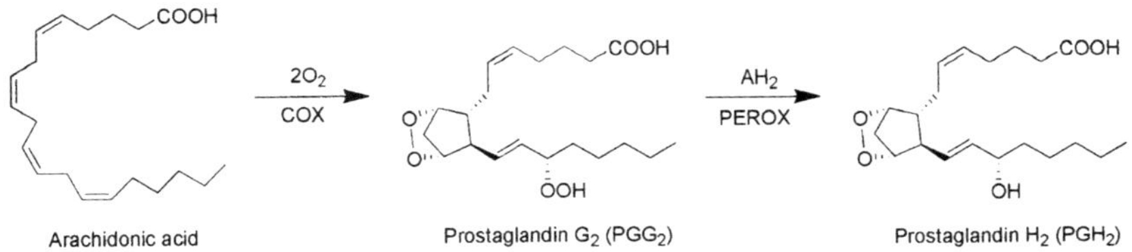

Figure 1. Biosynthesis of prostaglandin H_2.

potency and selectivity data for NSAIDs and selective COX-2 inhibitors from different laboratories has been somewhat of a challenge. Recombinant enzyme systems [4–6], microsomal membranes and cultured intact cells [7], insect cells [8], human whole blood [9–11], human monocytes [12], and various other mammalian cells and platelets [13–15] are routinely used in the screening of COX inhibitors. Examination of the COX-1/COX-2 selectivity ratio for indomethacin under various assay conditions reveals that this ratio varies from 2.0 to 14 [10, 14, 16, 17]. This highlights the influence different assay conditions can have and is discussed in more detail in Chapter 3.

Diarylheterocycles

Research within this class of compounds has culminated in the FDA approval and marketing of celecoxib, rofecoxib and valdecoxib [18–20]. Other agents from this class of compounds in various phases of clinical trials include etoricoxib and JTE-522 [21].

Historical perspective

The chemical structures of diarylheterocycles can be traced back to clinical candidates discovered over 30–40 years ago (Fig. 2). Examination of the structures of commercial selective COX-2 inhibitors from the diarylheterocycle series (e.g. celecoxib, rofecoxib, valdecoxib, and etoricoxib) reveals a striking resemblance to the older COX inhibitor phenylbutazone [22, 23]. The exceptional anti-inflammatory properties of phenylbutazone [24] undoubtedly provided much of the impetus that led chemists to exploit this type of structure. The discovery of the diphenyloxazole derivative oxaprozin as a clinical candidate for the treatment of inflammation is one example of the utility of the diarylheterocycle scaffold [25, 26]. Likewise, the 2,3-diarylindole indoxole was disclosed in the early 1960s as a potent anti-inflammatory agent; however, the drug was suspended from clinical development since it caused photosensitivity in humans [27–29]. Throughout the period from late 1960s to the early 1980s, a plethora of diarylheterocycles with variations on the central heterocyclic ring were evaluated as anti-inflammatory agents [30] and several were advanced into clinical trials based on their superior pharmacological and gastrointestinal profile in preclinical species compared with traditional NSAIDs [31, 32]. An example of such a compound is the 4,5-diarylthiazole derivative itazigrel that was shown to be less ulcerogenic than aspirin in subsequent clinical trials [33]. Relative to the unsubstituted congeners, structure-activity relationship (SAR) studies in the prototype diarylheterocycles indicated significant enhancement in anti-inflammatory properties when methoxy or halogen groups were incorporated in the *para* position on the two pendant phenyl rings (see Fig. 3) [34].

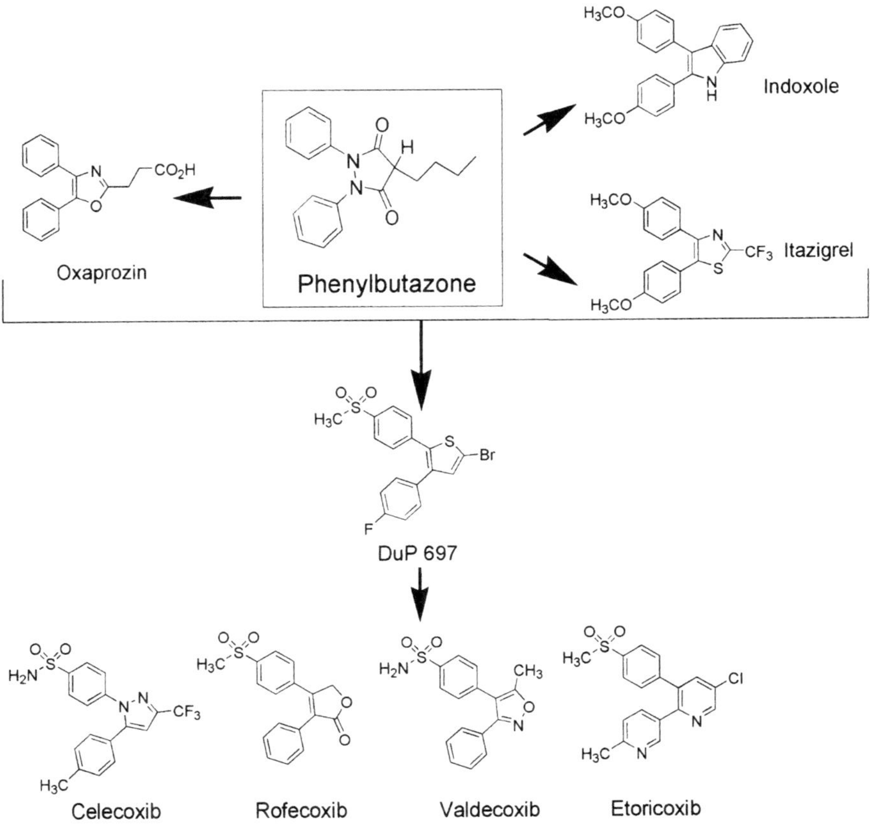

Figure 2. Evolution of selective COX-2 inhibitors.

General determinants of selective COX-2 inhibition by diarylheterocycles

The 4-methylsulfonyl (CH_3SO_2)-diarylheterocycle combination, commonly found in the most selective COX-2 inhibitors, was first observed in the structure of the 2,3-diarylthiophene, DuP 697, disclosed in the late 1980s (see Fig. 2) [35]. SAR studies on potential methylsulfone replacements in 2,3-diarylthiophenes indicated that only the 4-sulfonamido (SO_2NH_2) group maintained COX-2 inhibitory potency and selectivity observed with DuP697 [36, 37]. Incorporating the 4-sulfonamido group also results in increased COX-2 inhibitory potency, albeit with a loss of COX-2 selectivity. This SAR trend appears to be a common theme in all of the diarylheterocycles studied to date and is illustrated in Figure 3 with the 1,5-diarylpyrazole-based methylsulfone SC 58125 (COX-1/COX-2 selectivity > 1000) and the sulfonamide, celecoxib (COX-1/COX-2 selectivity ~325) [38]. Loss in COX-2 selectivity, however, is often compensated by an overall improvement in physiochemical and pharmacokinetic properties [38]. The oxidation state of the sulfur in the

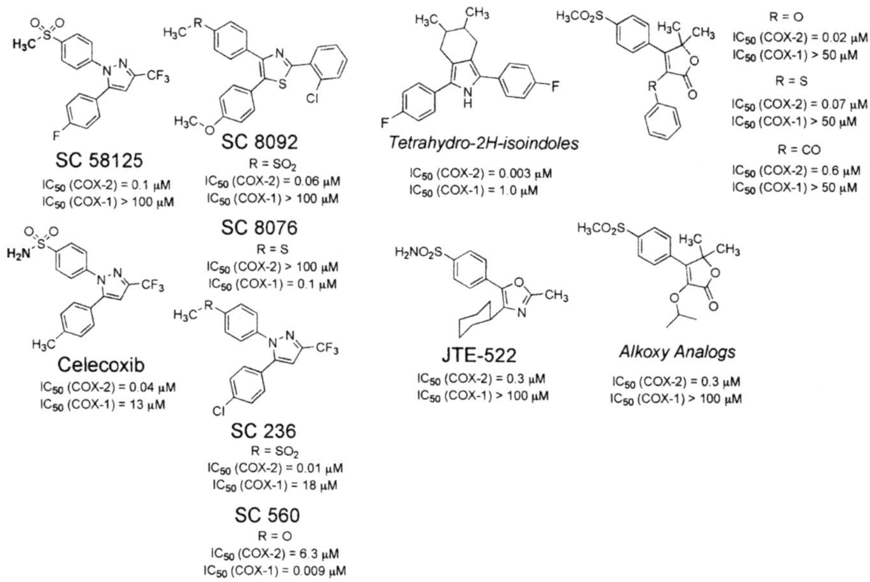

Figure 3. Structural requirements for COX-2 inhibitory potency and selectivity.

4-methylsulfone-containing compounds is extremely crucial for selective COX-2 inhibition; its reduction to sulfoxide or sulfide reverses isozyme selectivity. Thus, methylsulfone-containing 4,5-diarylthiazole (SC-8092) is a selective COX-2 inhibitor, whereas the corresponding methylthioether analog (SC-8076) exhibits COX-1-selective inhibitory properties (see Fig. 3) [17]. Likewise, replacement of the methylsulfonyl group in the selective COX-2 inhibitor SC-236 with a methoxy substituent also reverses isozyme specificity as highlighted with SC-560 [39].

In contrast with these observations, a series of novel 1,3-diaryltetrahydro-2H-isoindoles that lack the essential sulfonyl group were recently disclosed as potent and selective COX-2 inhibitors (IC_{50} = 0.6–100 nM for COX-2, 100–1000 nM for COX-1). SAR studies established that a bicyclic ring appended to the pyrrole nucleus and 4,4'-difluoro substitution on the phenyl rings was optimal for COX-2 potency (see Fig. 3) [40]. Interestingly, incorporation of the methylsulfonyl group in these compounds proved detrimental for COX-2 potency. The structural basis for COX-2 selectivity in this series remains unclear at the present time.

SAR studies also indicate that the phenyl substituent that does not contain the methylsulfonyl or the sulfonamide group can be replaced with a cyclohexyl substituent as illustrated with oxazole derivative JTE-522 (see Fig. 3). JTE-522 is currently under clinical evaluation as an anti-inflammatory agent [41, 42]. Introduction of a heteroatom (O and S) or a carbonyl link between the central ring system and the phenyl ring (lacking the methylsulfone or sulfonamide) also results in compounds that retain COX-2 potency and selectivity [43]. This is illustrated in Figure 3 with the 5,5-dimethylfuranone derivatives. The aryloxy moiety can also be replaced with alkoxy groups without any deleterious effect on potency and selectivity [43–46].

Overall, the common structural features essential for inhibition by most diarylheterocycles can be summarized as follows: (1) the presence of two lipophilic substituents (one of which has to be a phenyl ring) on adjacent positions within the central unsaturated ring system and (2) the presence of a methylsulfone/sulfonamide group on the essential phenyl ring. That the two lipophilic substituents need to be adjacent to one another is highlighted with 2,3- and the 3,4-diarylthiophenes, which are potent and selective COX-2 inhibitors [47], whereas the corresponding 2,4- or 2,5-diarylthiophenes suffer dramatic losses in COX-2 potency (Fig. 4).

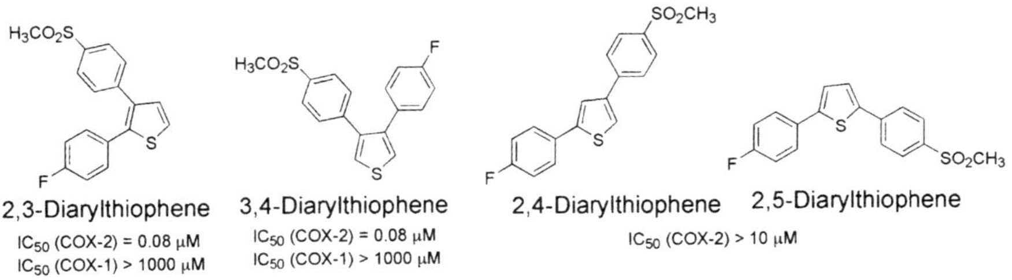

Figure 4. Effect of the phenyl group positioning on COX-2 inhibition by diarylthiophene isomers.

Variations on the central ring system

Numerous SAR studies have been focused on the alteration of the central heterocyclic ring and to date, virtually every imaginable carbocyclic (Fig. 5), heterocyclic (Fig. 6), and bicycloheterocyclic (Fig. 7) ring system has been incorporated.

Carbocycles

Examples of diarylcarbocycles include four-membered (e.g., cyclobutene and cyclobutenone) [48], five-membered (e.g., cyclopentene and cyclopentenone) [49–51], spirocyclic [52, 53] and substituted phenyl ring (terphenyl) systems [54, 55]. Although preliminary SAR studies suggested that the central ring was essential for favorable spatial orientation of the two pendant phenyl rings, a series of 1,2- and 1,1-diarylstilbenes were recently disclosed as potent and selective COX-2 inhibitors [56–58].

Heterocycles

Examples of central heterocycles include 5- or 6-membered rings that contain one (thiophenes [35–37], pyrroles [59–61], furans [62, 63] and pyridines

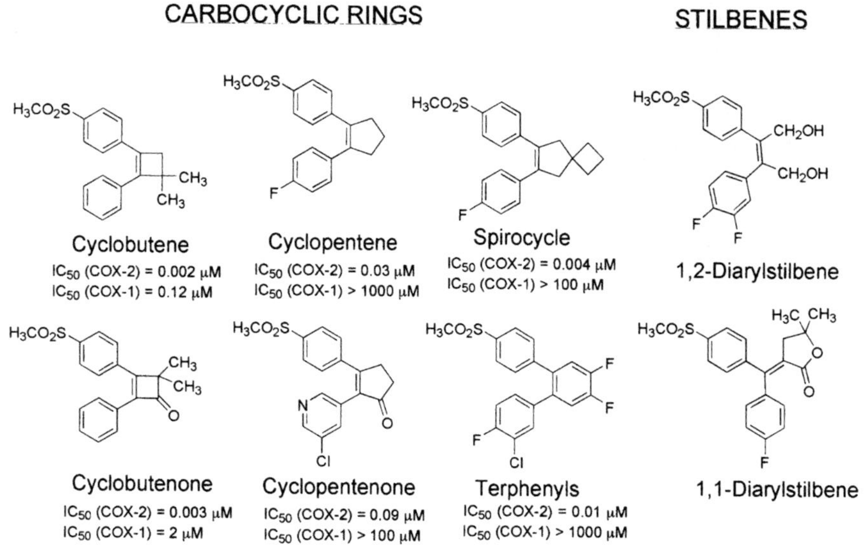

Figure 5. Diarylcarbocycles as selective COX-2 inhibitors.

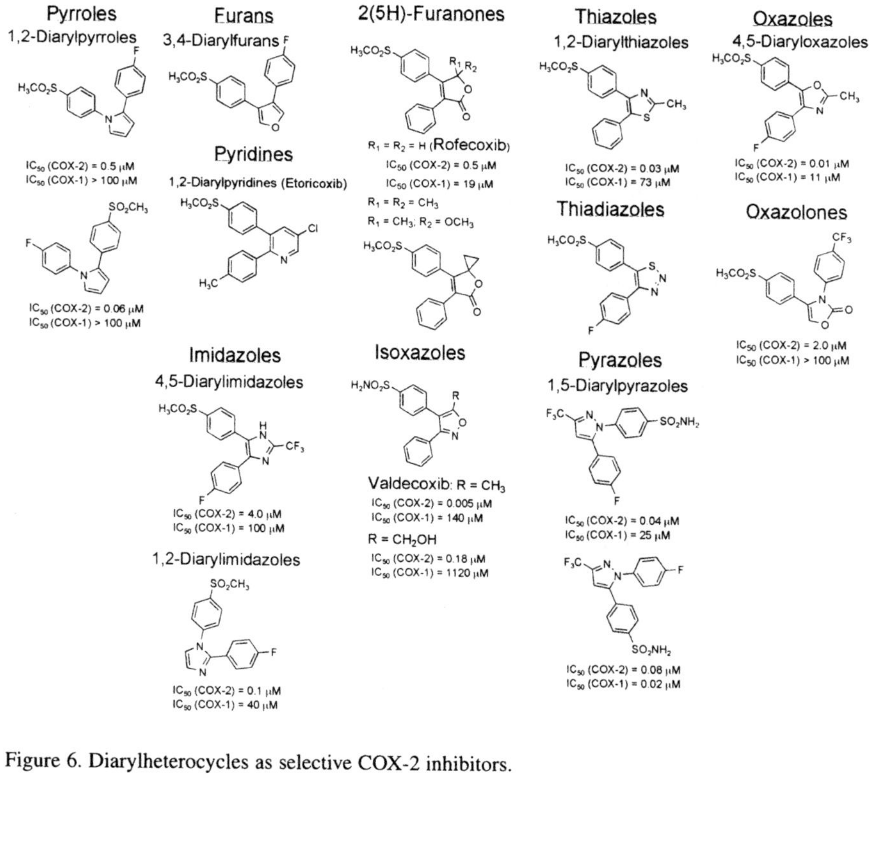

Figure 6. Diarylheterocycles as selective COX-2 inhibitors.

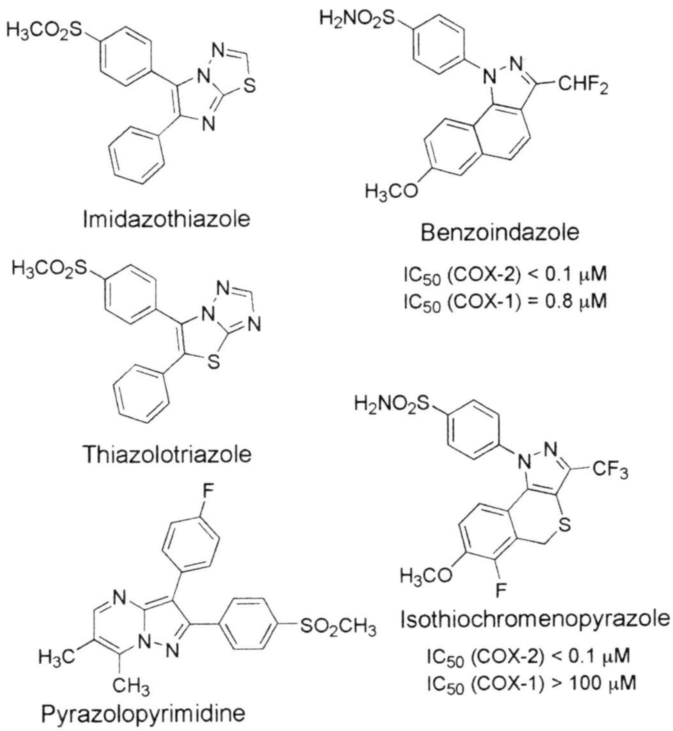

Figure 7. Diarylbicycloheterocycles as selective COX-2 inhibitors.

[64–66]), two (thiazoles [67, 68], oxazoles and the corresponding oxazolones [69, 70], imidazoles [71–73], isoxazoles [74–76], pyrazoles [38, 77–79]), or three (thiadiazole [80]) heteroatoms (Fig. 6).

As observed with thiophenes, both the 2,3- and 3,4-diarylfuran isomers demonstrate selective COX-2 inhibition [62, 63]. The inhibitory potency of diarylfurans, however, is inferior when compared with the corresponding thiophene derivatives. Structural alteration of the central furan ring to the 2(5H)-furanone scaffold however, results in substantial increases in COX-2 potency and selectivity as illustrated with the furanone rofecoxib (Fig. 6) [81, 82]. SAR analysis indicates that disubstitution at the 5-position in rofecoxib with alkyl, alkoxy or cycloalkyl groups leads to metabolically more stable compounds (see Fig. 6) with retention of COX-2 inhibitory potency and selectivity [45, 83–85].

Comparison of the SAR for selective COX-2 inhibition by terphenyls (see Fig. 5) and pyridines reveals that substitution other than in the 1- or 2-position in the central benzene ring in terphenyls with groups as small as fluorine affords inactive compounds, whereas such replacements are not problematic in the pyridine series as indicated with etoricoxib in Figure 6. SAR studies on diarylpyrroles indicate that COX-2 can be selectively inhibited by all possible isomers (1,2-diaryl, 2,3-diaryl and 3,4-diaryl) [59–61]. Furthermore, regardless of the 4-methylsulfonylphenyl-substitution pattern (N1 *versus* C-2), the 1,2-diarylpyrrole regioisomers maintained selective COX-2 inhibitory properties (see Fig. 6). In contrast, selective COX-2 inhibition by 1,2-diarylimidazoles and 1,5-diarylpyrazoles was only discernible with compounds in which the 4-methylsulfonylphenyl- or 4-sulfonamidophenyl groups were attached to the nitrogen atom. Imidazole or pyrazole analogs, in which the 4-methylsulfonylphenyl- or 4-sulfonamidophenyl groups were attached at the C-2 or the C-5 positions respectively, were nonselective or selective COX-1 inhibitors [38, 71–73]. This is depicted with the 1,5-diarylpyrazole derivatives in Figure 6.

Bicyclic or fused heterocycles

Examples of bicyclic or fused-ring systems as the central template have also been disclosed. Although many compounds demonstrate COX-2 potency and selectivity, they suffer from poor physiochemical and pharmacokinetic properties, which ultimately result in poor oral absorption. Examples of orally active analogs are few and include the imidazothiazole-, thiazolotriazole- and the pyrazolopyrimidine-containing diarylbicycloheterocycle derivatives (Fig. 7) [86–88]. Other variations in this class are highlighted with the benzindazole and the isothiochromenopyrazole ring system in which one of the pendant aryl rings was fused to the bicyclic central core (see Fig. 7) [89–91].

Alkylsulfonanilide derivatives

Investigations on the anti-inflammatory properties of alkylsulfonanilides such as nimesulide (Fig. 8) began in the 1960s [92]. Details on the pharmacologi-

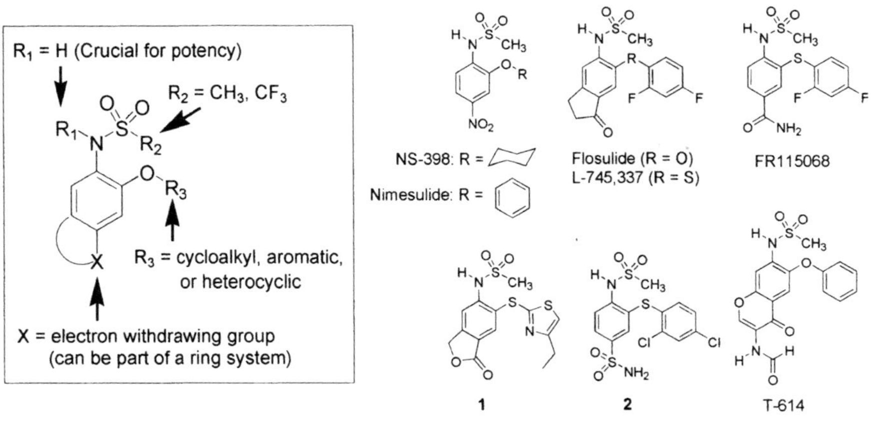

Figure 8. SAR studies on akylsulfonanilides as selective COX-2 inhibitors.

cal [93] and clinical [94] evaluation of nimesulide clearly demonstrated the excellent anti-inflammatory properties of this compound. The recent finding on the selective COX-2 inhibition by the structurally related sulfonanilide NS-398 (see Fig. 8) led to a renewed interest in this general class of anti-inflammatory agents [16, 95, 96]. Likewise, the nonulcerogenic anti-inflammatory activity of nimesulide [94] and the close structural analog flosulide [97] had been previously established, but re-evaluation of both compounds as COX inhibitors revealed their selective COX-2 inhibitory properties [98]. The common structural features of the alkylsulfonanilides are depicted in Figure 8. The alkyl substituent is typically methyl, but halogenated methyl substituents such as trifluoromethyl have also been reported [99]. N-Methylation of the sulfonamide nitrogen generates inactive compounds [99]. The *ortho*-substituent typically includes aryl (e.g., nimesulide), heterocyclic (thiazole analog **1**) [100, 101], or cycloalkyl ethers (e.g., NS-398) and thioethers (L-745,337 [99] and FR115068 [102], respectively). The thiazolyl-thioether analog **1** is reported to possess greater COX-2 potency and selectivity, superior *in vivo* pharmacokinetic and pharmacological profile, and a greater gastrointestinal safety profile than flosulide [100]. The *para*-substituent invariably bears an electron-withdrawing group that may be incorporated as part of a ring. A variety of methanesulfonanilides with different *para*-electron-withdrawing groups have been evaluated as selective COX-2 inhibitors and as orally active anti-inflammatory agents. Substituents include 4-acetyl, 4-cyano, 4-carboxamido (FR115068), nitro (NS-398 and nimesulide), trifluoromethyl, sulfonamido (compound **2**), sulfonyl and several others [102]. Another structural variant is the incorporation of the *para*-electron-withdrawing group as part of a ring system. For instance, the principal difference between nimesulide/NS-398 and flosulide/L-745,337 is the incorporation of the electron-withdrawing group as part of a five-membered cyclopentanone ring system in the latter compounds. The electron withdrawing substituent also may be incorporated as part of a heterocyclic ring as seen in T-614 [103–105]. Although the purpose of the electron-withdrawing group remains unclear, its presence is crucial for COX-2 inhibitory potency. Molecular modeling studies suggest that the electron-withdrawing group may function as an H-bond acceptor from the active site Tyr355 residue [106].

Structural modifications of NSAIDs into selective COX-2 inhibitors

Modifying nonselective NSAIDs into selective COX-2 inhibitors represents an interesting strategy and has met with variable degrees of success. Thus far, aryl acetic acid- (e.g., indomethacin, zomepirac, sulindac, diclofenac, and etodolac), fenamic acid- (e.g., meclofenamic acid), propionic acid- (e.g., flurbiprofen), enol-carboxamide- (e.g., piroxicam), and benzoic acid- (e.g., aspirin) containing NSAIDs have been successfully elaborated into potent and selective COX-2 inhibitors.

Aryl acetic acids

Indomethacin

Modifications on the N-substituted indole-3-acetic acid framework found in indomethacin have yielded tremendous success in the search for novel structural types as selective COX-2 inhibitors (Fig. 9). Some of the changes in indomethacin structure that imparts COX-2 potency and selectivity include: (1) replacement of the 4-chlorobenzoyl group with a 2,4,6-trichlorobenzoyl- or a 4-bromobenzyl group (examples **3** and **4**) [107, 108], (2) increasing the chain length of the 3-alkanoic acid group (example **5**) [108], (3) neutralization of the carboxylic acid group via conversion to the corresponding esters (compd. **6**) or amides (compd. **7**) [109–111] as well as the corresponding reverse esters/amides (compds. **8** and **9**) [112], and (4) replacing the carboxylic acid group with appropriate bioisosteres as exemplified with the iminoxy (**10**) or substituted thiazole (**11**) analogs, respectively [113, 114]. It is noteworthy to point out that the 2-methyl group in indomethacin and the neutral amide and ester derivatives is crucial for inhibitory potency, since its replacement with hydrogen results in compounds that are devoid of inhibitory potency against the COX isozymes [109, 110]. Exchanging the methoxy substituent in indomethacin with the methylsulfone group common in diarylheterocycles and

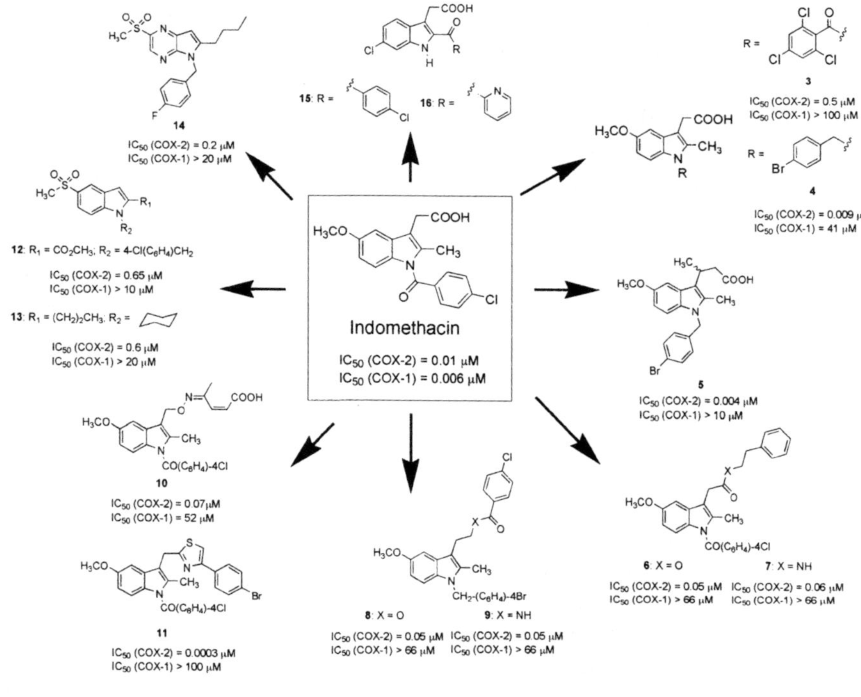

Figure 9. Strategies implemented in the conversion of indomethacin to a COX-2 selective inhibitor.

moving the alkanoic ester group from the 3- to the 2-position on the indole ring also results in reasonably selective and potent COX-2 inhibitors as highlighted with compound **12** [115]. Additional improvements in the COX-2 potency and selectivity of **12** have also been described as exemplified with the mono- and diazaindolyl derivatives **13** and **14** [116–118]. These latter compounds lack the carboxylic acid group that is a key determinant for COX inhibition by indomethacin and instead possess a methylsulfone group, an essential criterion for selective COX-2 inhibition by diarylheterocycles. Whether these analogs bind in the same region of the COX-2 enzyme, as diarylheterocycles remains unclear. An interesting variation on the utilization of the indomethacin template in the design of COX-2 inhibitors is shown with 2,3-disubstituted indolyl derivatives (compds. **15** and **16**) [119]. These analogs contain an aroyl group on the 2-position of the indole ring, unlike indomethacin, in which the aroyl group is attached to the nitrogen on the indole ring.

Zomepirac and sulindac

As observed with indomethacin, replacement of the carboxylic acid group in zomepirac and sulindac with certain bioisosteres also confers COX-2 selectivity in the resulting compounds (Fig. 10). For instance, exchanging the carboxylic group in sulindac with a tetronic acid moiety (compd. **17**) or in

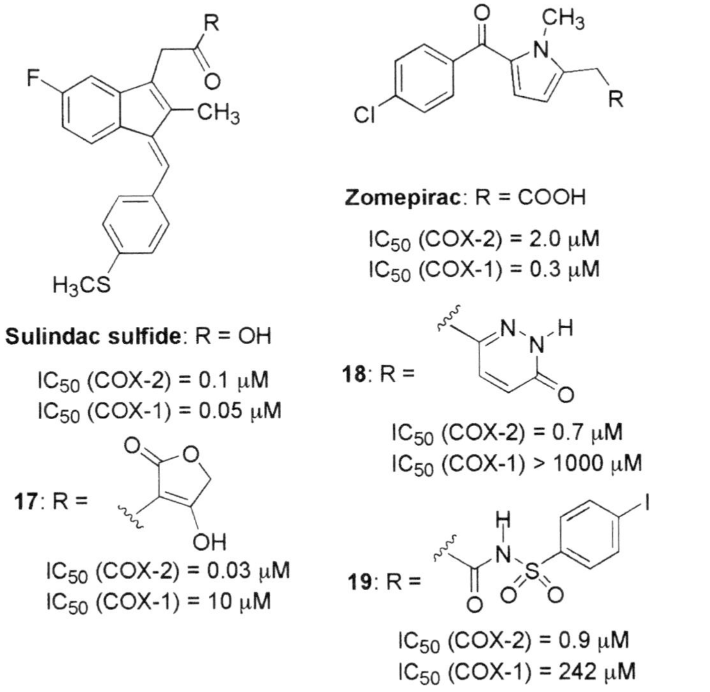

Figure 10. Carboxylic acid bioisosteres of sulindac and zomepirac as selective COX-2 inhibitors.

zomepirac with pyridazinone (compd. **18**) or an acylsulfonamide group (compd. **19**) affords potent and selective COX-2 inhibitors [120–122].

Diclofenac

Carboxylic acid bioisosteres of diclofenac also demonstrate potent and selective COX-2 inhibition. This is highlighted with the iminoxy analog **20** (Fig. 11) [123]. The observation that the active site in COX-2 is larger than in COX-1 has led to the hypothesis that introducing steric constraints in the NSAID scaffold should potentially generate compounds with COX-2 selectivity. This proposal has been validated in the case of diclofenac. Incorporation of *meta*-alkyl substituents in the arylacetic acid ring in diclofenac reverses isozyme specificity as shown with analog **21** [124]. Replacement of a chloro substituent from the dichlorophenyl ring in **21** with a fluoro group resulted in the discovery of Lumiracoxib (see Fig. 11) [125].

Etodolac

Evaluation of the COX isozyme selectivity for the NSAID etodolac (Fig. 12) led to the finding on the moderately selective COX-2 inhibitory properties of

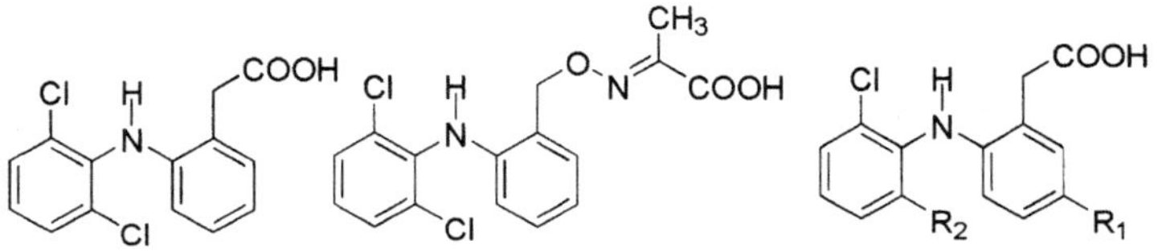

Diclofenac

20
IC_{50} (COX-2) = 0.1 μM
IC_{50} (COX-1) > 1 μM

21: $R_1 = CH_3$; $R_2 = Cl$
IC_{50} (COX-2) = 0.01 μM
IC_{50} (COX-1) > 10 μM
Lumiracoxib ($R_1 = CH_3$; $R_2 = F$)
IC_{50} (COX-2) = 0.007 μM
IC_{50} (COX-1) > 10 μM

Figure 11. Structural modifications on diclofenac that result in selective COX-2 inhibition.

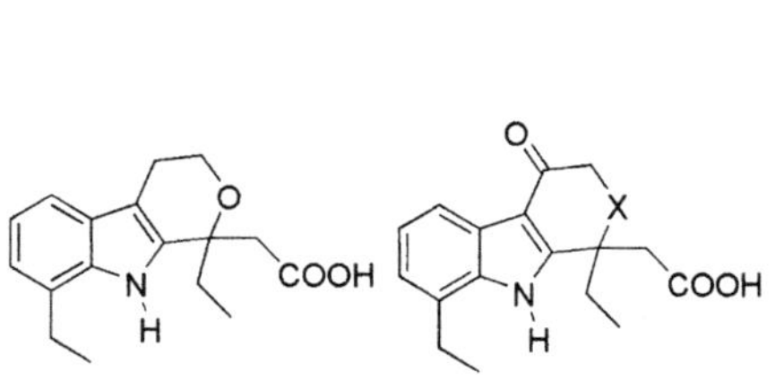

Etodolac
IC_{50} (COX-2) = 1.4 μM
IC_{50} (COX-1) = 15 μM

22: X = O
IC_{50} (COX-2) = 1.9 μM
IC_{50} (COX-1) = 270 μM

23: X = CH_2
IC_{50} (COX-2) = 2.0 μM
IC_{50} (COX-1) > 90 μM

Figure 12. Etodolac and related analogs as selective COX-2 inhibitors.

this NSAID [126]. These findings were consistent with the better gastrointestinal safety of etodolac compared with other NSAIDs. Adjustments in the oxidation state from a methylene at C-4 in the pyran ring of etodolac to the corresponding keto derivative **22** and removal of the oxygen atom in the pyran ring of **22** to afford **23** led to dramatic improvement in the COX-2 selectivity of the resulting compounds (see Fig. 12) [127, 128].

Fenamates

As observed with indomethacin, secondary amide derivatives of the fenamic acid NSAID, meclofenamic acid, also demonstrated potent and selective COX-2 inhibition [109, 129]. The 2-phenoxyethylamide derivative **24** was the most selective inhibitor in the series, with a COX-2 selectivity ratio of ~440 (Fig. 13). Unlike indomethacin SAR, only the amide derivatives of meclofenamic acid demonstrated COX-2 selectivity. The esters were either inactive or nonselective COX inhibitors. The reason(s) for this discrepancy remains unclear.

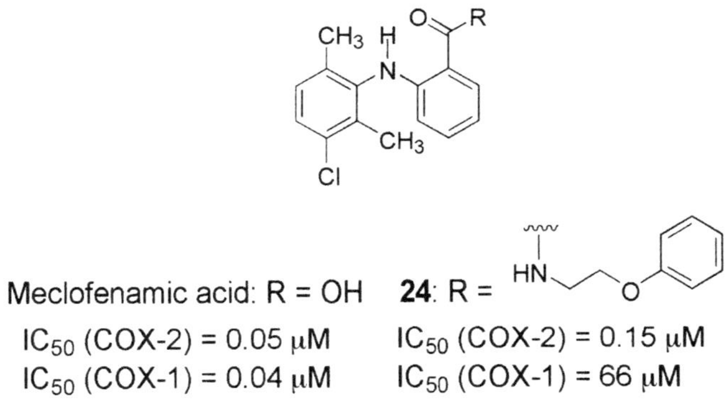

Figure 13. Meclofenamic acid amides as selective COX-2 inhibitors.

Propionates

Flurbiprofen is the only example in this structural class that has been successfully elaborated into a selective COX-2 inhibitor. Examination of the flurbiprofen-COX-1 and -COX-2 cocrystal structure revealed that introducing steric constraints in the 4-phenyl ring in flurbiprofen should in theory afford compounds devoid of COX-1 inhibitory properties without affecting COX-2 potency [94]. The finding that introduction of diethoxy substituents in the 4-phenyl ring of flurbiprofen led to selective COX-2 inhibition (compd. **25**, Figure 14) validated this hypothesis [130].

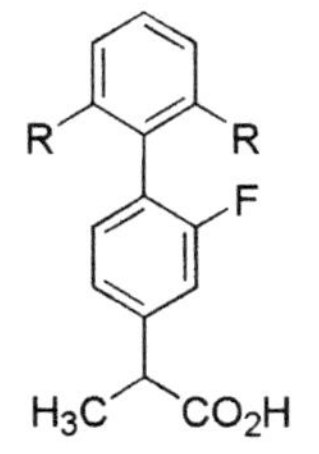

IC_{50} (COX-2) = 0.01 μM
IC_{50} (COX-1) = 0.01 μM

25: R = OC_2H_5

IC_{50} (COX-2) = 0.1 μM
IC_{50} (COX-1) = 7.7 μM

Flurbiprofen: R = H

Figure 14. Modification of the propionic acid NSAID flurbiprofen into a selective COX-2 inhibitor.

Enol-carboxamides

Replacement of the pyridine ring in the nonselective enolcarboxamide NSAID piroxicam with a 5-methylthiazoyl group resulted in the discovery of meloxicam, which demonstrated reasonable selectivity as a COX-2 inhibitor (Fig. 15) [131]. Further structural modifications on meloxicam resulted in isoquinolinediones such as **26** (see Fig. 15) that represent a new class of orally active COX-2-selective inhibitors [132, 133].

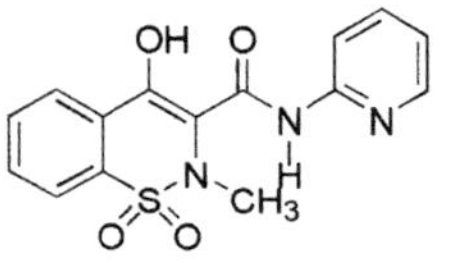

Piroxicam

IC_{50} (COX-2) = 0.5 μM
IC_{50} (COX-1) = 0.4μM

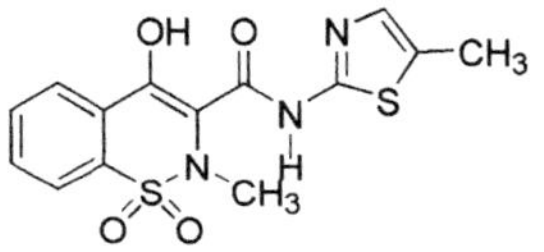

Meloxicam

IC_{50} (COX-2) = 0.49 μM
IC_{50} (COX-1) = 37 μM

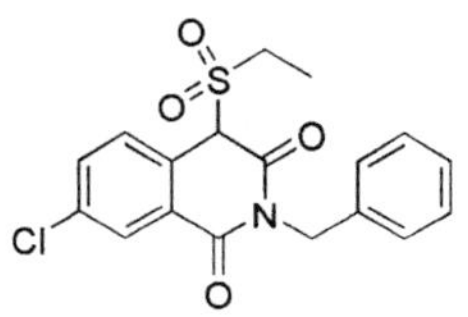

26

IC_{50} (COX-2) = 0.06 μM
IC_{50} (COX-1) = 1.4 μM

Figure 15. Structural modifications on the enol-carboxamide piroxicam that lead to selective COX-2 inhibition.

Benzoic acid

Recently, we demonstrated the structural alteration of aspirin to APHS, a selective COX-2 inhibitor (Fig. 16). Aspirin is the only NSAID that covalently modifies COX isozymes [134]. Covalent bond formation is achieved through a regioselective acetylation of the active site Ser530 residue in the two isozymes [135]. Replacement of the carboxylic acid group in aspirin with alkyl- or alkynylsulfide groups reverses isozyme specificity as shown with APHS [136, 137]. SAR studies indicated that the presence of the sulfur is crucial for COX inhibition; its replacement with oxygen, nitrogen, and carbon generates inactive compounds [137]. Furthermore, the acetoxy group in APHS

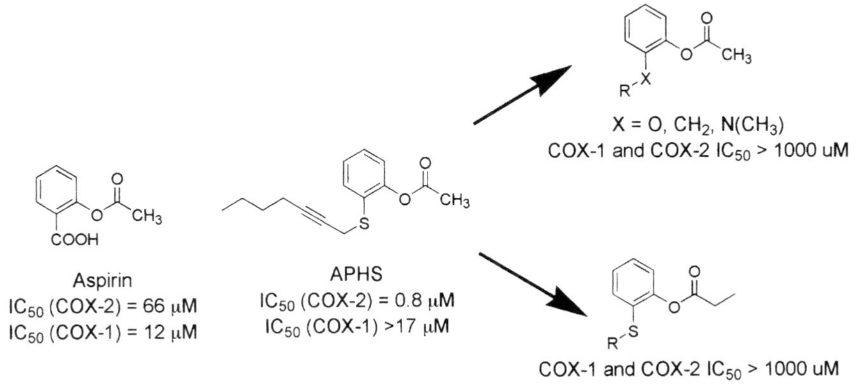

Figure 16. Transformation of aspirin to the selective COX-2 inhibitor APHS.

was optimal for inhibition since the corresponding propionoxy derivative was inactive as a COX inhibitor [137]. APHS acetylates the same serine residue on COX-2 (Ser530) that is acylated by aspirin [137]. Site-directed mutagenesis studies have indicated that the hept-2-ynyl chain in APHS probably binds in the same region of the protein as the ω-end of the fatty acid substrate arachidonic acid [138]. This binding mode of APHS contrasts with aspirin in which the carboxylate moiety is involved in electrostatic interactions with the positively charged Arg120 residue in the mouth of the active site [139]. Whether these differences in binding are also responsible for the COX-2 selectivity of APHS remain unclear.

Di-tert-butylphenols and dihydrobenzofurans

Certain di-*tert*-butylphenolic analogs containing five-membered heterocyclic rings *para* to the phenolic OH group were recently disclosed as a novel class of potent and selective COX-2 inhibitors [140–142]. In addition to inhibiting COX-2, these compounds also possess 5-lipoxygenase inhibitory properties [140]. Examples of five-membered cyclic heterocyclic substituent included the thiazolone, oxazolone, 1,3,4-thiadiazole, and 1,3,4-oxadiazole ring systems (Fig. 17). Thus, the thiazolone analog PD138387 indicated good *in vitro* COX-2 potency and selectivity (IC_{50} (COX-1) > 100 μM; IC_{50} (COX-2) = 0.98 μM). Upon oral administration, PD138387 demonstrated anti-inflammatory activity in the mouse footpad edema model (ED_{40} = 16 mg/kg) as well as analgesic properties in the acetic acid writhing test (ED_{40} = 0.1 mg/kg). The bioisosteric replacement of the sulfur atom in PD138387 with an oxygen atom resulted in the oxazolone **27**, which demonstrated selective COX-1 inhibition (IC_{50} (COX-1) = 7.3 μM; IC_{50} (COX-2) > 100 μM). Direct attachment of the di-*tert*-butylphenolic portion to heterocyclic rings resulted in improved COX-2 potency and selectivity as exemplified with the 1,3,4-thiadiazole PD164387 (IC_{50}

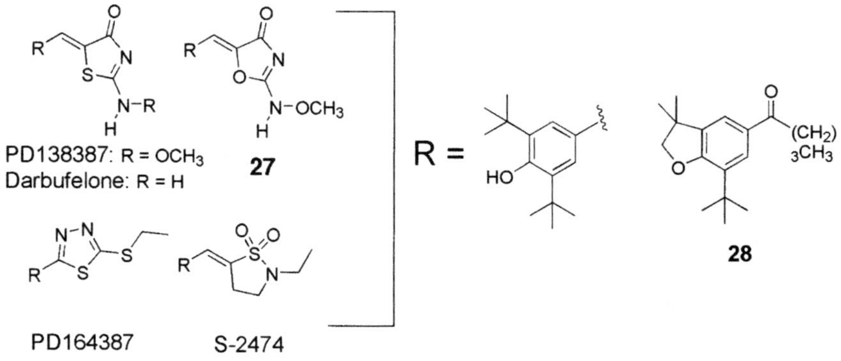

Figure 17. Di-*tert*-butylphenols and dihydrobenzofurans as selective COX-2 inhibitors.

(COX-1) = 0.14 μM; IC_{50} (COX-2) > 100 μM) [142]. The C-4 5-membered cyclic sulfonamide moiety in S-2474 serves as an additional example of structural diversity in the series. In addition to inhibiting COX-2 and 5-lipoxygenase, S-2474 also inhibited the production of IL-1 *in vitro*. S-2474 has also proven to be effective in several animal models of inflammation and is devoid of ulcerogenic liability [143]. Based on these attributes, S-2474 is under clinical investigation as an anti-inflammatory agent [143]. The recent disclosure on the selective COX-2 inhibition by the structurally related dihydrobenzofurans such as **28** (IC_{50} (COX-1) = 10 μM; IC_{50} (COX-2) = 0.09 μM) broadens the scope of the di-*tert*-butylphenol template [144–146]. As observed in the case of diarylheterocycles, di-*tert*-butylphenols such as darbufelone (see Fig. 17) also demonstrate slow, tight-binding inhibition of COX-2 [147]. Structural information is not yet available on the binding of this class of compounds to COX-2.

Chromenes

The discovery of novel structural types continues as evident from the potent and selective COX-2 inhibition by certain chromene derivatives (Fig. 18) [148]. *In vitro* inhibition data suggest that COX-2 potency and selectivity of

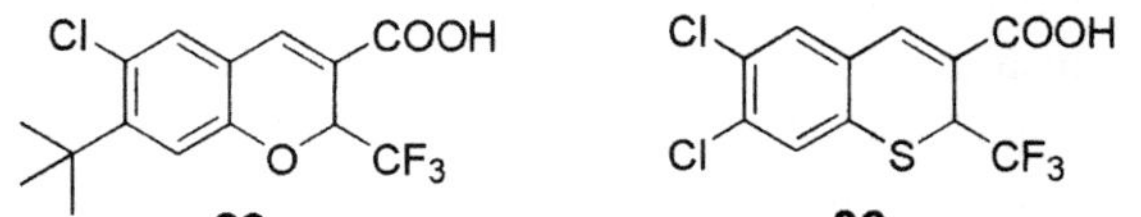

29

IC_{50} (COX-2) < 0.1 μM
IC_{50} (COX-1) = 61 μM

30

IC_{50} (COX-2) < 0.1 μM
IC_{50} (COX-1) = 0.8 μM

Figure 18. Chromene derivatives as selective COX-2 inhibitors.

chromenes is comparable to that of diarylheterocycles. The carboxylic acid and the trifluoromethyl substituents are major determinants of potency in the series as highlighted with analog **29**, whereas COX-2 selectivity is dependent on the presence of the oxygen atom in the chromene ring; its replacement with a sulfur atom results in dramatic loss of selectivity as shown with **30**.

Kinetic basis for selective COX-2 inhibition

COX-1 and COX-2 exist as homodimers with a molecular mass of ~70 kDa per monomer. COX-1 and COX-2 are bifunctional enzymes that carry out two sequential chemical reactions in spatially distinct but mechanistically coupled active sites. The cyclooxygenase activity *bis*oxygenates arachidonic acid to PGG_2 and the peroxidase activity reduces the hydroperoxyendoperoxide PGG_2 to PGH_2 (see Fig. 1) [149]. Both the cyclooxygenase and the peroxidase active sites are located in the catalytic domain, which accounts for 80% of the protein. All NSAIDs, as well as selective COX-2 inhibitors, bind in the COX active site but not the peroxidase site [3]. A general model for COX inhibition is emerging in which multiple equilibria are established between free enzyme, inhibitor and two or three enzyme-inhibitor complexes (Eq. 1) [150, 151]. The most potent COX inhibitors are slow, tight-binding inhibitors that form very stable complexes. The structural differences between COX-inhibitor complexes have been the subject of considerable debate and speculation. Crystallographic data on the slow, tight-binding NSAIDs (e.g., flurbiprofen and indomethacin) bound to COX-1 and COX-2 are assumed to reflect the most stable complexes [152–154]. However, flurbiprofen and the competitive inhibitor ibuprofen exhibit similar conformations in the COX active site and the structures of these cocrystals are virtually superimposible [155]. Thus, no structural differences are apparent in complexes of COX-1 bound to a slow, tight-binding inhibitor or a competitive inhibitor. Many selective COX-2 inhibitors from the diarylheterocycle or alkylsulfonanilide class appear to be competitive inhibitors of COX-1, but display time-dependent inhibition of COX-2 that is responsible for their selectivity [151, 156–158]. Fluorescence quenching analysis indicates that diarylheterocycles or neutral indomethacin amides binds to both enzymes and the kinetics of fluorescence decay reveal sequential bimolecular and unimolecular steps that occur at comparable rates [159, 160]. A third unimolecular step is only observed with COX-2. The first two steps in the binding are reversible, whereas the third step in the binding is pseudo irreversible. Recovery of fluorescence in the presence of a substantial excess of the nonfluorescent inhibitor flurbiprofen takes on the order of ~3 h with COX-2, whereas it is complete in ~40 s with COX-1. Of some interest in this context are the recent biochemical observations on the slow, tight-binding inhibition of COX-2 by the di-*tert*-butylphenol darbufelone [147]. Although, the biochemical mechanism for the selective inhibition of COX-2 is similar to that displayed by diarylheterocycles, darbufelone appears to be a selective noncompetitive inhibitor of COX-2, since arachidonic acid and various other COX inhibitors do not alter the binding affinity of this agent. It has been proposed that di-*tert*-butylphenols inhibit COX-2 by exploiting a previously unrecognized binding site on the enzyme [147].

$$E + I \underset{k_{-1}}{\overset{k_1}{\rightleftharpoons}} [EI] \underset{k_{-2}}{\overset{k_2}{\rightleftharpoons}} [EI^*] \underset{k_{-3}}{\overset{k_{intact}}{\rightleftharpoons}} [EX]$$

Equation 1. Kinetic scheme for COX inhibition. k_1, k_2, and k_{inact} represent forward rate constants for the three discrete steps of the overall reaction, whereas k_{-1}, k_{-2}, and k_{-3} represent the backward rate constants. E and I represent the free enzyme and inhibitor. Three distinct enzyme-inhibitor complexes, identified from enzyme kinetics, are shown by [EI], [EI*], and [EX]. The precise structure of each of these complexes remain unclear at the present time.

Concluding remarks

COX-2 is one of the most attractive and highly pursued targets in the history of the pharmaceutical industry. The depth and breadth of discovery efforts is attested to by the structural diversity of the inhibitor candidates summarized above. The clinical and commercial successes of celecoxib and rofecoxib validate the "COX-2 hypothesis" and guarantee that additional candidates will be advanced to the clinic. It is also likely that accumulating structural information on COX-inhibitor complexes will begin to have an impact on drug discovery and optimization. Despite the fact that COX enzymes have been the targets of enormous investigation over the past 30 years and that considerable structural information is available on COX-inhibitor complexes, remarkably little is known about the molecular details of the slow, tight binding that is responsible for the potency and selectivity of most selective or non-selective COX inhibitors. This suggests that functional studies must be performed in combination with structural studies to uncover the discrete interactions that lead to this these most fundamental of interactions responsible for COX inhibition.

Acknowledgements
Research in the Marnett laboratory is supported by research grants from the National Institutes of Health (CA89450 and GM15431).

References

1 Brune K (2002) Next generation of everyday analgesics. *Am J Ther* 9: 215–223
2 Kuhnert N (2000) A century of aspirin. The history of the most successful drug of the last century. *Pharmazie in unserer Zeit* 1: 32–39
3 Marnett LJ, Rowlinson SW, Goodwin DC, Kalgutkar AS, Lanzo CA (1999) Arachidonic acid oxygenation by COX-1 and COX-2. Mechanisms of catalysis and inhibition. *J Biol Chem* 274: 22903–22906
4 Barnett J, Chow J, Ives D, Chiou M, Mackenzie R, Osen E, Nguyen B, Tsing S, Bach C, Freire J

et al (1994) Purification, characterization and selective inhibition of human prostaglandin G/H synthase 1 and 2 expressed in the baculovirus system. *Biochim Biophys Acta* 1209: 130–139
5 Laneuville O, Breuer DK, Dewitt DL, Hla T, Funk CD, Smith WL (1994) Differential inhibition of human prostaglandin endoperoxide H synthases-1 and -2 by nonsteroidal anti-inflammatory drugs. *J Pharmacol Exp Ther* 271: 927–934
6 Meade EA, Smith WL, DeWitt DL (1993) Differential inhibition of prostaglandin endoperoxide synthase (cyclooxygenase) isozymes by aspirin and other non-steroidal anti-inflammatory drugs. *J Biol Chem* 268: 6610–6614
7 Mitchell JA, Akarasereenont P, Thiemermann C, Flower RJ, Vane JR (1993) Selectivity of nonsteroidal antiinflammatory drugs as inhibitors of constitutive and inducible cyclooxygenase. *Proc Natl Acad Sci USA* 90: 11693–11697
8 Cromlish WA, Kennedy BP (1996) Selective inhibition of cyclooxygenase-1 and -2 using intact insect cell assays. *Biochem Pharmacol* 52: 1777–1785
9 Warner TD, Giuliano F, Vojnovic I, Bukasa A, Mitchell JA, Vane JR (1999) Nonsteroid drug selectivities for cyclo-oxygenase-1 rather than cyclo-oxygenase-2 are associated with human gastrointestinal toxicity: a full *in vitro* analysis. *Proc Natl Acad Sci USA* 96: 7563–7568
10 Patrignani P, Panara MR, Greco A, Fusco O, Natoli C, Iacobelli S, Cipollone F, Ganci A, Creminon C, Maclouf J (1994) Biochemical and pharmacological characterization of the cyclooxygenase activity of human blood prostaglandin endoperoxide synthases. *J Pharmacol Exp Ther* 271: 1705–1712
11 Panara MR, Greco A, Santini G, Sciulli MG, Rotondo MT, Padovano R, di Giamberardino M, Cipollone F, Cuccurullo F, Patrono C et al (1995) Effects of the novel anti-inflammatory compounds, N-[2-(cyclohexyloxy)-4-nitrophenyl] methanesulphonamide (NS-398) and 5-methanesulphonamido-6-(2,4-difluorothio-phenyl)-1-inda none (L-745,337), on the cyclo-oxygenase activity of human blood prostaglandin endoperoxide synthases. *Br J Pharmacol* 116: 2429–2434
12 Kato M, Nishida S, Kitasato H, Sakata N, Kawai S (2001) Cyclooxygenase-1 and cyclooxygenase-2 selectivity of non-steroidal anti-inflammatory drugs: investigation using human peripheral monocytes. *J Pharm Pharmacol* 53: 1679–1685
13 Chulada PC, Langenbach R (1997) Differential inhibition of murine prostaglandin synthase-1 and -2 by nonsteroidal anti-inflammatory drugs using exogenous and endogenous sources of arachidonic acid. *J Pharmacol Exp Ther* 280: 606–613
14 Grossman CJ, Wiseman J, Lucas FS, Trevethick MA, Birch PJ (1995) Inhibition of constitutive and inducible cyclooxygenase activity in human platelets and mononuclear cells by NSAIDs and Cox 2 inhibitors. *Inflamm Res* 44: 253–257
15 Kargman S, Wong E, Greig GM, Falgueyret JP, Cromlish W, Ethier D, Yergey JA, Riendeau D, Evans JF, Kennedy B et al (1996) Mechanism of selective inhibition of human prostaglandin G/H synthase-1 and -2 in intact cells. *Biochem Pharmacol* 52: 1113–1125
16 Futaki N, Takahashi S, Yokoyama M, Arai I, Higuchi S, Otomo S (1994) NS-398, a new anti-inflammatory agent, selectively inhibits prostaglandin G/H synthase/cyclooxygenase (COX-2) activity *in vitro*. *Prostaglandins* 47: 55–59
17 Kalgutkar AS, Crews BC, Marnett LJ (1996) Kinetics of the interaction of nonsteroidal antiinflammatory drugs with prostaglandin endoperoxide synthase-1 studied by limited proteolysis. *Biochemistry* 35: 9076–9082
18 McKenna F, Arguelles L, Burke T, Lefkowith J, Geis GS (2002) Upper gastrointestinal tolerability of celecoxib compared with diclofenac in the treatment of osteoarthritis and rheumatoid arthritis. *Clin Exp Rheumatol* 20: 35–43
19 Weaver AL (2001) Rofecoxib: clinical pharmacology and clinical experience. *Clin Ther* 23: 1323–1338
20 Ormrod D, Wellington K, Wagstaff AJ (2002) Valdecoxib. *Drugs* 62: 2059–2071
21 Gottesdiener K, Schnitzer T, Fisher C, Bockow B, Markenson J, Ko A, DeTora L, Curtis S, Geissler L, Gertz BJ (2002) Results of a randomized, dose-ranging trial of etoricoxib in patients with osteoarthritis. *Rheumatology (Oxford)* 41: 1052–1061
22 Flower R, Gryglewski R, Herbaczynska-Cedro K, Vane JR (1972) Effects of anti-inflammatory drugs on prostaglandin biosynthesis. *Nat New Biol* 238: 104–106
23 Lombardino JG (1974) In: RA Sherrer, MW Whitehouse (eds): *Antiinflammatory agents, chemistry and pharmacology*. Academic Press, New York, 131–164
24 Frey HH, Nielsen CK (1966) Contributions to the evaluation of antiphlogistic test results. 2. Comparison of the effects of phenylbutazone with various models of inflammation.

Arzneimittelforschung 16: 87–89
25 Brown K, Cavalla JF, Green D, Wilson AB (1968) Diaryloxazole and diarylthiazolealkanoic acids: two novel series of non-steroidal anti-inflammatory agents. *Nature* 219: 164
26 Rosenthale ME, Begany AJ, Dervinis A, Malis JL, Shriver DA, Datko SL, Gluckman MI (1974) Anti-inflammatory properties of 4,5-diphenyl-2-oxazolepropionic acid (oxaprozin). *Agents Actions* 4: 151–159
27 Szmuszkovicz J, Glenn EM, Heinzelman RV, Hester JB Jr, Youngdale GA (1966) Synthesis and antiinflammatory activity of 2,3-bis(p-methoxyphenyl)indole and related compounds. *J Med Chem* 9: 527–536
28 Glenn EM, Bowman BJ, Kooyers W, Koslowske T, Myers ML (1967) The pharmacology of 2,3-bis-(p-methoxyphenyl)-indole (indoxole). *J Pharmacol Exp Ther* 155: 157–166
29 Kaiser DG, Glenn EM, Johnson RH, Johnston RL (1967) Correlation of serum indoxole [2,3-bis(p-methoxyphenyl)-indole] levels with biologic activity. *J Pharmacol Exp Ther* 155: 174–180
30 Talley JJ (1999) Selective inhibitors of cyclooxygenase-2 (COX-2). *Prog Med Chem* 36: 201–234
31 Tanaka A, Sakai H, Motoyama Y, Ishikawa T, Takasugi H (1994) Antiplatelet agents based on cyclooxygenase inhibition without ulcerogenesis. Evaluation and synthesis of 4,5-bis(4-methoxyphenyl)-2-substituted-thiazoles. *J Med Chem* 37: 1189–1199
32 Hsyu PH, Koets MD, Luderer JR (1994) Pharmacokinetics and cyclooxygenase inhibition of itazigrel in normal volunteers after single oral doses. *J Pharm Sci* 83: 1747–1750
33 Demke DM, Luderer JR, Wakefield LK, Euler AR, Brouwer RD, Metzler CM (1987) The effect of itazigrel and aspirin on the mucosa of the esophagus, stomach, and duodenum of normal subjects. *J Clin Pharmacol* 27: 916–920
34 Sharpe TR, Cherkofsky SC, Hewes WE, Smith DH, Gregory WA, Haber SB, Leadbetter MR, Whitney JG (1985) Preparation and antiarthritic and analgesic activity of 4,5-diaryl-2-(substituted thio)-1H-imidazoles and their sulfoxides and sulfones. *J Med Chem* 28: 1188–1194
35 Gans KR, Galbraith W, Roman RJ, Haber SB, Kerr JS, Schmidt WK, Smith C, Hewes WE, Ackerman NR (1990) Anti-inflammatory and safety profile of DuP 697, a novel orally effective prostaglandin synthesis inhibitor. *J Pharmacol Exp Ther* 254: 180–187
36 Leblanc Y, Gauthier JY, Ethier D, Guay J, Mancini J, Riendeau D, Tagari P, Vickers P, Wong E, Prasit P (1995) Synthesis and biological evaluation of 2,3-diarylthiophenes as selective Cox-2 and Cox-1 inhibitors. *Bioorg Med Chem Lett* 5: 2123–2128
37 Bertenshaw SR, Talley JJ, Rogier DJ, Graneto MJ, Rogers RS, Kramer SW, Penning TD, Koboldt CM, Veenhuizen AW, Zhang Y et al (1995) 3,4-diarylthiophenes are selective COX-2 inhibitors. *Bioorg Med Chem Lett* 5: 2919–2922
38 Penning TD, Talley JJ, Bertenshaw SR, Carter JS, Collins PW, Docter S, Graneto MJ, Lee LF, Malecha JW, Miyashiro JM et al (1997) Synthesis and biological evaluation of the 1,5-diarylpyrazole class of cyclooxygenase-2 inhibitors: identification of 4-[5-(4-methylphenyl)-3-(trifluoromethyl)-1H-pyrazol-1-yl]benzenesulfonamide (SC-58635, celecoxib). *J Med Chem* 40: 1347–1365
39 Carter JS (2000) Inhibitors of cyclooxygenase-2. *Exp Opin Ther Pat* 10: 1011–1020
40 Portevin B, Tordjman C, Pastoureau P, Bonnet J, De Nanteuil G (2000) 1,3-Diaryl-4,5,6,7-tetrahydro-2H-isoindole derivatives: a new series of potent and selective COX-2 inhibitors in which a sulfonyl group is not a structural requisite. *J Med Chem* 43: 4582–4593
41 Hashimoto H, Imamura K, Haruta J, Wakitani K (2002) 4-(4-cycloalkyl/aryl-oxazol-5-yl) benzenesulfonamides as selective cyclooxygenase-2 inhibitors: enhancement of the selectivity by introduction of a fluorine atom and identification of a potent, highly selective, and orally active COX-2 inhibitor JTE-522. *J Med Chem* 45: 1511–1517
42 Hashimoto H, Maeda K, Ozawa K, Haruta J, Wakitani K (2002) 4-Aryl/cycloalkyl-5-phenyloxazole derivatives as selective COX-2 inhibitors. *Bioorg Med Chem Lett* 12: 65–68
43 Lau CK, Brideau C, Chan CC, Charleson S, Cromlish WA, Ethier D, Gauthier JY, Gordon R, Guay J, Kargman S et al (1999) Synthesis and biological evaluation of 3-heteroaryloxy-4-phenyl-2(5H)-furanones as selective COX-2 inhibitors. *Bioorg Med Chem Lett* 9: 3187–3192
44 Leblanc Y, Roy P, Boyce S, Brideau C, Chan CC, Charleson S, Gordon R, Grimm E, Guay J, Leger S et al (1999) SAR in the alkoxy lactone series: the discovery of DFP, a potent and orally active COX-2 inhibitor. *Bioorg Med Chem Lett* 9: 2207–2212
45 Leblanc Y, Roy P, Wang Z, Li CS, Chauret N, Nicoll-Griffith DA, Silva JM, Aubin Y, Yergey JA, Chan CC et al (2002) Discovery of a potent and selective COX-2 inhibitor in the alkoxy lactone

series with optimized metabolic profile. *Bioorg Med Chem Lett* 12: 3317–3321

46 Li CS, Black WC, Brideau C, Chan CC, Charleson S, Cromlish WA, Claveau D, Gauthier JY, Gordon R, Greig G et al (1999) A new structural variation on the methanesulfonylphenyl class of selective cyclooxygenase-2 inhibitors. *Bioorg Med Chem Lett* 9: 3181–3186

47 Pinto DJP, Copeland RA, Covington MB, Pitts WJ, Batt DG, Orwat MJ, Lam GN, Joshi A, Chan Y-C, Wang S et al (1996) Chemistry and pharmacokinetics of diarylthiophenes and terphenyls as selective COX-2 inhibitors. *Bioorg Med Chem Lett* 6: 2907–2912

48 Friesen RW, Dubé D, Fortin R, Frenette R, Prescott S, Cromlish W, Greig GM, Kargman S, Wong E, Chan CC et al (1996) Novel 1,2-diarylcyclobutenes: selective and orally active cox-2 inhibitors. *Bioorg Med Chem Lett* 6: 2677–2682

49 Reitz DB, Li JJ, Norton MB, Reinhard EJ, Collins JT, Anderson GD, Gregory SA, Koboldt CM, Perkins WE, Seibert K et al (1994) Selective cyclooxygenase inhibitors: novel 1,2-diarylcyclopentenes are potent and orally active COX-2 inhibitors. *J Med Chem* 37: 3878–3881

50 Li JJ, Anderson GD, Burton EG, Cogburn JN, Collins JT, Garland DJ, Gregory SA, Huang HC, Isakson PC, Koboldt CM et al (1995) 1,2-Diarylcyclopentenes as selective cyclooxygenase-2 inhibitors and orally active anti-inflammatory agents. *J Med Chem* 38: 4570–4578

51 Black WC, Brideau C, Chan CC, Charleson S, Chauret N, Claveau D, Ethier D, Gordon R, Greig G, Guay J et al (1999) 2,3-Diarylcyclopentenones as orally active, highly selective cyclooxygenase-2 inhibitors. *J Med Chem* 42: 1274–1281

52 Reitz DB, Huang HC, Li JJ, Garland DJ, Manning RE, Anderson GD, Gregory SA, Koboldt CM, Perkins WE, Seibert K et al (1995) Selective cyclooxygenase inhibitors: novel 4-spiro 1,2-diarylcyclopentenes are potent and orally active COX-2 inhibitors. *Bioorg Med Chem Lett* 5: 867–872

53 Huang HC, Li JJ, Garland DJ, Chamberlain TS, Reinhard EJ, Manning RE, Seibert K, Koboldt CM, Gregory SA, Anderson GD et al (1996) Diarylspiro[2.4]heptenes as orally active, highly selective cyclooxygenase-2 inhibitors: synthesis and structure-activity relationships. *J Med Chem* 39: 253–266

54 Pinto DJ, Batt DG, Pitts WJ, Petraitis JJ, Orwat MJ, Wang S, Jetter JW, Sherk SR, Houghton GC, Copeland RA et al (1999) Terphenyl cyclooxygenase-2 (COX-2) inhibitors: optimization of the central ring and o-biphenyl analogs. *Bioorg Med Chem Lett* 9: 919–924

55 Li JJ, Norton MB, Reinhard EJ, Anderson GD, Gregory SA, Isakson PC, Koboldt CM, Masferrer JL, Perkins WE, Seibert K et al (1996) Novel terphenyls as selective cyclooxygenase-2 inhibitors and orally active anti-inflammatory agents. *J Med Chem* 39: 1846–1856

56 Atkinson J, Wang Z (1998) Stilbene derivatives useful as cyclooxygenase-2 inhibitors. Merck Frosst Canada Inc US5849943.

57 Nicolai E, Teulon J-M (2000) Preparation of diarylmethylidine furan derivatives as anti-inflammatory, analgesic and chemopreventive agents. Lab UPSA PCT WO 2000014082.

58 Nicolai E, Teulon J-M (1997) Preparation of 3-diarylmethylidene-2-furanones and analogs as cyclooxygenase-2 inhibitors. Lab UPSA PCT WO9737984.

59 Khanna IK, Weier RM, Yu Y, Collins PW, Miyashiro JM, Koboldt CM, Veenhuizen AW, Currie JL, Seibert K, Isakson PC (1997) 1,2-Diarylpyrroles as potent and selective inhibitors of cyclooxygenase-2. *J Med Chem* 40: 1619–1633

60 Wilkerson WW, Galbraith W, Gans-Brangs K, Grubb M, Hewes WE, Jaffee B, Kenney JP, Kerr J, Wong N (1994) Antiinflammatory 4,5-diarylpyrroles: synthesis and QSAR. *J Med Chem* 37: 988–998

61 Wilkerson WW, Copeland RA, Covington M, Trzaskos JM (1995) Antiinflammatory 4,5-diarylpyrroles. 2. Activity as a function of cyclooxygenase-2 inhibition. *J Med Chem* 38: 3895–3901

62 Bertenshaw SR, Collins PW, Penning TD, Reitz DB, Rogers RS (1994) Novel 3,4-diaryl thiophenes and analogs thereof having use as antiinflammatory agents. GD Searle & Co PCT WO9415932.

63 Ducharme Y, Gauthier JY, Prasit P, Leblanc Y, Wang Z, Leger S, Therien M (1995) Phenyl heterocycles as COX-2 inhibitors. Merck Frosst Canada Inc US5474995.

64 Dube D, Brideau C, Deschenes D, Fortin R, Friesen RW, Gordon R, Girard Y, Riendeau D, Savoie C, Chan CC (1999) 2-heterosubstituted-3-(4-methylsulfonyl) phenyl-5-trifluoromethyl pyridines as selective and orally active cyclooxygenase-2 inhibitors. *Bioorg Med Chem Lett* 9: 1715–1720

65 Friesen RW, Brideau C, Chan CC, Charleson S, Deschenes D, Dube D, Ethier D, Fortin R, Gauthier JY, Girard Y et al (1998) 2-Pyridinyl-3-(4-methylsulfonyl) phenylpyridines: selective and orally active cyclooxygenase-2 inhibitors. *Bioorg Med Chem Lett* 8: 2777–2782

66 Riendeau D, Percival MD, Brideau C, Charleson S, Dube D, Ethier D, Falgueyret JP, Friesen RW, Gordon R, Greig G et al (2001) Etoricoxib (MK-0663): preclinical profile and comparison with other agents that selectively inhibit cyclooxygenase-2. *J Pharmacol Exp Ther* 296: 558–566
67 Carter JS, Kramer S, Talley JJ, Penning T, Collins P, Graneto MJ, Seibert K, Koboldt CM, Masferrer J, Zweifel B (1999) Synthesis and activity of sulfonamide-substituted 4,5-diaryl thiazoles as selective cyclooxygenase-2 inhibitors. *Bioorg Med Chem Lett* 9: 1171–1174
68 Carter JS, Rogier DJ, Graneto MJ, Seibert K, Koboldt CM, Zhang Y, Talley JJ (1999) Design and synthesis of sulfonyl-substituted 4,5-diarylthiazoles as selective cyclooxygenase-2 inhibitors. *Bioorg Med Chem Lett* 9: 1167–1170
69 Talley JJ, Bertenshaw SR, Brown DL, Carter JS, Graneto MJ, Koboldt CM, Masferrer JL, Norman BH, Rogier DJ Jr, Zweifel BS et al (1999) 4,5-Diaryloxazole inhibitors of cyclooxygenase-2 (COX-2). *Med Res Rev* 19: 199–208
70 Puig C, Crespo MI, Godessart N, Feixas J, Ibarzo J, Jimenez JM, Soca L, Cardelus I, Heredia A, Miralpeix M et al (2000) Synthesis and biological evaluation of 3,4-diaryloxazolones: A new class of orally active cyclooxygenase-2 inhibitors. *J Med Chem* 43: 214–223
71 Barta TE, Stealey MA, Collins PW, Weier RM (1998) Antiinflammatory 4,5-diarylimidazoles as selective cyclooxygenase inhibitors. *Bioorg Med Chem Lett* 8: 3443–3448
72 Khanna IK, Yu Y, Huff RM, Weier RM, Xu X, Koszyk FJ, Collins PW, Cogburn JN, Isakson PC, Koboldt CM et al (2000) Selective cyclooxygenase-2 inhibitors: heteroaryl modified 1,2-diarylimidazoles are potent, orally active antiinflammatory agents. *J Med Chem* 43: 3168–3185
73 Khanna IK, Weier RM, Yu Y, Xu XD, Koszyk FJ, Collins PW, Koboldt CM, Veenhuizen AW, Perkins WE, Casler JJ et al (1997) 1,2-Diarylimidazoles as potent, cyclooxygenase-2 selective, and orally active antiinflammatory agents. *J Med Chem* 40: 1634–1647
74 Talley JJ, Brown DL, Carter JS, Graneto MJ, Koboldt CM, Masferrer JL, Perkins WE, Rogers RS, Shaffer AF, Zhang YY et al (2000) 4-[5-Methyl-3-phenylisoxazol-4-yl]-benzenesulfonamide, valdecoxib: a potent and selective inhibitor of COX-2. *J Med Chem* 43: 775–777
75 Talley JJ, Bertenshaw SR, Brown DL, Carter JS, Graneto MJ, Kellogg MS, Koboldt CM, Yuan J, Zhang YY, Seibert K (2000) N-[[(5-methyl-3-phenylisoxazol-4-yl)-phenyl]sulfonyl]propanamide, sodium salt, parecoxib sodium: A potent and selective inhibitor of COX-2 for parenteral administration. *J Med Chem* 43: 1661–1663
76 Chavatte P, Yous S, Marot C, Baurin N, Lesieur D (2001) Three-dimensional quantitative structure-activity relationships of cyclo-oxygenase-2 (COX-2) inhibitors: a comparative molecular field analysis. *J Med Chem* 44: 3223–3230
77 Tsuji K, Nakamura K, Konishi N, Tojo T, Ochi T, Senoh H, Matsuo M (1997) Studies on anti-inflammatory agents. IV. Synthesis and pharmacological properties of 1,5-diarylpyrazoles and related derivatives. *Chem Pharm Bull* 45: 987–995
78 Ochi T, Yamane-Sugiyama A, Ohkubo Y, Sakane K, Tanaka H (2001) The anti-inflammatory effect of FR188582, a highly selective inhibitor of cyclooxygenase-2, with an ulcerogenic sparing effect in rats. *Jpn J Pharmacol* 85: 175–182
79 Seibert K, Zhang Y, Leahy K, Hauser S, Masferrer J, Perkins W, Lee L, Isakson P (1994) Pharmacological and biochemical demonstration of the role of cyclooxygenase 2 in inflammation and pain. *Proc Natl Acad Sci USA* 91: 12013–12017
80 Lau CK (1997) Diphenyl-1,2,3-thiadiazoles as anti-inflammatory agents. Merck Frosst Canada Inc US5677318.
81 Prasit P, Wang Z, Brideau C, Chan CC, Charleson S, Cromlish W, Ethier D, Evans JF, Ford-Hutchinson AW, Gauthier JY et al (1999) The discovery of rofecoxib, [MK 966, Vioxx, 4-(4'-methylsulfonylphenyl)-3-phenyl-2(5H)-furanone], an orally active cyclooxygenase-2-inhibitor. *Bioorg Med Chem Lett* 9: 1773–1778
82 Chan CC, Boyce S, Brideau C, Charleson S, Cromlish W, Ethier D, Evans J, Ford-Hutchinson AW, Forrest MJ, Gauthier JY et al (1999) Rofecoxib [Vioxx, MK-0966; 4-(4'-methylsulfonylphenyl)-3-phenyl-2-(5H)-furanone]: a potent and orally active cyclooxygenase-2 inhibitor. Pharmacological and biochemical profiles. *J Pharmacol Exp Ther* 290: 551–560
83 Shin SS, Noh MS, Byun YJ, Choi JK, Kim JY, Lim KM, Ha JY, Kim JK, Lee CH, Chung S (2001) 2,2-Dimethyl-4,5-diaryl-3(2H)furanone derivatives as selective cyclo-oxygenase-2 inhibitors. *Bioorg Med Chem Lett* 11: 165–168
84 Riendeau D, Percival MD, Boyce S, Brideau C, Charleson S, Cromlish W, Ethier D, Evans J, Falgueyret JP, Ford-Hutchinson AW et al (1997) Biochemical and pharmacological profile of a tetrasubstituted furanone as a highly selective COX-2 inhibitor. *Br J Pharmacol* 121: 105–117

85 Yergey JA, Trimble LA, Silva J, Chauret N, Li C, Therien M, Grimm E, Nicoll-Griffith DA (2001) *In vitro* metabolism of the COX-2 inhibitor DFU, including a novel glutathione adduct rearomatization. *Drug Metab Dispos* 29: 638–644
86 Therien M, Brideau C, Chan CC, Cromlish WA, Gauthier JY, Gordon R, Greig G, Kargman S, Lau CK, Leblanc Y et al (1997) Synthesis and biological evaluation of 5,6-diarylimidazo[2.1-b]thiazole as selective COX-2 inhibitors. *Bioorg Med Chem Lett* 7: 47–52
87 Roy P, LeBlanc Y, Ball RG, Brideau C, Chan CC, Chauret N, Cromlish W, Ethier D, Gauthier JY, Gordon R et al (1997) A new series of selective COX-2 inhibitors: 5,6-diarylthiazolo[3,2-b][1,2,4]triazoles. *Bioorg Med Chem Lett* 7: 57–62
88 Almansa C, de Arriba AF, Cavalcanti FL, Gomez LA, Miralles A, Merlos M, Garcia-Rafanell J, Forn J (2001) Synthesis and SAR of a new series of COX-2-selective inhibitors: pyrazolo[1,5-a]pyrimidines. *J Med Chem* 44: 350–361
89 Farrerons Gallemi C, Miquel Bono I-J, Fernandez Serrat AM, Monserrat Vidal C, Lagunas Arnal C, Gimenez Guasch F, Fernandez Garcia A (2000) Substituted imidazo[1,2a]azines as selective inhibitors of COX-2. Lab SALVAT SA PCT WO200008024.
90 Talley JJ, Bartenshaw SR, Graneto MJ, Rogier DJ (1996) Preparation of cytokine-inhibiting benzopyranopyrazole antiinflammatories and analgesics. GD Searle & Co PCT WO9609304.
91 Talley JJ, Bertenshaw SR, Rogers RS (1997) Benz[G]indazolyl derivatives for the treatment of inflammation. GD Searle & Co US5696143.
92 Harrington J, Robertson JE, Kvam DC, Hamilton RR, McGurran KT, Trancik RJ, Swingle KF, Moore GG, Gerster JF (1970) Antiinflammatory agents. I. 3-Benzoylfluoroalkanesulfonamides. *J Med Chem* 13: 137
93 Swingle KF, Moore GG, Grant TJ (1976) 4-Nitro-2-phenoxymethanesulfonanilide (R-805): a chemically novel anti-inflammatory agent. *Arch Int Pharmacodyn Ther* 221: 132–139
94 Davis R, Brogden RN (1994) Nimesulide. An update of its pharmacodynamic and pharmacokinetic properties, and therapeutic efficacy. *Drugs* 48: 431–454
95 Masferrer JL, Zweifel BS, Manning PT, Hauser SD, Leahy KM, Smith WG, Isakson PC, Seibert K (1994) Selective inhibition of inducible cyclooxygenase 2 *in vivo* is antiinflammatory and nonulcerogenic. *Proc Natl Acad Sci USA* 91: 3228–3232
96 Futaki N, Takahashi S, Kitagawa T, Yamakawa Y, Tanaka M, Higuchi S (1997) Selective inhibition of cyclooxygenase-2 by NS-398 in endotoxin shock rats *in vivo*. *Inflamm Res* 46: 496–502
97 Wiesenberg-Bottcher I, Schweizer A, Green JR, Seltenmeyer Y, Muller K (1989) The pharmacological profile of CGP 28238, a highly potent anti-inflammatory compound. *Agents Actions* 26: 240–242
98 Vago T, Bevilacqua M, Norbiato G (1995) Effect of nimesulide action time dependence on selectivity towards prostaglandin G/H synthase/cyclooxygenase activity. *Arzneimittelforschung* 45: 1096–1098
99 Li CS, Black WC, Chan CC, Ford-Hutchinson AW, Gauthier JY, Gordon R, Guay D, Kargman S, Lau CK, Mancini J et al (1995) Cyclooxygenase-2 inhibitors. Synthesis and pharmacological activities of 5-methanesulfonamido-1-indanone derivatives. *J Med Chem* 38: 4897–4905
100 Ouimet N, Chan CC, Charleson S, Claveau D, Gordon R, Guay D, Li CS, Ouellet M, Percival DM, Riendeau D et al (1999) Substituted heterocyclic analogs as selective COX-2 inhibitors in the flosulide class. *Bioorg Med Chem Lett* 9: 151–156
101 Guay D, Li C-S, Quimet N (1996) Preparation of 5-methanesulfonamido-3H-isobenzofuran-1-one antiinflammatory inhibitors of cyclooxygenase-2. Merck Frosst Canada Inc PCT WO9623786.
102 Tsuji K, Nakamura K, Konishi N, Okumura H, Matsuo M (1992) Studies on antiinflammatory agents. I. Synthesis and pharmacological properties of 2'-phenoxymethanesulfonanilide derivatives. *Chem Pharm Bull* 40: 2399–2409
103 Tanaka K, Shimotori T, Makino S, Aikawa Y, Inaba T, Yoshida C, Takano S (1992) Pharmacological studies of the new antiinflammatory agent 3-formylamino-7-methylsulfonylamino-6-phenoxy-4H-1-benzopyran-4-one. 1st communication: antiinflammatory, analgesic and other related properties. *Arzneimittelforschung* 42: 935–944
104 Berg J, Fellier H, Christoph T, Kremminger P, Hartmann M, Blaschke H, Rovensky F, Towart R, Stimmeder D (2000) Pharmacology of a selective cyclooxygenase-2 inhibitor, HN-56249: a novel compound exhibiting a marked preference for the human enzyme in intact cells. *Naunyn-Schmiedeberg's Archives of Pharmacology* 361:363-372.
105 Kawasaki H, Tanaka K (1996) Cyclooxygenase 2 inhibitors containing phenoxybenzopyranone

derivative. Toyama Kagaku Kogyo KK JP8157361.
106 Park CH, Siomboing X, Yous S, Gressier B, Luyckx M, Chavatte P (2002) Investigations of new lead structures for the design of novel cyclooxygenase-2 inhibitors. *Eur J Med Chem* 37: 461–468
107 Black WC, Bayly C, Belley M, Chan CC, Charleson S, Denis D, Gauthier JY, Gordon R, Guay D, Kargman S et al (1996) From indomethacin to a selective COX-2 inhibitor: Development of indolalkanoic acids as potent and selective cyclooxygenase-2 inhibitors. *Bioorg Med Chem Lett* 6: 725–730
108 Lau CK, Black WC, Belley M, Chan CC, Charleson S, Denis D, Gauthier JY, Gordon R, Guay D, Hamel P et al (1997) From indomethacin to a selective COX-2 inhibitor. Development of indolalkanoic acids as potent and selective cyclooxygenase-2 inhibitors. *Adv Exp Med Biol* 407: 73–78
109 Kalgutkar AS, Crews BC, Rowlinson SW, Marnett AB, Kozak KR, Remmel RP, Marnett LJ (2000) Biochemically based design of cyclooxygenase-2 (COX-2) inhibitors: facile conversion of nonsteroidal antiinflammatory drugs to potent and highly selective COX-2 inhibitors. *Proc Natl Acad Sci USA* 97: 925–930
110 Kalgutkar AS, Marnett AB, Crews BC, Remmel RP, Marnett LJ (2000) Ester and amide derivatives of the nonsteroidal antiinflammatory drug, indomethacin, as selective cyclooxygenase-2 inhibitors. *J Med Chem* 43: 2860–2870
111 Kozak KR, Prusakiewicz JJ, Rowlinson SW, Marnett LJ (2002) Enantiospecific, selective cyclooxygenase-2 inhibitors. *Bioorg Med Chem Lett* 12: 1315–1318
112 Kalgutkar AS, Marnett LJ (2001) Esters derived from indolealkanols and novel amides derived from indolealkylamides that are selective COX-2 inhibitors. Vanderbilt University US6306890.
113 Woods KW, McCroskey RW, Michaelides MR (1998) Heterocyclic compounds as COX-2 inhibitors. Abbott Laboratories US5811425.
114 Woods KW, McCroskey RW, Michaelides MR, Wada CK, Hulkower KI, Bell RL (2001) Thiazole analogues of the NSAID indomethacin as selective COX-2 inhibitors. *Bioorg Med Chem Lett* 11: 1325–1328
115 Palomer A, Cabre F, Pascual J, Campos J, Trujillo MA, Entrena A, Gallo MA, Garcia L, Mauleon D, Espinosa A (2002) Identification of novel cyclooxygenase-2 selective inhibitors using pharmacophore models. *J Med Chem* 45: 1402–1411
116 Matsuoka H, Kato N, Takahashi T, Maruyama N, Ishizawa T, Suzuki Y (1999) Preparation of heterocyclic indole derivatives and mono- or diazaindole derivatives as cyclooxygenase-2 (COX-2) inhibitors. Chugai Seiyaku Kabushiki Kaisha PCT WO9961436.
117 Matsuoka H, Maruyama N, Kashiwagi H (1998) Preparation of indole, monoazaindole, and diazaindole derivatives as cyclooxygenase inhibitors. Chugai Seiyaku Kabushiki Kaisha (1998) PCT WO9851667.
118 Matsuoka H, Maruyama N, Kato Y (1998) Preparation of indene derivatives as COX 2 inhibitors. Chugai Seiyaku Kabushiki Kaisha PCT WO9857924.
119 Nakao K, Stevens RW, Kawamura K, Uchida C, Koike H, Caron S (1999) Preparation of 2,3-substituted indoles as COX-2 inhibitors. Pfizer Pharmaceuticals, Japan PCT WO9935130.
120 Failli AA (1999) Indene inhibitors of COX-2. American Home Products Corp US5869524.
121 Barnett JW, Dunn JP, Kertesz DJ, Miller AB, Morgans DJ Jr, Ramesha CS, Sigal CE, Sjogren EB, Smith DB, Talamas FX (1996) Pyrrole derivatives. F Hoffmann-La Roche AG EP-714895.
122 Luong C, Miller A, Barnett J, Chow J, Ramesha C, Browner MF (1996) Flexibility of the NSAID binding site in the structure of human cyclooxygenase-2. *Nature Struct Biol* 3: 927–933
123 Brooks CDW, Craig RA, Kolasa T, Stewart AO (1999) Iminoxyderivatives of fenamates as inhibitors of prostaglandin biosynthesis. Abbott Laboratories US5863946.
124 Fujimoto RA, Mcquire LW, Mugrage BB, Van Duzer JH, Xu D (1999) Preparation of 5-alkyl-2-arylaminophenylacetic acids as COX-2 cyclooxygenase inhibitors. Novartis AG PCT WO9911605.
125 Sorbera LA, Castaner J, Bayes M, Silvestre JS (2002) Lumiracoxib. *Drugs of the Future* 27: 740–747
126 Glaser K, Sung ML, O'Neill K, Belfast M, Hartman D, Carlson R, Kreft A, Kubrak D, Hsiao CL, Weichman B (1995) Etodolac selectively inhibits human prostaglandin G/H synthase 2 (PGHS-2) *versus* human PGHS-1. *Eur J Pharmacol* 281: 107–111
127 Kreft AF, Caufield CE, Failli AA, Caggiano TJ, Greenfield AA, Kubrak DM (1998) Pyranoindole inhibitors of COX-2. American Home Products Corp US5776967.

128 Kreft AF, Caufield CE, Failli AA, Caggiano TJ, Greenfield AA, Kubrak DM (1998) Carbazole inhibitors of COX-2. American Home Products Corp US5824699.
129 Kalgutkar AS, Rowlinson SW, Crews BC, Marnett LJ (2002) Amide derivatives of meclofenamic acid as selective cyclooxygenase-2 inhibitors. *Bioorg Med Chem Lett* 12: 521–524
130 Bayly CI, Black WC, Leger S, Quimet N, Quellet M, Percival MD (1999) Structure-based design of COX-2 selectivity into flurbiprofen. *Bioorg Med Chem Lett* 9: 307–312
131 Ogino K, Hatanaka K, Kawamura M, Katori M, Harada Y (1997) Evaluation of pharmacological profile of meloxicam as an anti-inflammatory agent, with particular reference to its relative selectivity for cyclooxygenase-2 over cyclooxygenase-1. *Pharmacology* 55: 44–53
132 Lazer ES, Sorcek R, Cywin CL, Thome D, Possanza GJ, Graham AG, Churchill L (1998) Antiinflammatory 2-benzyl-4-sulfonyl-4H-isoquinoline-1,3-diones: novel inhibitors of COX-2. *Bioorg Med Chem Lett* 8: 1181–1186
133 Lazer ES, Miao CK, Cywin CL, Sorcek R, Wong HC, Meng Z, Potocki I, Hoermann M, Snow RJ, Tschantz MA et al (1997) Effect of structural modification of enol-carboxamide-type nonsteroidal antiinflammatory drugs on COX-2/COX-1 selectivity. *J Med Chem* 40: 980–989
134 Lecomte M, Laneuville O, Ji C, DeWitt DL, Smith WL (1994) Acetylation of human prostaglandin endoperoxide synthase-2 (cyclooxygenase-2) by aspirin. *J Biol Chem* 269: 13207–13215
135 Roth GJ, Stanford N, Majerus PW (1975) Acetylation of prostaglandin synthase by aspirin. *Proc Natl Acad Sci USA* 72: 3073–3076
136 Kalgutkar AS, Crews BC, Rowlinson SW, Garner C, Seibert K, Marnett LJ (1998) Aspirin-like molecules that covalently inactivate cyclooxygenase-2. *Science* 280: 1268–1270
137 Kalgutkar AS, Kozak KR, Crews BC, Hochgesang GP, Marnett LJ (1998) Covalent modification of cyclooxygenase-2 (COX-2) by 2-acetoxyphenyl alkyl sulfides, a new class of selective COX-2 inactivators. *J Med Chem* 41: 4800–4818
138 Hochgesang GP, Nemeth-Cawley JF, Rowlinson SW, Caprioli RM, Marnett LJ (2003) Functional analysis of the molecular determinants of cyclooxygenase-2 acetylation by 2-acetoxyphenyl-hept-2-ynyl sulfide. *Arch Biochem Biophys* 409: 127–133
139 Hochgesang GP, Rowlinson SW, Marnett LJ (2000) Tyrosine 385 is critical for acetylation of cyclooxygenase-2 by aspirin. *J Am Chem Soc* 122: 6514–6515
140 Mullican MD, Wilson MW, Connor DT, Kostlan CR, Schrier DJ, Dyer RD (1993) Design of 5-(3,5-di-tert-butyl-4-hydroxyphenyl)-1,3,4-thiadiazoles, -1,3,4-oxadiazoles, and -1,2,4-triazoles as orally-active, nonulcerogenic antiinflammatory agents. *J Med Chem* 36: 1090–1099
141 Song Y, Connor DT, Doubleday R, Sorenson RJ, Sercel AD, Unangst PC, Roth BD, Gilbertsen RB, Chan K, Schrier DJ et al (1999) Synthesis, structure-activity relationships, and *in vivo* evaluations of substituted di-tert-butylphenols as a novel class of potent, selective, and orally active cyclooxygenase-2 inhibitors. 1. Thiazolone and oxazolone series. *J Med Chem* 42: 1151–1160
142 Song Y, Connor DT, Sercel AD, Sorenson RJ, Doubleday R, Unangst PC, Roth BD, Beylin VG, Gilbertsen RB, Chan K et al (1999) Synthesis, structure-activity relationships, and *in vivo* evaluations of substituted di-tert-butylphenols as a novel class of potent, selective, and orally active cyclooxygenase-2 inhibitors. 2. 1,3,4- and 1,2,4-thiadiazole series. *J Med Chem* 42: 1161–1169
143 Inagaki M, Tsuri T, Jyoyama H, Ono T, Yamada K, Kobayashi M, Hori Y, Arimura A, Yasui K, Ohno K et al (2000) Novel antiarthritic agents with 1,2-isothiazolidine-1,1-dioxide (gamma-sultam) skeleton: cytokine suppressive dual inhibitors of cyclooxygenase-2 and 5-lipoxygenase. *J Med Chem* 43: 2040–2048
144 Janusz JM, Young PA, Ridgeway JM, Scherz MW, Enzweiler K, Wu LI, Gan L, Darolia R, Matthews RS, Hennes D et al (1998) New cyclooxygenase-2/5-lipoxygenase inhibitors. 1. 7-tert-butyl-2,3-dihydro-3,3-dimethylbenzofuran derivatives as gastrointestinal safe antiinflammatory and analgesic agents: discovery and variation of the 5-keto substituent. *J Med Chem* 41: 1112–1123
145 Janusz JM, Young PA, Scherz MW, Enzweiler K, Wu LI, Gan L, Pikul S, McDow-Dunham KL, Johnson CR, Senanayake CB et al (1998) New cyclooxygenase-2/5-lipoxygenase inhibitors. 2. 7-tert-butyl-2,3-dihydro-3,3-dimethylbenzofuran derivatives as gastrointestinal safe antiinflammatory and analgesic agents: variations of the dihydrobenzofuran ring. *J Med Chem* 41: 1124–1137
146 Janusz JM, Young PA, Ridgeway JM, Scherz MW, Enzweiler K, Wu LI, Gan L, Chen J, Kellstein DE, Green SA et al (1998) New cyclooxygenase-2/5-lipoxygenase inhibitors. 3. 7-tert-butyl-2,3-dihydro-3,3-dimethylbenzofuran derivatives as gastrointestinal safe antiinflammatory and anal-

gesic agents: variations at the 5 position. *J Med Chem* 41: 3515–3529
147 Johnson AR, Marletta MA, Dyer RD (2001) Slow-binding inhibition of human prostaglandin endoperoxide synthase-2 with darbufelone, an isoform-selective antiinflammatory di-tert-butyl phenol. *Biochemistry* 40: 7736–7745
148 Carter JS, Obukowicz MG, Devadas B, Talley JJ, Brown DL, Graneto MJ, Bertenshaw SR, Rogier DJ Jr, Nagarajan SR, Hanau CE, Hartmann SJ, Ludwig CL, Metz S (1998) Substituted benzopyran derivatives for the treatment of inflammation. G D Searle & Co PCT WO9847890.
149 Kurumbail RG, Kiefer JR, Marnett LJ (2001) Cyclooxygenase enzymes: catalysis and inhibition. *Curr Op Struct Biol* 11: 752–760
150 Rome LH, Lands WE (1975) Structural requirements for time-dependent inhibition of prostaglandin biosynthesis by anti-inflammatory drugs. *Proc Natl Acad Sci USA* 72: 4863–4865
151 Walker MC, Kurumbail RG, Kiefer JR, Moreland KT, Koboldt CM, Isakson PC, Seibert K, Gierse JK (2001) A three-step kinetic mechanism for selective inhibition of cyclo-oxygenase-2 by diarylheterocyclic inhibitors. *Biochem J* 357: 709–718
152 Picot D, Loll PJ, Garavito RM (1994) The X-ray crystal structure of the membrane protein prostaglandin H2 synthase-1. *Nature* 367: 243–249
153 Kurumbail RG, Stevens AM, Gierse JK, McDonald JJ, Stegeman RA, Pak JY, Gildehaus D, Miyashiro JM, Penning TD, Seibert K et al (1996) Structural basis for selective inhibition of cyclooxygenase-2 by anti-inflammatory agents. *Nature* 384: 644–648
154 Loll PJ, Picot D, Ekabo O, Garavito RM (1996) Synthesis and use of iodinated nonsteroidal anti-inflammatory drug analogs as crystallographic probes of the prostaglandin H2 synthase cyclooxygenase active site. *Biochemistry* 35: 7330–7340
155 Selinsky BS, Gupta K, Sharkey CT, Loll PJ (2001) Structural analysis of NSAID binding by prostaglandin H2 synthase: time-dependent and time-independent inhibitors elicit identical enzyme conformations. *Biochemistry* 40: 5172–5180
156 Copeland RA, Williams JM, Giannaras J, Nurnberg S, Covington M, Pinto D, Pick S, Trzaskos JM (1996) Mechanism of selective inhibition of the inducible isoform of prostaglandin G/H synthase. *Proc Natl Acad Sci USA* 91: 11202–11206
157 Gierse JK, Koboldt CM, Walker MC, Seibert K, Isakson PC (1999) Kinetic basis for selective inhibition of cyclo-oxygenases. *Biochem J* 339: 607–614
158 Lanzo CA, Sutin J, Rowlinson S, Talley J, Marnett LJ (2000) Fluorescence quenching analysis of the association and dissociation of a diarylheterocycle to cyclooxygenase-1 and cyclooxygenase-2: dynamic basis of cyclooxygenase-2 selectivity. *Biochemistry* 39: 6228–6234
159 Timofeevski SL, Prusakiewicz JJ, Rouzer CA, Marnett LJ (2002) Isoform-selective interaction of cyclooxygenase-2 with indomethacin amides studied by real-time fluorescence, inhibition kinetics, and site-directed mutagenesis. *Biochemistry* 41: 9654–9662
160 Kiefer JR, Pawlitz JL, Moreland KT, Stegeman RA, Hood WF, Gierse JK, Stevens AM, Goodwin DC, Rowlinson SW, Marnett LJ et al (2000) Structural insights into the stereochemistry of the cyclooxygenase reaction. *Nature* 405: 97–101

COX-2 Inhibitors
Edited by M. Pairet and J. van Ryn

The molecular and biological basis for COX-2 selectivity

Guenter Trummlitz[1], Joanne van Ryn[1] and Timothy D. Warner[2]

[1] *Boehringer Ingelheim Pharma GmbH & Co. KG, Birkendorfer Str. 65, D-88397 Biberach, Germany*
[2] *TheWilliam Harvey Research Institute, Bart's and the London Queen Mary's School of Medicine and Dentistry, London EC1M 6BQ, United Kingdom*

Introduction

Before the development of the first three-dimensional structure of cyclooxygenase (COX) [1], it was difficult to understand why very chemically diverse inhibitors acted via similar mechanisms at this same enzyme. The information from X-ray analyses and from mutagenesis experiments now allow us to use molecular modelling approaches to gain insight into the molecular mechanism of selective COX-2 inhibition by at least eight different structural classes of inhibitors. The particular advantage of molecular modelling approaches is that they allow us to visualize and rationalize the recognition process of substrates and inhibitors by COX isoenzymes [2].

In addition to the differential binding of NSAIDs the realisation by three independent groups that two COX isozymes exist (see Chapter 1 for details) further progressed understanding of the different efficacy/side-effect profiles of different NSAIDs. COX-1/-2 selectivity measurements have since become standard methods to determine the *in vitro* COX-2 selectivity of already known compounds, and to assist in the discovery of new compounds. Additionally, it is possible to correlate the *in vitro* COX-2 selectivity data with information from *in vivo* studies and so to build a continuum from molecular biology through to the clinic.

COX genes

Prostaglandin biosynthesis and the existence of COX isoenzymes have been well characterised in vertebrates. In mammals at least one COX isoenzyme is found in most cell types and organs. Selective synthesis of prostaglandins by one COX isoenzyme over the other has been observed in cells that express both isoforms. This indicates that the COX-1 and COX-2 pathways are segregated within cells permitting products derived from the two pathways to be channelled to different downstream signalling targets. This appears to be explained both by particular co-localisation of COX isoenzymes and prostanoid synthetic enzymes and by the preferential utilisation of low concentrations of free arachidonic acid by COX-2 [3]. Prostaglandins formed following the actions of COX-1 and COX-2 largely signal through G-protein linked receptors present on the cell surface. In addition, prostanoids, particularly COX-2 derived products, may signal through nuclear receptors. For example, prostacyclin can activate nuclear PPAR-δ receptors and so promote apoptosis [4]. Further refinement of these differences in production is supplied by the tissue differential expression of the COX-1 and COX-2 genes.

COX-1 is constitutively expressed in most tissues and produces prostaglandins to regulate physiological processes in response to hormonal and other stimuli. In contrast, COX-2 expression is highly regulated and increases in response to pathological stresses, such as inflammation. It must be noted, however, that this distinction is now appreciated to be somewhat blurred as it is clear that in certain tissues, e.g., the kidney and brain, COX-2 is constitutively expressed. Consistent with the general idea of COX-1 being constitutive and COX-2 inducible are the different natures of the COX-1 and COX-2 genes. The COX-1 gene, which is 22 kB in size and located on chromosome 9 in humans, has few binding sites for inducible transcription factors. The COX-2 gene in comparison is a so-called immediate-early gene – 8.3 kB in size and located on chromosome 1 – the expression of which is tightly controlled. The COX-2 gene is responsive to many regulators from various signalling pathways, such as growth factors and mediators of inflammation such as IL-1, TNF-α and LPS [3, 5].

COX enzyme structures

COX-1 and COX-2 are monotropic enzmyes bound on the membranes of the inner surface of the endoplasmic reticulum and on the inner and outer surface of the nuclear envelope. Both isoenzymes are homodimer proteins with similar structure and catalytic activity. The amino acid sequences of COX-1 and COX-2 from numerous mammalian, bird and fish species are known. COX-1 and COX-2 amino acid sequences are 60 to 65% identical and each isoform is conserved by 85% and 65% within mammalian and vertebrate species, respectively.

Both isoforms contain signal peptides of varying lengths, 22–26 amino acids in COX-1 and 16–20 in COX-2. After cleavage of the signal peptides the mature COX-1 and COX-2 proteins contain 576 amino acids and 587 amino acids, respectively in mammalians. Major structural differences between these two isoforms occur: (a) in the EGF-like domain, which is shorter in COX-2; (b) in the membrane binding domain of COX-2, in which a proline is inserted after Tyr106 and (c) in the catalytic domain of COX-2, in which 18 amino acids are inserted close to the C-terminus (Fig. 1). The high degree of sequence identity between the mature isoforms and between species allows one-to-one

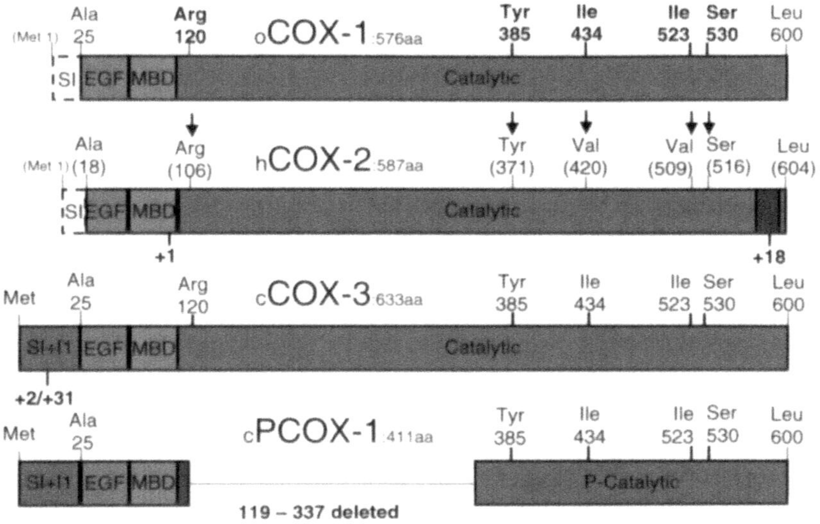

Figure 1. Amino acid sequences of ovine(o) COX-1, human(h) COX-2, canine(c) COX-3 and cPCOX-1a. SI = signal peptides, I1 = Intron 1, EGF = Epidermal growth factor-like domain, MBD = membrane binding domain. For cross-species and cross-isoform comparisons the numbering of oCOX-1 is used.

comparisons of the known sequences to be made. To aid in these cross-isoform and cross-species comparisons, the common consensus is to use the ovine COX-1 sequence numbering as the reference [5].

Recently it has been reported that there could be additional variant protein products from the COX-1 gene [6]. COX-3 and PCOX-1a (partial COX-1a), are derived from the COX-1 gene, but their mRNAs retain intron 1 yielding an insertion of 30–34 amino acids, depending on the species, into the signal peptide, which is not cleaved off. PCOX-1 proteins additionally contain a deletion of exons 5–8 of the COX-1 mRNA resulting in deletion of amino acids 119–337 in the protein (Fig. 1). The initial report also showed that COX-3, but not PCOX-1a, possesses COX activity. Most recently, however, it has been shown that the human COX-1 mRNA is such that COX-3 could not be transcribed [7].

The three-dimensional structure of ovine(o) COX-1 has been resolved by X-ray analyses in two crystal modifications in the presence of four different substrates and nine inhibitors. In addition, three other groups have used recombinant human(h) and murine(m) COX-2 complexed with eight different ligands to make structural determinations of the binding modes of COX-2 inhibitors (Tab. 1, [1, 8–20]).

Given the high degree of amino acid conservation between the two isoenzymes, it is not surprising that the structures of COX-1 and COX-2 are very similar. COX-1 and COX-2 crystallise as dimers, with each monomer consisting of three independent folding units: (1) an EGF-like region that forms a portion of

Table 1. X-ray structures of COX isoenzyme/ligand complexes and critical interaction sites

Ligand	COX	Reference	Critical interaction sites
Unliganded:	mCOX-2	Kiefer [8]	
Substrates:			Arachidonic acid:
Arachidonic acid	oCOX-1	Malkowski [9]	Tyr385: 13-pro-S-hydrogen abstraction
	mCOX-2	Kiefer [8]	Tyr348, Gly533: important positioning
	oCOX-1	Thuresson [10]	Arg120: ionic binding COX-1
Dihomo-γ-linolenic acid	oCOX-1	Malkowski [11]	Val349, Trp387, Leu534: steric position
Eicosapentaenoic acid	oCOX-1	Malkowski [11]	11 amino acids: catalysis optimization
Linoleic acid			
Salicylic acids:			Bromoaspirin:
Bromo-aspirin	oCOX-1	Loll [12]	Ser530: acetylation site
O-Acetylsalicylhydroxamic acid	oCOX-1	Loll [13]	Arg120: ionic binding to carboxylate
Arylacetic acids:			Indomethacin:
Indomethacin	mCOX-2	Kurumbail [14]	Arg120: ionic binding
Iodo-indomethacin	oCOX-1	Loll [15]	Ser530: hydrogen bonding
L-758048 (Indomethacin-analog)	hCOX-2	Merck & Co [16]	Leu384: steric interaction
Alclofenac	oCOX-1	Selinsky [17]	RS-104897:
RS-57067 (Zomepirac-analog)	hCOX-2	Luong [18]	Ser530: hydrogen bond
RS-104897 (Zomepirac-analog)	hCOX-2	Luong [18]	Arg120, Tyr355, (Gln524): hydrogen bonding

(Continued on next page)

Table 1. (Continued)

Ligand	COX	Reference	Critical interaction sites
Arylpropionic acids:			Flurbiprofen:
Flurbiprofen	oCOX-1	Picot [1]; Garavito [19]	Arg120: ionic + hydrogen interaction
	mCOX-2	Kurumbail [14]	Tyr355: hydrogen bond
Methylflurbiprofen	oCOX-1	Selinsky [17]	Arg120,Glu524,Tyr355,His90,H_2O:
Ibuprofen	oCOX-1	Selinsky [17]	ionic + hydrogen network
Iodosuprofen	oCOX-1	Loll [18]	Ser530: contact to distal phenyl
Enolcarboxamides:			
Meloxicam	oCOX-1	Trummlitz [20]	Phe518: steric interaction with 5'-methyl
			Arg120, Tyr355,Ser530:hydrogen bonds
Diarylheterocycles:			
SC-558 (Celecoxib-analog)	mCOX-2	Kurumbail [14]	Val523, Phe518: access to side pocket,
			Arg513, His90: hydrogen bond to sulfonamide)

the dimer interface and has a role in protein folding; (2) a membrane binding domain that is incorporated into the inner side of the membrane bilayer and (3) the catalytic domain (Fig. 2). The hydrophobic amino acid residues at the membrane binding domain are associated with the membrane monolayers, thereby allowing direct access from the interior of the bilayer to the active site or "channel" of COX. In this way the enzyme substrate, arachidonic acid, enters the COX channel and this is also the route of entry for competitive enzyme inhibitors. The catalytic domain contains both the COX and the peroxidase active sites. In order to pass into the upper part of the COX channel, substrate and inhibitors need to pass the constriction formed by Arg120 and Tyr355. The residue Tyr385 is located at the top of the channel and represents the active centre.

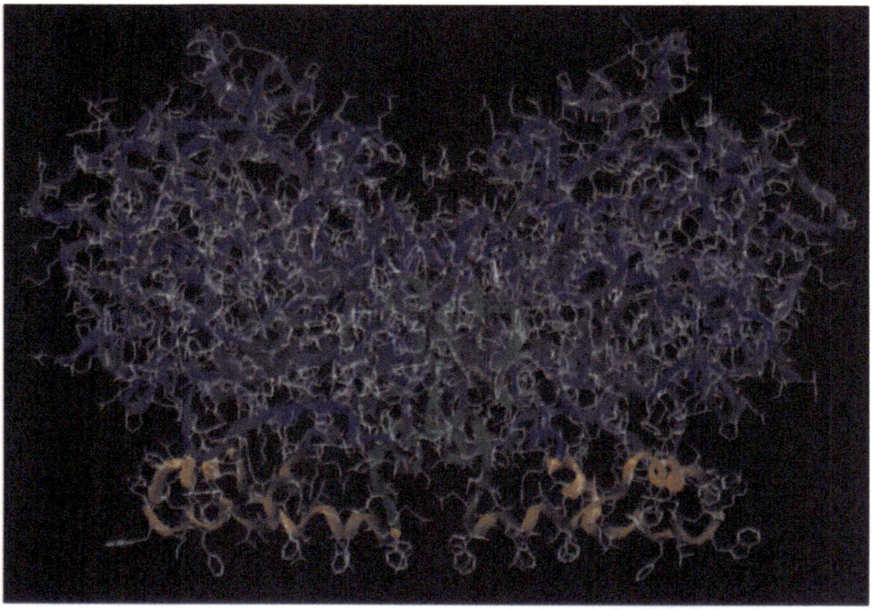

Figure 2. Three dimensional structure of the oCOX-1 dimer. Ribbon representation of the amino acid backbone is used. EGF-like domain: green; MBD: orange; catalytic domain: blue. Amino acid residues are represented as stick models. The coordinates are taken from Malkowski et al. [9].

Substrate binding

Recently, Malkowski et al. prepared a native-like, but functionally inert, Co^{3+} protoporphyrin reconstituted oCOX-1 [21] and determined the structure of arachidonic acid bound within the active site of COX [9]. Arachidonic acid is bound in an extended L-shaped conformation and makes two hydrophilic and a total of 48 hydrophobic contacts with the enzyme, involving 19 different amino acids. The interaction of arachidonic acid with the COX channel has been visualized (Fig. 3a) using an algorithm that represents channel and cavity surfaces by colour coding local hydrophobicity [2]. The carboxylate part of arachidonic acid binds with Arg 120 with the carbon 13 being oriented near the Tyr385, where the pro-S-hydrogen can be abstracted in order to initiate the COX reaction. Additionally, the binding to COX of other substrates has been studied (Tab. 1) and it has been shown, for instance, that the selective oxidation of 2-arachidonylglycerol by COX-2 has as a critical determinant interaction with Arg513 [22–24].

Selective COX-2 inhibitors

NSAIDs display a high degree of structural diversity (Tab. 2) and numerous structural classes with a wide range of physicochemical properties were already available when COX-2 was discovered [25]. To date, all newly discovered selective COX-2 inhibitors belong to one of these classes. Interestingly very slight modifications to a compound can drastically change its selectivity for one of the COX isoenzymes (see Chapter 2 for details). X-ray structure analyses (Tab. 1), assays of binding to COX mutants (Tab. 3, [26–42]) and molecular modelling of COX-2 inhibitors have supplied us with a tremendous amount of knowledge that allow us to rationalise this selectivity based upon the molecular features of the selective binding in each structural class. (See reviews: [2, 3, 5, 43]).

Aspirin is the only marketed COX inhibitor that covalently modifies the enzyme. Detailed structural data taken from the X-ray structure of the bromo-aspirin/COX-1 adduct [12] show that aspirin acetylates Ser530. When COX-2 is acetylated by aspirin arachidonic acid can still be oxidised but 15*R*-HPETE is produced [26, 44]. Conversely, when Ser530 is acetylated in COX-1 the enzyme is completely inactivated. Structural alterations arising from mutations of Ser 530 confirmed the critical role of this amino acid (Tab. 3).

Replacement of the carboxylate moiety of aspirin with alkyl- and alkynyl-sulfide moieties results in COX-2 selective, irreversible inhibitors [30]. Mutational experiments revealed that the lead compound of this group, 2-(acetoxyphenyl)hept-2-ynyl sulphide (APHS), also acetylates Ser530, but does not interact with the side pocket (Val523, Arg513). Mutations that introduce steric

---

Figure 3a–d. (see pp 48, 49) The inside surface of the active side (COX channel) has been identified using the channel surface calculation of MOLCAD (Tripos Inc), and the local hydrophobicity is mapped on this surface (from brown: lipophilic areas to green-blue for hydrophilic areas). The COX coordinates used for modelling are taken from the PDB-files 1DIH, 6COX, 1PRH. For clarity, only the most important amino acid residues are shown as stick models and labelled: catalytic side: Tyr365, ASA acetylation side Ser530, hydrogen network at the gate: Arg120 and Tyr355, side pocket: Val/Ile523 and Arg/His513, flexible extra space: Phe518 and Val/Ile 434, extra space: Leu384 and Leu/Phe503. Arachidonic acid and inhibitors are shown as stick models with white carbon atoms and colored heteroatoms. (a) Arachidonic acid bound in the COX-1 channel, H-abstraction at C13 in yellow. (b) Flurbiprofen bound to the centre of the COX-1 channel. (c) Meloxicam bound at COX-2; the 5'-methyl group of meloxicam occupies the flexible extra space at the top of the active site. (d) Celecoxib bound to COX-2, the phenylsulfamoyl group occupies the side pocket. Figures 3a–d are adapted from [2].

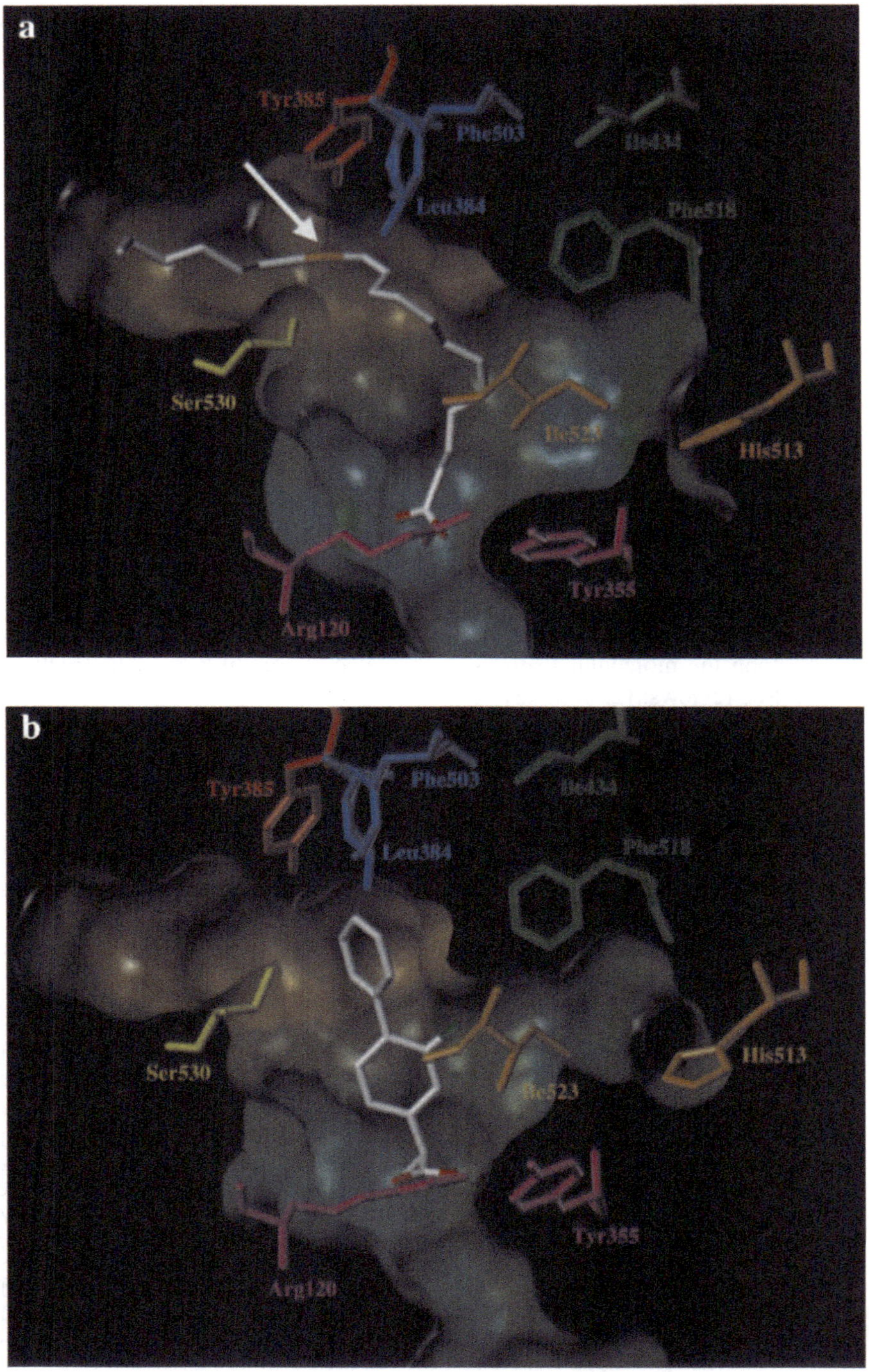

Figure 3 a, b (legend see p. 47)

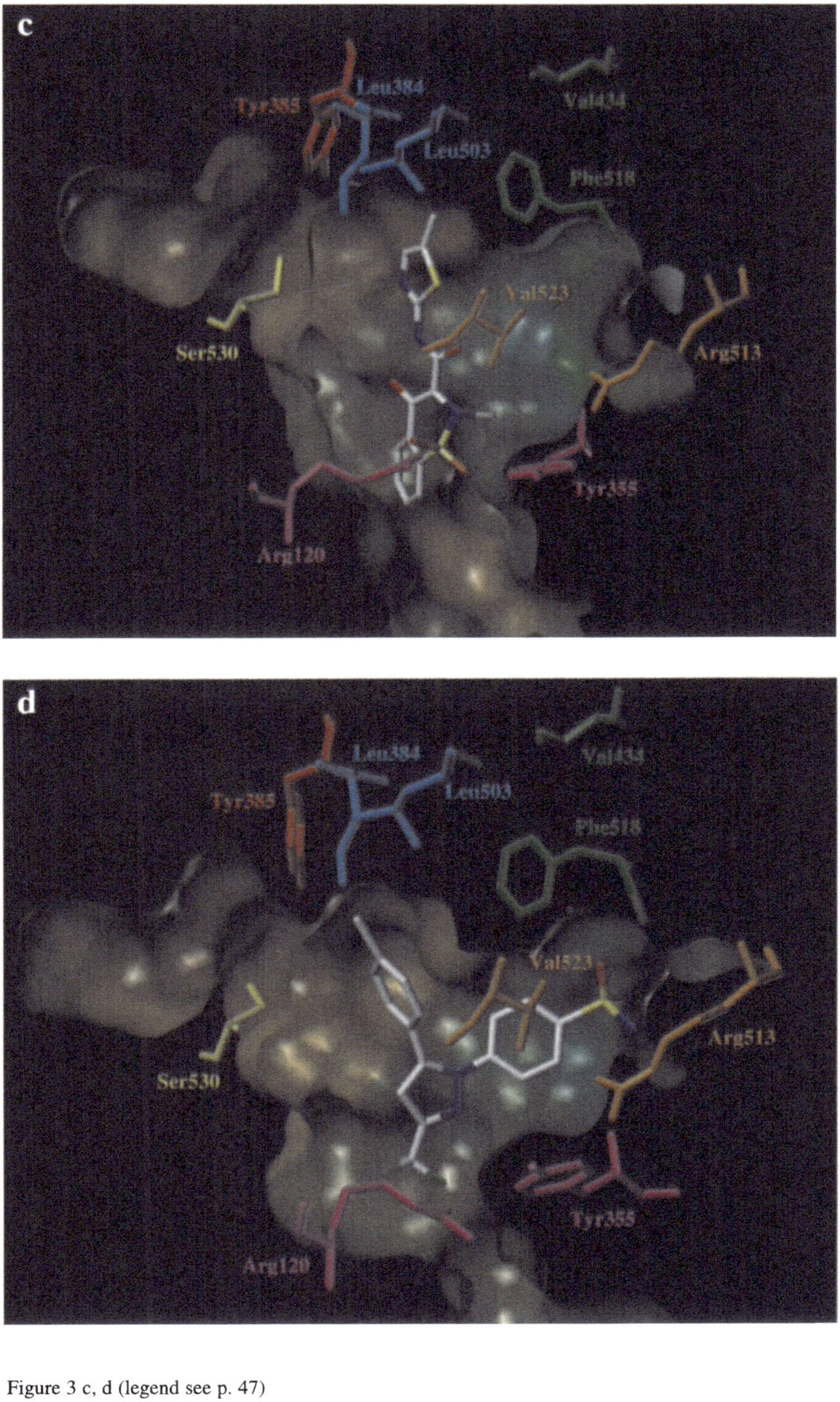

Figure 3 c, d (legend see p. 47)

Table 2. Structures and properties of selective COX-2 inhibitors and structurally related NSAIDs

Structural class (Non-selective inhibitors: *italics*; selective COX-2 inhibitors: **bold**)		Solubility in water (pH 7, 25°C)	Ionization (pKa value)	Lipophilicity (logP at pH 7.4)	Protein binding	Volume of distribution (Vd, ss)
1. Acetoxyphenyl-derivatives (aspirin):						
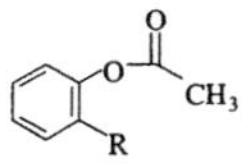	*aspirin* (R=COOH)	*~10 g/L*	*3.5 (acid)*	*–2.8*	*(~85%)*[1]	*(12 L)*[1]
	APHS (R=heptinylsulfide)	–	–	–	–	–
2. Anthranilic acids (fenamates):						
	ester of meclofenamic acid	–	–	–	–	–
3. Arylacetic acids (fenacs):						
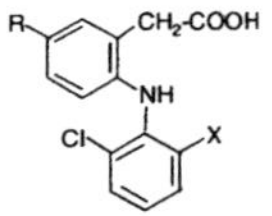	*diclofenac* (R=H, X=Cl)	*~5 g/L*	*4 (acid)*	*1.2*	*>99%*	*~10 L*
	COX-189 (R=CH_3, X=F)	**280 mg/L**	**4.3 (acid)**	**1.5**	**>99%**	**~10 L**
	others: etodolac, RS57067, L-761066					
4. Arylpropionic acids (profens):						
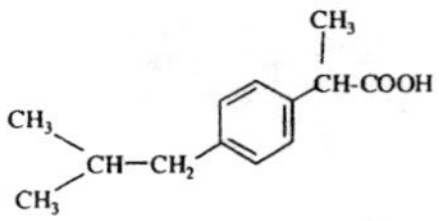	*ibuprofen*	*~5 g/L*	*4.5*	*1.5*	*>99%*	*~10 L*
	bisethoxyflurbiprofen	–	–	–	–	–

(Continued on next page)

Table 2. (Continued)

Structural class (Non-selective inhibitors: *italics*; selective COX-2 inhibitors: **bold**)	Solubility in water (pH 7, 25°C)	Ionization (pKa value)	Lipophilicity (logP at pH 7.4)	Protein binding	Volume of distribution (Vd, ss)
5. Enolcarboxamides (oxicams):					
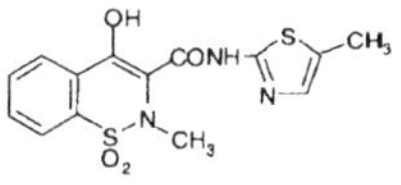 **meloxicam**	**250 mg/L**	**1.1 + 4.2 (zwitterion)**	**0.1**	**>99%**	**~10 L**
6. Sulfanilides (sulides):					
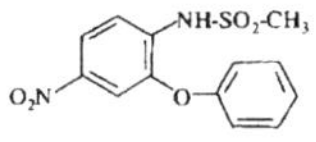 **nimesulide**	**20 mg/L**	**6.9 (weak acid)**	**2.6**	**99%**	**~20 L**
others: **NS-398, flosulide, L-745337**					
7. Diarylheterocycles (coxibs):					
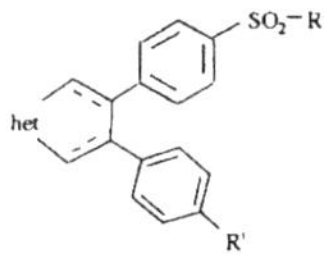 **celecoxib** (R=NH_2, R'=CH_3, het=CF_3-pyrazol)	**5 μg/L**	**11.1 (neutral)**	**>3**	**97%**	**~400 L (app VdSS)**
rofecoxib (R=CH_3, R'=H, het=furanone)	**4 μg/L**	**12.5 (neutral)**	**>3**	**87%**	**100 L (app VdSS)**
others: **DUP697, SC58125, DFP, DFU, valdecoxib, etoricoxib**					
8. o,o-Dialkylphenols:					
BF398, CI-1004, PD-138387, PD-136095	–	–	–	–	–

Physicochemical values were determined according to the methods described by Luger et al. [25]. Pharmacokinetic values were taken from the product information sheets. [1]Rapid transformation of aspirin to salicylic acid, values given are concentration dependent and are given for low doses. -: no data available

Table 3. Overview of the interactions of COX inhibitors with COX mutants

Ligand	COX-mutants [Reference]
Salicylic acids:	
Aspirin	*hCOX-2:* Ser530Met [26]; *hCOX-1:* Arg120Glu [27] *mCOX-2:* Tyr385Phe; Ser530Ala; Arg120Ala,Gln; Tyr348Phe; VRV-IHI [28] *mCOX-2:* Val434Ile; Arg 513His; Val523Ile; VRV-IHI; Val228Phe; Gly533Ala [29]
2-Acetoxyphenylhept-2-ynyl sulfide (APHS)	*mCOX-2:* Arg120Gln; Tyr355Ala; VRV-IHI [30] *mCOX-2:* Tyr385Phe; Ser530Ala; Arg120Gln; Gly533Ala,Val; Tyr348Phe; VRV-IHI [31]
O-Acetylsalicylhydroxamic acid (AcSHA)	*hCOX-1:* Ser530Ala, Arg120Gln [13]
Arylacetic acids:	
Indomethacin	*hCOX-1:* Arg120Glu [27]; *hCOX-2*: Arg120Gln [32] *hCOX-2:* Tyr355Phe [33]; *hCOX-2:* Ser530Met [34] *hCOX-1:* His513Arg;Ile523Val; HI-AV [36]; *hCOX-2:* Val523Ile [35]
Diclofenac	*hCOX-1:* Arg120Glu [27]; *hCOX-2:* Arg120Gln [32] *hCOX-2:* Ser530Met [34] *hCOX-1:* His513Arg;Ile523Val; HI-AV [36]; *hCOX-2:* Val523Ile [35]
Sulindac	*hCOX-2:* Ser530Met [34]; *hCOX-1:* His513Arg;Ile523Val; HI-AV [36]
RS-57067 (Zomepirac analog)	*hCOX-2:* Tyr355Phe [33]
Indomethacin analogs:	
L-588,983; L-761,066	*hCOX-2:* Arg120Gln [32]
LM-4143	*mCOX-2:* Arg120Gln,Ala; Tyr355Phe,Ala; Glu524Leu [37]
Arylpropionic acids:	
Ibuprofen	*hCOX-1:* Arg120Glu [27]; *oCOX-1:* Arg120Gln,Lys,Glu; Tyr355Phe D- and L-ibuprofen [38]
Flurbiprofen	*hCOX-1:* Arg120Glu [27]; *oCOX-1:* Arg120Gln,Lys,Glu [38]; *hCOX-2:* Arg120Gln [32, 39]; *hCOX-2:* Ser530Met [34]; *hCOX-2:* Val523Ile [35]; *hCOX-2:* Tyr355Phe [33]
Ketoprofen	*hCOX-1:* Arg120Glu [27]; *hCOX-2:* Ser530Met [34]
Naproxen	*hCOX-2:* Val523Ile [35]
Flufenamic acids:	
Meclofenamic acid	*hCOX-1:* Arg120Glu [27]; *hCOX-2:* Arg120Gln [32]; *hCOX-2:* Ser530Met [34]
N-Phenoxyethylamid of Meclofenamic acid	*mCOX-2:*Arg120Gln,Ala; Tyr355Phe, Ser530Ala; Glu524Leu; Leu503Phe; Val523Ile, VRV–IHI [40]
Mefenamic acid	*hCOX-2:* Val523Ile [35]

(Continued on next page)

Table 3. (Continued)

Ligand	COX-mutants [Reference]
Sulfanilides:	
NS 398	*hCOX-2:* Arg120Gln [32, 39]; *hCOX-1:* His513Arg; Ile523Val; HI-AV [36]; *hCOX-2:* Val523Ile [35]; Val523Ile,Ala,Glu,Lys [40]
Nimesulide	*hCOX-2:* Val523Ile,Ala,Glu,Lys [41]
Flosulide	*h-COX-2:* Arg120Gln [32]
L-745,337	*hCOX-2:* Ser530Met [34]
Diarylheterocycles:	
DUP-697	*hCOX-1:* Arg120Glu [27]; *hCOX-2:* Arg120Gln [32, 39]; *hCOX-2*: Ser530Met [34]); *hCOX-2:* Val523Ile [35]; Val523Ile,Ala,Glu,Lys [41]; *hCOX-1:* His513Arg; Ile523Val; HI-AV [36]
SC 58125	*hCOX-2:* Arg120Gln [39]; *hCOX-2:* Val523Ile; TSP-LLL [35]; *hCOX-2:* Val523Ile,Ala,Glu,Lys [41]; *hCOX-1:* His513Arg; Ile523Val; HI-AV [36]
Celecoxib, Valdecoxib	*mCOX-2:* Arg120Ala,Gln,Asn; Tyr355Ala; Val523Ile; VRV-IHI [42]
L-746,483 (DUP-697 analog)	*hCOX-1:* Arg120Glu [27]; *hCOX-2:* Arg120Gln [34]
L-745,296 (DUP-697 analog)	*hCOX-2:* Arg120Gln [34]; *hCOX-1:* His513Arg; Ile523Val; HI-AV [36]
DFU	*hCOX-1:* His513Arg; Ile523Val; HI-AV [36]

HI-AV: double mutant *COX-1* His513Arg + Ile523Val; VRV-IHI: triple mutant *COX-2* Val434Ile + Arg513His + Val523Ile; TSP-LLL: triple mutant *COX-2* Tyr101Leu + Ser105Leu + Phe343Leu.

bulk into the COX channel at the place where the ω-end of arachidonic acid binds (Gly533) lead to reduced COX activity [31].

Before the introduction of selective COX-2 inhibitors arylacetic acids, arylpropionic acids and fenamic acids represented the largest group of marketed NSAIDs. Detailed X-ray analyses (Tab. 1), site-directed mutagenesis experiments (Tab. 3) and comparative molecular modelling studies [2] provide us with a large amount of information as to the critical amino acid residues in COX-1 and COX-2 responsible for the selective interaction with compounds of these classes. In Figure 3b the interaction of the nonselective flurbiprofen within the COX channel has been visualized using an algorithm that represents channel and cavity surfaces by color coding local hydrophobicity.

Meloxicam was characterised during development as being COX-2 selective, and it was therefore launched as the first selective COX-2 inhibitor. By applying molecular modelling methods [45], it has subsequently been established that the enolate anion of meloxicam does not interact with Arg120 via an ionic bond but that the thiazine-sulfone can form hydrogen bridges to Arg120 and Tyr355. More importantly, the 5-methyl group of the thiazole ring exploits the "flexible extra space" at the top of the COX-2 channel. The substitution of the second-shell amino acid residue Ile434 in COX-1 for Val in COX-2 allows the side chain of Phe518 (a residue at the active side) to open an "extra space", which favours the binding of meloxicam to COX-2 (Fig. 3c). The binding conformation of meloxicam is less optimal in COX-1; for example, a steric interaction with Phe518 is most notable. X-ray analyses and newer molecular modelling studies have confirmed these interactions and have allowed even more precise evaluations to be made [20].

Nimesulide and the experimental compounds L-745,337 (Merck & Co), flosulide (Novartis AG) and NS-398 (Taisho Pharmaceutical Co) are members of the sulfanilide group of NSAIDs. Mutational analyses [35, 39] suggest that the methylsulfone group of these compounds interact particularly with Arg120 at the constriction of the COX channel.

Our knowledge about the molecular basis of the selective action of compounds such as celecoxib and rofecoxib is based on the X-ray analysis of SC 558 [14], a structural analogue of celecoxib, together with mutagenesis [42], kinetic [46, 47] and modelling experiments [48]. The phenylsulfonamide of celecoxib and the phenylmethylsulfone moiety of rofecoxib occupy the side pocket in COX-2, made accessible by the substitution of Val523 for Ile523, and are within binding distance of Arg513. One other amino acid substitution, Leu503 in COX-2 replacing Phe503 in COX-1, also extends the volume of the active site in COX-2 and this is exploited by the coxibs. In Figure 3d the binding of celecoxib into the COX-2 channel based on the X-ray data [14] is demonstrated.

Because of their structural distinctions, the COX-2 inhibitors have markedly different molecular properties, as demonstrated by their solubility, protonation behaviour and lipophilicities (Tab. 2). These different properties lead to differences in the pharmacokinetic behaviours of the COX-2 inhibitors, such as their protein binding, tissue distribution and metabolic pathways. For this reason adverse event profiles of selective COX-2 inhibitors should take into account not only the advantages of COX-2 selectivity (i.e., COX-1 sparing) but also differences in drug distribution and metabolism as compared to standard NSAIDs.

Enzyme kinetics and inhibitor complexes

Enzyme kinetics play an important role in the activation and inhibition of COX. Selective COX-2 inhibition is correlated with a time-dependent COX-2 inhibition and a time-independent reversible COX-1 inhibition [49]. Kinetic aspects of binding interactions are discussed in this chapter based on work from crystallographic studies that have led to insight into the enzyme-inhibitor complexes. Further kinetic studies are also described in Chapter 2.

Prior to the work of Selinsky et al. [17], it was believed that competitive inhibitors formed only a freely reversible complex (EI) and that time-dependent inhibitors could additionally slowly form tightly bound irreversible complexes (EI*):

$$E + I_{rev} \rightleftharpoons EI$$

$$E + I_{irr} \rightleftharpoons EI \longrightarrow EI^*$$

It had been postulated, therefore, that selectivity of COX-2 inhibitors may occur by the formation of EI* with COX-2 and of reversible EI with COX-1 [48, 46]. Selinsky et al. [17] proposed that both reversible time-dependent inhibitors and irreversible time-independent inhibitors bind by the same mechanism and differ by the activation energy (E^a) of the EI $\longrightarrow$ EI* transition:

$$E + I_{rev} \rightleftharpoons EI \overset{\text{low } E^a}{\rightleftharpoons} EI^*$$

$$E + I_{irr} \rightleftharpoons EI \overset{\text{high } E^a}{\rightleftharpoons} EI^*$$

With a low E^a separating EI and EI*, the transition to EI* will be rapid. One should observe only the presence of EI* in kinetic measurements and so crystal analyses will be of COX in the EI* state. The same conformational status (EI*) will be observed in X-ray experiments for slow, tightly-bound inhibitors, and due to the high E^a, the first step to EI could also be measured kinetically. COX-2 selective inhibitors, like meloxicam might gain their selectivity by having a low E^a for COX-1 and a high E^a for COX-2.

More recently Walker et al. [47] proposed that some selective COX-2 inhibitors of the diarylheterocyclic type (i.e., valdecoxib) could generate a three-step time-dependent irreversible COX-2 inhibition:

$$E + I_{irr} \rightleftharpoons EI \rightleftharpoons EI^* \xrightarrow{\text{low } E^a} EI^{**}$$

COX assays

In vitro *assays*

Before the availability of enzyme modelling many *in vitro* assay systems were developed to investigate the COX-2 selectivity of NSAIDs. These have been reviewed in detail elsewhere [50] but in broad terms these tests systems can be classified into three groups: assays using animal enzymes, cells, or cell lines, which were the first to be developed; assays using human recombinant enzymes, cell lines, or blood cells (mainly platelets and monocytes), which are the current standards; and models using those human cells that are targets for the anti-inflammatory and adverse effects of NSAIDs, e.g., human gastric mucosa cells, chondrocytes, and synoviocytes.

Apart from variations in the sources of the COX enzymes there are wide variations in the experimental conditions in these various assays. For example, arachidonic acid can come from both endogenous (often in cell based assays) and exogenous sources (in isolated enzyme or broken cell assays). It is important to remember that the potencies of some NSAIDs, particularly those that act as competitive substrate inhibitors [46], are influenced by the supply of arachidonic acid. Also, expression systems and cell transfection targets vary considerably, including even non-mammalian cells such as those from insects. Other experimenters have used cells that express the COX enzymes constitutively, following stable transfection with the recombinant enzymes, to avoid the problems associated with variability in the types and concentrations of inducing agents used. As well as substrate, the time of incubation of the test drug with the COX enzyme systems can have a strong influence on the apparent potencies of drugs. In particular, a number of NSAIDs and COX-2 selective agents inhibit COX-2 in a time-dependent manner [46, 51]. Another important factor is the protein concentrations present in the medium. This is a critical issue for NSAIDs, which bind avidly to plasma proteins such that within the circulation, for instance, the concentration of free drug is usually only a few percent of the total drug concentration. In general terms it appears that an ideal assay of use in trying to predict the clinical effects of these agents should use native human enzymes that are present in whole cells; that the cells used should be target cells for the anti inflammatory and adverse effects of NSAIDs; that COX-2 should be induced, thereby simulating an inflammatory process, rather than being constitutively expressed; that prostaglandin synthesis should be measured from arachidonic acid released from endogenous stores rather than from exogenously added arachidonic acid; and, that the protein concentration in the medium should also closely mimic plasma protein concentrations.

Exploiting data from in vitro *assays*

The IC_{50} values of COX-1 and COX-2 inhibition obtained in *in vitro* assays are often used to calculate selectivity ratios. However, this is a very simplistic approach. It relies upon concentration curves being parallel, and takes little account of the plasma concentrations of drugs achieved following therapeutic dosing. Furthermore, if two drugs have similar selectivity ratios but very different slopes to their inhibitor curves the conclusions we can draw must be very different. In simple terms, steep curves allow a better dissociation between COX-1 and COX-2 inhibition than flat curves [50]. So, even though COX-2 selectivity can be expressed by selectivity ratios derived from *in vitro* assays we must be very cautious before extrapolating to clinical conditions.

Human whole-blood assay (hWBA)

For many of the reasons discussed above the human whole blood assay (hWBA) appears to offer many advantages over other *in vitro* tests. In this assay, COX-1 activity is assessed by the release of prostanoids from platelets following stimulation of human whole blood clotting, while COX-2 activity is assessed by the release over about 18 h of prostanoids from white blood cells (principally monocytes) following incubation of the whole blood with bacterial lipopolysaccharide (LPS). The hWBA has many advantages: intact human cells are used, which are target cells for the anti inflammatory effects (monocytes) and side-effects (platelets) of NSAIDs; plasma proteins are present; the whole blood used for both assays is taken from the same volunteer (or patient) at the same time, allowing a direct comparison of the results from each assay; the assay can be performed using blood from volunteers (or patients) who have been treated previously with NSAIDs (*ex vivo* assay). Results obtained by different laboratories, using the hWBA have demonstrated similar rank orders for COX-2 selectivity [50]. Standard NSAIDs range from being COX-1 selective to being approximately equally effective on both isoenzymes with diclofenac having the most favourable profile. Compounds such as etodolac, nimesulide, and meloxicam followed by celecoxib, flosulide, rofecoxib, etoricoxib, valdecoxib and COX-189 show increasing selectivity for COX-2. Another way to express this data, if COX-2 is inhibited by >80% (i.e., the therapeutic target) how much COX-1 inhibition can one expect *in vivo*. This is illustrated in Figure 4 for both traditional NSAIDs and COX-2 selective compounds. It indicates that traditional NSAIDs will inhibit COX-1 to the same extent as COX-2, while more selective inhibitors have less effect on COX-1. The new generation of COX-2 inhibitors do not inhibit COX-1 at concentrations that maximally inhibit COX-2.

A further development of the hWBA has been to mix with aliquots of human whole blood human A549 cells already induced to express COX-2 (human modified whole blood assay, hmWBA) [52]. This permits the COX-1 and COX-2 assays to be conducted over identical time courses. As for the standard hWBA the hmWBA has been used to generate ratios for the COX-1/COX-2 selectivities of a wide range of NSAIDs and newer COX-2 selective agents.

Other assays

Alternative and newer *in vitro* assays are always being developed. On the face of it test systems that appear the most interesting are those using human cells such as gastric mucosa cells, chondrocytes or synoviocytes which are target cells for the anti inflammatory or adverse effects of NSAIDs. These generate similar data to those seen with human recombinant enzymes and the hWBA [53–56] and it can be shown that the data from such assays correlates with the

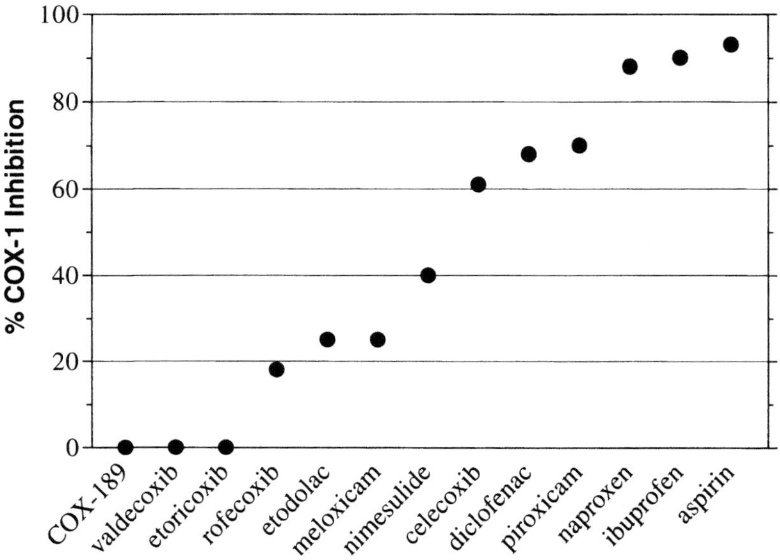

Figure 4. The degree of inhibition of COX-1 when COX-2 is inhibited by 80% (i.e., IC_{80}). The data were obtained from [52] using IC_{80} determination *in vitro* and unpublished results (valdecoxib, etoricoxib and COX-189).

idea that increased COX-2 selectivity is associated with reduced gastrointestinal toxicity [57] but still these assays have the problem that they do not mimic well the drug binding to proteins seen *in vivo*. Thus, one must be careful when trying to draw conclusions about the effects on tissue prostaglandin formation by comparing potencies against tissue and whole blood prostanoid formation [58].

In vivo relevance of *in vitro* results

Clearly the purpose of *in vitro* tests is to predict the usefulness of NSAIDs *in vivo*. In particular, following the idea that inhibition of COX-2 explains efficacy and inhibition of COX-1 explains side-effects, will a drug be efficacious without causing side-effects? Using the whole blood assays we can map blood therapeutic concentrations derived by pharmacokinetic studies (taking into account the differences between plasma and whole blood concentrations) and compare them to the concentrations of drug required to inhibit COX-1 and COX-2 in *in vitro* assays. Such composite figures have been derived from the hmWBA, examples being those for diclofenac (Fig. 5a), meloxicam (Fig. 5b) and rofecoxib (Fig. 5c). This modelling shows us that these three drugs (diclofenac, meloxicam and rofecoxib), which are established to be efficacious at standard doses, should produce similar inhibition of COX-2 when applied at

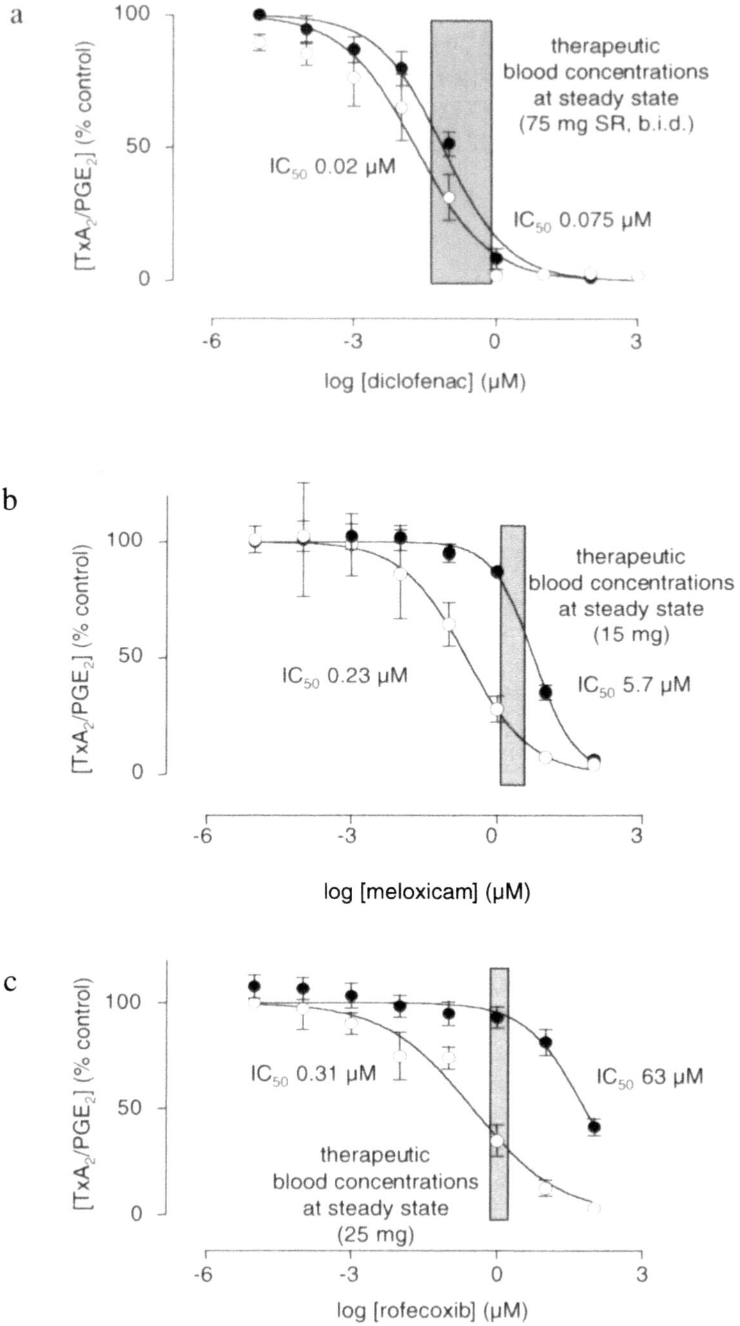

Figure 5a–c. Inhibition of COX-1 (●) and COX-2 (○) by diclofenac (a), meloxicam (b), and rofecoxib (c) in hmWBA [52]. Greyed bar indicates concentrations achieved *in vivo* following ingestion of therapeutic doses.

standard doses. However, the level of COX-1 inhibition varies greatly (from approximately 70% to approximately 10%). Once again, this supports the idea that it is inhibition of COX-2 that underlies the efficacious effects of COX inhibitors; i.e., if inhibition of COX-1 was required for drugs to be efficacious diclofenac would be considerably more efficacious than either meloxicam or rofecoxib.

Human pharmacology studies

The most relevant markers of COX-1 and COX-2 activity *in vivo* would be PGE_2 or PGI_2 production by the gastric mucosa, and PGE_2 production by sites of inflammation, most notably the inflamed synovium. Indeed, similar models have been developed for use in experimental animals. However, for practical reasons, such studies are much more difficult to perform in humans, since target tissues are not easily obtainable. Furthermore, prostaglandin production can be stimulated when tissues are harvested. Despite these reservations there are a number of studies that appear promising [59, 60]. Generally the differential inhibition of COX-l and COX-2 by NSAIDs is studied in humans by *in vivo* pharmacological studies. The effects of repeated administration of anti-inflammatory doses on COX-2 activity measured in a hWBA *ex vivo*, or on markers of COX-1 activity, such as platelet aggregation, plasma thromboxane B_2 (TxB_2) concentrations, or urinary excretion of PGE_2, are usually investigated. Of these serum TxB_2 appears to be the most sensitive measure of COX-1 inhibition by a drug, as platelet aggregation does not appear affected until plasma TxB_2 is reduced by more than 95% [61]. This is illustrated with meloxicam, which was given in doses of 7.5, 15 and 30 mg to healthy volunteers [62]. *Ex vivo* serum TxB_2 was inhibited dose-dependently by ~40, 60 and 75%, respectively, however, there was no effect on platelet aggregation or bleeding time at these therapeutic and supratherapeutic doses. Nimesulide, is also without effect upon bleeding time or *ex vivo* aggregation [63]. Studies performed with several non-selective "standard" NSAIDs showed a close relationship between plasma drug concentrations and inhibition of serum TxB_2 [64–68] (most likely to be COX-1 inhibition) and inhibition of synovial PGE_2 (most likely to be COX-2 inhibition) [59, 60]. Naproxen and ibuprofen both inhibit serum TxB_2 by more than 95% at therapeutic doses and inhibit platelet aggregation *ex vivo* [69, 70].

Celecoxib did not inhibit collagen, ADP, or arachidonate-induced platelet aggregation, and also had no effect on TxB_2 levels, 2 and 4 h after the last administration of 600 mg twice daily for 6 days [71]. Similarly, more studies using celecoxib at doses of up to 1200 mg in comparison to 800 mg ibuprofen or 1000 mg naproxen have shown that the traditional NSAIDs, but not celecoxib, inhibited *ex vivo* platelet aggregation induced by agonists such as collagen. Also, while naproxen but not celecoxib depressed circulating thromboxane A_2 levels, both ibuprofen and celecoxib inhibited LPS-induced COX-2 expression *ex vivo* [72]. Interestingly, however, the plasma TxB_2 concentration, a more sensitive measure of COX-1 inhibition than platelet aggregation (see above), was significantly reduced by approximately 30% after a single administration of the supratherapeutic dose of 800 mg celecoxib [73]. In the same studies, both ibuprofen and celecoxib were found to inhibit urinary excretion of the PGI_2 metabolite 2,3-dinor-6-keto-$PGF_{1\alpha}$ [73]. This may indicate that COX-2 is responsible for the formation of PGI_2 in the circulation. Indeed, further studies with rofecoxib have shown that like its comparator indomethacin this COX-2-selective inhibitor also reduces the urinary excretion of the PGI_2 metabolite 2,3-dinor-6-keto-prostaglandin $F_{1\alpha}$ [74]. At the same time even supra-therapeutic doses of rofecoxib of up to 375 mg for 14 days produces no inhibition of COX-1 activity *ex vivo*, as assessed by TxB_2 generation in clotting blood, whereas there is profound inhibition of *ex vivo* COX-2 activity [75]. Consistent with these results, the new generation of COX-2 selective agents such as etoricoxib [76, 77], valdecoxib [78], and COX-189 [79] all do not inhibit serum TxB_2 at therapeutic doses and do not affect platelet aggregation or bleeding time at therapeutic or supratherapeutic doses.

Acknowledgements
The authors wish to thank U. Werthmann for determining the physiochemical values of various NSAIDs. We also wish to thank H. Wittneben and C. Klein for performing the 3D-molecular modelling and to E. Haaksma and H. Nar for their helpful suggestions.

References

1 Picot D, Loll PJ, Garavito RM (1994) The X-ray crystal structure of the membrane protein prostaglandin H2 synthase-1. *Nature* 367: 243–249
2 Trummlitz G, van Ryn J (2002) Designing selective COX-2 inhibitors: Molecular modeling approaches. *Curr Opin Drug Discov Devel* 5: 550–561
3 Smith, WL, DeWitt, DL, Garavito, RM (2000) Cyclooxygenases: structural, cellular and molecular biology. *Annu Rev Biochem* 69: 145–182
4 Hatae T, Wada M, Yokoyama C, Shimonishi M, Tanabe T (2001) Prostacyclin-dependent apoptosis mediated by PPAR delta. *J Biol Chem* 276: 46260–46267
5 Garavito RM (2001) Prostaglandin endoperoxide H2 synthases-1 and -2. In: A Messerschmidt, R Huber, T Poulos (eds): *Handbook of Metalloproteins*. John Wiley and Sons Ltd, Chichester, UK, 245–263
6 Chandrasekharan NV, Dai H, Roos LT, Evanson NK, Tomsik J, Elton TS, Simmons DL (2002) COX-3, a cyclooxygenase-1 variant inhibited by acetaminophen and other analgesic/antipyretic drugs: Cloning, structure, and expression. *Proc Natl Acad Sci USA* 99: 13926–13931
7 Dinchuk JE, Liu RQ, Trzaskos JM (2003) COX-3: in the wrong frame in mind. *Immunol Lett* 86: 121
8 Kiefer JR, Pawlitz JL, Moreland KT, Stegeman RA, Hood WF, Gierse JK, Stevens AM, Goodwin DC, Rowlinson SW, Marnett LJ et al (2000) Structural insights into the stereochemistry of the cyclooxygenase reaction. *Nature* 405: 97–101
9 Malkowski MG, Ginell SL, Smith WL, Garavito RM (2000) The productive conformation of arachidonic acid bound to prostaglandin synthase. *Science* 289: 1933–1937
10 Thuresson ED, Malkowski MG, Lakkides KM, Rieke CJ, Mulichak AM, Ginell SL, Garavito RM, Smith WL (2001) Mutational and X-ray crystallographic analysis of the interaction of dihomo-gamma-linolenic acid with prostaglandin endoperoxide H syntheses. *J Biol Chem* 276: 10358–10365

11 Malkowski MG, Thuresson ED, Lakkides KM, Rieke CJ, Micielli R, Smith WL, Garavito RM (2001) Structure of eicosapentaenoic and linoleic acids in the cyclooxygenase site of prostaglandin endoperoxide H synthase-1. *J Biol Chem* 276: 37547–37555

12 Loll PJ, Picot D, Garavito RM (1995) The structural basis of aspirin activity inferred from the crystal structure of inactivated prostaglandin H2 synthase. *Nature Struct Biol* 2: 637–643

13 Loll PJ, Sharkey CT, O'Connor SJ, Dooley CM, O'Brien E, Devocelle M, Nolan KB, Selinsky BS, Fitzgerald DJ (2001) O-Acetylsalicylhydroxamic acid, a novel acetylating inhibitor of prostaglandin H2 Synthase: Structural and functional characterization of enzyme-inhibitor interactions. *Mol Pharmacol* 60: 1407–1413

14 Kurumbail RG, Stevens AM, Gierse JK, McDonald JJ, Stegeman RA, Pak JY, Gildehaus D, Miyashiro JM, Penning TD, Seibert K et al (1996) Structural basis for selective inhibition of cyclooxygenase-2 by anti-inflammatory agents. *Nature* 384: 644–648

15 Loll PJ, Picot D, Ekabo O, Garavito RM (1996) Synthesis and use of iodinated nonsteroidal anti-inflammatory drug analogs as crystallographic probes of the prostaglandin H2 synthase cyclooxygenase active site. *Biochemistry* 35: 7330–7340

16 Merck and Co. Inc. (Bayly CI, McKeever BM, Ouellet M, Pandrya SR, Percival MD, Roy P) (1998) Crystals of human cyclooxygenase-2 (hCOX-2). GB2319032A

17 Selinsky BS, Gupta K, Sharkey CT, Loll PJ (2001) Structural analysis of NSAID binding by prostaglandin H2 synthase: Time-dependent and time-independent inhibitors elicit identical enzyme conformations. *Biochemistry* 40: 5172–5180

18 Luong C, Miller A, Barnett J, Chow J, Ramesha C, Browner MF (1996) Flexibility of the NSAID binding site in the structure of human cyclooxygenase-2. *Nature Struct Biol* 3: 927–933

19 Garavito RM, Picot D, Loll PJ (1995) The 3.1 A X-ray crystal structure of the integral membrane enzyme prostaglandin H2 synthase-1. *Adv Prostaglandin Thromboxane Leukot Res* 23: 99–103

20 Trummlitz G, Wittneben H, van Ryn J, Harman C, Clayton G, Garavito RM (2003) X-Ray crystallographic analysis and computational docking studies of cyclooxygenase isoform inhibitors. *Ann Rheum Dis 62 (Suppl)*: Abs FRI0230

21 Malkowski MG, Theissen MJ, Scharmen A, Garavito RM (2000) The formation of stable fatty acid substrate complexes in prostaglandin H_2 synthetase-1. *Arch Biochem Biophys* 380: 39–45

22 Ringbom T, Huss U, Stenholm A, Flock S, Skattebol L, Perera P, Bohlin L (2001) COX-2 inhibitory effects of naturally occurring and modified fatty acids. *J Nat Prod* 64: 745–749

23 Kozak KR, Crews BC, Ray JL, Tai HH, Morrow JD, Marnett LJ (2001) Metabolism of prostaglandin glycerol esters and prostaglandin ethanolamides *in vitro* and *in vivo*. *J Biol Chem* 276: 36993–36998

24 Kozak KR, Prusakiewicz JJ, Rowlinson SW, Schneider C, Marnett LJ (2001) Amino acid determinants in cyclooxygenase-2 oxygenation of the endocannabinoid 2-aracidonylglycerol. *J Biol Chem* 276: 30072–30077

25 Luger P, Daneck K, Engel W, Trummlitz G, Wagner K (1996) Structure and physicochemical properties of meloxicam, a new NSAID. *Eur J Pharmac Sci* 4: 75–187

26 Mancini JA, O'Neill GP, Bayly C, Vickers PJ (1994) Mutation of serine-516 in human prostaglandin G/H synthase-2 to methionine or aspirin acetylation of this residue stimulates 15-*R*-HETE synthesis. *FEBS Letters* 342: 33–37

27 Mancini JA, Riendeau D, Falgueyret JP, Vickers PJ, O'Neill GP (1995) Arginine 120 of prostaglandin G/H synthase-1 is required for the inhibition by nonsteroidal anti-inflammatory drugs containing carboxylic acid moiety. *J Biol Chem* 270: 29372–29377

28 Hochgesang GP, Rowlinson SW, Marnett LJ (2000) Tyrosine-85 is critical for acetylation of cyclooxgenase-2 by aspirin. *J Amer Chem Soc* 122: 6514–6515

29 Rowlinson SW, Crews BC, Goodwin DC, Schneider C, Gierse JK, Marnett LJ (2000) Spatial requirements for 15-(*R*)-Hydroxy-5Z,8Z,11*Z*,13*E*-eicosatertraenoic acid synthesis within the cyclooxygenase active site of murine COX-2. *J Biol Chem* 275: 6586–6591

30 Kalgutkar AS, Crews BC, Rowlinson SW, Garner C, Seibert K, Marnett LJ (1998) Aspirin-like molecules that covalently inactivate cyclooxygenase-2. *Science* 280: 1268–1270

31 Hochgesang GP, Nemeth-Cawley JF, Rowlinson SW, Caprioli RM, Marnett LJ (2003) Functional analysis of the molecular determinants of cyclooxygenase-2 acetylation by 2-acetoxyphenylhept-2-ynyl sulfide. *Arch Biochem Biophys* 409: 127–133

32 Greig GM, Francis DA, Falgueyret JP, Ouellet M, Percival MD, Roy P, Bayly C, Mancini JA, O'Neill GP (1997) The interaction of arginine 106 of human prostaglandin G/H synthase-2 with inhibitors is not a universal component of inhibition mediated by nonsteroidal anti-inflammatory

drugs. *Mol Pharmacol* 52: 829–838
33 So OY, Scarafia LE, Mak AY, Callan OH, Swinney DC (1998) The dynamics of prostaglandin H synthase 2 Y355F unmask mechanisms of time-dependent inhibition and allosteric activation. *J Biol Chem* 273: 5801–5807
34 Mancini JA, Vickers PJ, O'Neill GP, Boily C, Falgueyret JP, Riendeau D (1997) Altered sensitivity of aspirin-acetylated prostaglandin G/H Synthase to inhibition by nonsteroidal anti-inflammatory drugs. *Mol Pharmacol* 51: 52–60
35 Gierse JK, McDonald JJ, Hauser SD, Rangwala SH, Koboldt CM, Seibert K (1996) A single amino acid difference between cyclooxygenase-1 (COX-1) and -2 (COX-2) reverses the selectivity of COX-2 specific inhibitors. *J Biol Chem* 271: 15810–15814
36 Wong E, Bayly C, Waterman HL, Riendeau D, Mancini JA (1997) Conversion of prostaglandin G/H synthase-1 into an enzyme sensitive to PGHS-2-selective inhibitors by a double His513 → Arg and Ile523 → Val mutation. *J Biol Chem* 272: 9280–9286
37 Timofeevski SL, Prusakiewicz JJ, Rouzer CA, Marnett LJ (2002) Isoform-selective interaction of cyclooxygenase-2 with indomethacin amides studied by real-time fluorescence, inhibition kinetics, and site-directed mutagenesis. *Biochemistry* 41: 9654–9662
38 Bhattacharyya DK, Lecomte M, Rieke CJ, Garavito RM, Smith WL (1996) Involvement of arginine 120, Glutamate 524, and tyrosine 355 in the binding of arachidonate and 2-phenylpropionic acid inhibitors to the cyclooxygenase active site of ovine prostaglandin endoperoxide H synthase-1. *J Biol Chem* 271: 2179–2184
39 Rieke CJ, Mulichak AM, Garavito RM, Smith WL (1999) The role of arginine 120 of human prostaglandin endoperoxide H synthase in the interaction with fatty acids and inhibitors. *J Biol Chem* 274: 17109–17114
40 Kalgutkar AS, Crews BC, Rowlinson SW, Marnett AB, Kozak KR, Remmel RP, Marnett LJ (2000) Biochemically based design of cyclooxygenase-2 (COX-2) inhibitors: facile conversion of nonsteroidal antiinflammatory drugs to potent and highly selective COX-2 inhibitors. *Proc Natl Acad Sci USA* 97: 925–930
41 Guo QP, Wang LH, Ruan KH, Kulmacz RJ (1996) Role of Val509 in time-dependent inhibition of human prostaglandin H synthase-2 cyclooxygenase activity by isoform-selective agents. *J Biol Chem* 271: 9134–19139
42 Hood WF, Gierse JK, Isakson PC, Kiefer JR, Kurumbail RG, Seibert K, Monahan JB (2003) Characterization of celecoxib and valdecoxib binding to cyclooxygenase. *Mol Pharmacol* 63: 870–877
43 Kurumbail RG, Kiefer JR, Marnett LJ (2001) Cyclooxygenase enzymes: catalysis and inhibition. *Curr Opin Struct Biol* 11: 752–760
44 Lecomte M, Laneuville O, Ji C, DeWitt DL, Smith WL (1994) Acetylation of human prostaglandin endoperoxide synthase-2 (cyclooxygnase-2) by aspirin. *J Biol Chem* 269: 13207–13215
45 Trummlitz G, Wittneben H (2001) Insight into the structural basis of selective cyclooxygenase-2 inhibition. *Ann Rheum Dis* 60 (Suppl I): 234 Abstr SAT0080
46 Gierse JK, Koboldt CM, Walker MC, Seibert K, Isakson PC (1999) Kinetic basis for selective inhibition of cyclo-oxygenase. *Biochem J* 339: 607–614
47 Walker MC, Kurumbail RG, Kiefer JR, Moreland KT, Koboldt CM, Isakson PC, Seibert K, Gierse JK (2001) A three-step kinetic mechanism for selective inhibition of cyclo-oxygenase-2 by diarylheterocyclic inhibitors. *Biochem J* 357: 709–718
48 Price MLP, Jorgensen WL (2000) Analysis of binding affinities for celecoxib analogues with COX-1 and COX-2 from combined docking and Monte Carlo simulations and insight into the COX-2/COX-1 selectivity. *J Am Chem Soc* 122: 9455–9466
49 Copeland RA, Williams JM, Giannaras J, Nurnberg S, Covington M, Pinto D, Pick S, Trzaskos JM (1994) Mechanism of selective inhibition of the inducible isoform of prostaglandin G/H synthase. *Proc Natl Acad Sci USA* 91: 11202–11206
50 Pairet M, van Ryn J (1998) Experimental models used to investigate the differential inhibition of cyclooxygenase-1 and cyclooxygenase-2 by non-steroidal anti-inflammatory drugs. *Inflamm Res* 47 (Suppl 2): S93–S101
51 Lanzo CA, Sutin J, Rowlinson S, Talley J, Marnett LJ (2000) Fluorescence quenching analysis of the association and dissociation of a diarylheterocycle to cyclooxygenase-1 and cyclooxygenase-2: dynamic basis of cyclooxygenase-2 selectivity. *Biochemistry* 39: 6228–6234
52 Warner TD, Giuliano F, Vojnovic I, Bukasa A, Mitchell JA, Vane JR (1999) Nonsteroid drug selec-

tivities for cyclo-oxygenase-1 rather than cyclo-oxygenase-2 are associated with human gastrointestinal toxicity: a full *in vitro* analysis. *Proc Natl Acad Sci USA* 96: 7563–7568
53 Blanco F, Guitian R, Moreno 1, Hernandez A, Freire M, Atanes A, Grana J, Toro FJ de, Galdo F (1997) NSAIDs effects on COX 1 and COX 2 activity in human articular chondrocytes. *Arthritis Rheum* 40 (Suppl 9): 347
54 Kawai S, Nishida S, Kato M, Furumaya Y, Okamoto R, Koshino T, Mizushima Y (1998) Comparison of cyclooxygenase-1 and -2 inhibitory activities of various nonsteroidal anti-inflammatory drugs using human platelets and synovial cells. *Eur J Pharmacol* 347: 87–94
55 Tavares IA (2000) The effects of meloxicam, indomethacin or NS-398 on eicosanoid synthesis by fresh human gastric mucosa. *Aliment Pharmacol Ther* 14: 795–799
56 Blanco FJ, Guitian R, Moreno J, de Toro FJ, Galdo F (1999) Effect of antiinflammatory drugs on COX-1 and COX-2 activity in human articular chondrocytes. *J Rheumatol* 26: 1366–1373
57 Kawai S (1998) Cyclooxygenase selectivity and the risk of gastro-intestinal complications of various non-steroidal anti-inflammatory drugs: a clinical consideration. *Inflamm Res* 47 (Suppl 2): S102–106
58 Cryer B, Feldman M (1998) Cyclooxygenase-1 and cyclooxygenase-2 selectivity of widely used nonsteroidal anti-inflammatory drugs. *Am J Med* 104: 413–421
59 Day RO, Francis H, Vial J, Geisslinger G, Williams KM (1995) Naproxen concentrations in plasma and synovial fluid and effects on prostanoid concentrations. *J Rheumatol* 22: 2295–2303
60 Bertin P, Lapicque F, Payan E, Rigaud M, Bailleul F, Jaeger S, Treves R, Netter P (1994) Sodium naproxen: Concentration and effect on inflammatory response mediators in human rheumatoid synovial fluid. *Clin Pharmacol* 46: 3–7
61 Reilly IA, FitzGerald GA (1987) Inhibition of thromboxane formation *in vivo* and *ex vivo*: implications for therapy with platelet inhibitory drugs. *Blood* 69: 180–186
62 Rinder HM, Tracey JB, Souhrada M, Wang C, Gagnier RP, Wood CC (2002) Effects of meloxicam on platelet function in healthy adults: a randomized, double-blind, placebo-controlled trial. *J Clin Pharmacol* 42: 881–886
63 Marbet GA, Yasikoff Strub ML, Macciocchi A, Tsakiris DA (1998) The effect of nimesulide *versus* placebo on hemostasis in healthy volunteers. *Eur J Clin Pharmacol* 54: 383–387
64 Rane A, Oelz O, Frölich JC, Seyberth HW, Sweetman BJ, Watson J, Wilkinson GR, Oates JA (1978) Relationship between plasma concentrations of indomethacin and its effect on prostaglandin synthesis and platelet aggregation in man. *Clin Pharmacol Ther* 23: 658–668
65 Cronberg S, Wallmark E, Sodeberg I (1984) Effect on platelet aggregation of oral administration of 10 non steroidal analgesics to humans. *Scand J Hacmatol* 33: 155–159
66 Vinge E (1985) Arachidonic acid induced platelet aggregation and prostanoid formation in whole blood in relation to plasma concentration of indomethacin. *Eur J Clin Pharmacol* 28: 163–169
67 Cox SR, Vanderlugt JT, Gumbleton TJ, Smith RB (1987) Relationships between thromboxane production, platelet aggregability, and serum concentrations of ibuprofen and flurbiprofen. *Clin Pharmacol Ther* 41: 510–521
68 Schafer AI (1995) Effects of nonsteroidal antiinflammatory drugs on platelet function and systemic hemostasis. *J Clin Pharmacol* 35: 209–219
69 Van Hecken A, Schwartz JI, Depre M, De Lepeleire I, Dallob A, Tanaka W, Wynants K, Buntinx A, Arnout J, Wong PH et al (2000) Comparative inhibitory activity of rofecoxib, meloxicam, diclofenac, ibuprofen, and naproxen on COX-2 *versus* COX-1 in healthy volunteers. *J Clin Pharmacol* 40: 1109–1120
70 Catella-Lawson F, Reilly MP, Kapoor SC, Cucchiara AJ, DeMarco S, Tournier B, Vyas SN, FitzGerald GA (2001) Cyclooxygenase inhibitors and the antiplatelet effects of aspirin. *N Engl J Med* 345: 1809–1817
71 Mengle Gaw L, Hubbard R, Karim A, Yu S, Talwalker S, Isakson P (1997) A study of the platelet effects of SC 58635, a novel COX 2 selective inhibitor [abstract]. *Arthritis Rheum* 40 (Suppl 9): 374
72 Leese PT, Hubbard RC, Karim A, Isakson PC, Yu SS, Geiss GS (2000) Effects of celecoxib, a novel cyclooxygenase-2 inhibitor, on platelet function in healthy adults: a randomised, controlled trial. *J Clin Pharmacol* 40: 124–132
73 McAdam BF, Catella-Lawson F, Mardini IA, Kapoor S, Lawson JA, FitzGerald GA (1999) Systemic biosynthesis of prostacyclin by cyclooxygenase (COX)-2: the human pharmacology of a selective inhibitor of COX-2. *Proc Natl Acad Sci USA* 96: 272–277
74 Catella-Lawson F, McAdam B, Morrison BW, Kapoor S, Kujubu D, Antes L, Lasseter KC, Quan

H, Gertz BJ, FitzGerald GA (1999) Effects of specific inhibition of cyclooxygenase-2 on sodium balance, hemodynamics, and vasoactive eicosanoids. *J Pharmacol Exp Ther* 289: 735–741
75 Depre M, Ehrich E, Van Hecken A, De Lepeleire, I, Dallob A, Wong P, Porras A, Gertz BJ, De Schepper PJ (2000) Pharmacokinetics, COX-2 specificity, and tolerability of supratherapeutic doses of rofecoxib in humans. *Eur J Clin Pharmacol* 56: 167–174
76 Dallob A, Depré M, De Lepeleire I, Agrawal N, Musser B, Kher U, Van Hecken A, Wildonger L, Matthews C, De Schepper P et al (2000) MK-0663 a highly selective inhibitor of COX-2 in humans. *Ann Rheum Dis* 59 (Suppl 1): 331
77 Riendeau D, Percival MD, Brideau S, Charleson S, Dubé D, Ethier D, Falgueyret J-P, Friesen RW, Gordon R, Greig G et al (2001) Etoricoxib (MK-0663): Preclinical profile and comparison with other agents that selectively inhibit cyclooxygenase-2. *J Pharm Exp Ther* 296: 558–566
78 Leese PT, Talwalker S, Kent JD, Recker DP (2002) Valdecoxib does not impair platelet function. *Am J Emerg Med* 20: 275–281
79 Sorbera LA, Canstañer J, Bayés M, Silvestre JS (2002) Lumiracoxib. *Drugs of the Future* 27: 740–747

COX-2 Inhibitors
Edited by M. Pairet and J. van Ryn

Pharmacology and toxicology of COX-2 inhibitors

K.D. Rainsford

Biomedical Research Centre, Sheffield Hallam University, Howard Street, Sheffield S1 1WB, UK

Introduction

The pharmacological premise upon which the prostaglandin G/H synthase II (PGHS-II) or cyclo-oxygenase-2 (COX-2) inhibitors was developed was essentially that founded in the identification of two different genes coding for the cyclo-oxygenases-1 and –2 [1–3]. COX-1 is considered to be a constitutive enzyme that is responsible for "housekeeping" or physiological functions (e.g., some gastrointestinal (GI) mucosal, renal and haemostatic functions and protective effects) while COX-2, which is induced by inflammatory stimuli, is considered to be important in inflammation, pain and fever [1–5]. Comprehensive reviews on the regulation and roles of these two isoenzymes in inflammation are found in [1–9] and Figure 1 summarizes their essential features.

This review examines the published studies which support the claims for COX-2 selective inhibitors having comparable therapeutic actions to conventional NSAIDs, and in standard laboratory and clinical models under controlled conditions has improved GI safety compared with some of the conventional NSAIDs.

Determination of COX selectivity

The determination of COX-2 selectivity of NSAIDs has relied upon a variety of different assay methods that evolved after the discovery of the COX-isoforms about a decade ago. However, the clinical relevance of many of these *in vitro* assays has been questioned. The pharmacokinetic relationships, especially the plasma concentrations of drugs, have been related to their *in vitro* effects in an attempt to more accurately define the clinical significance of relative COX-1 and COX-2 inhibitory effects to clinical actions [10–12]. An attempt to arrive at consensus criteria for defining the clinical significance of COX-2/COX-1 selectivity was made at an international meeting sponsored by one of the major coxib drug companies (Merck) [11]. This has proven useful in bringing together the essential pharmacokinetic and clinical criteria.

Understanding of the role of intracellular kinetics and physico-chemical parameters underlying drug actions *in vitro* has been elegantly explored [13,

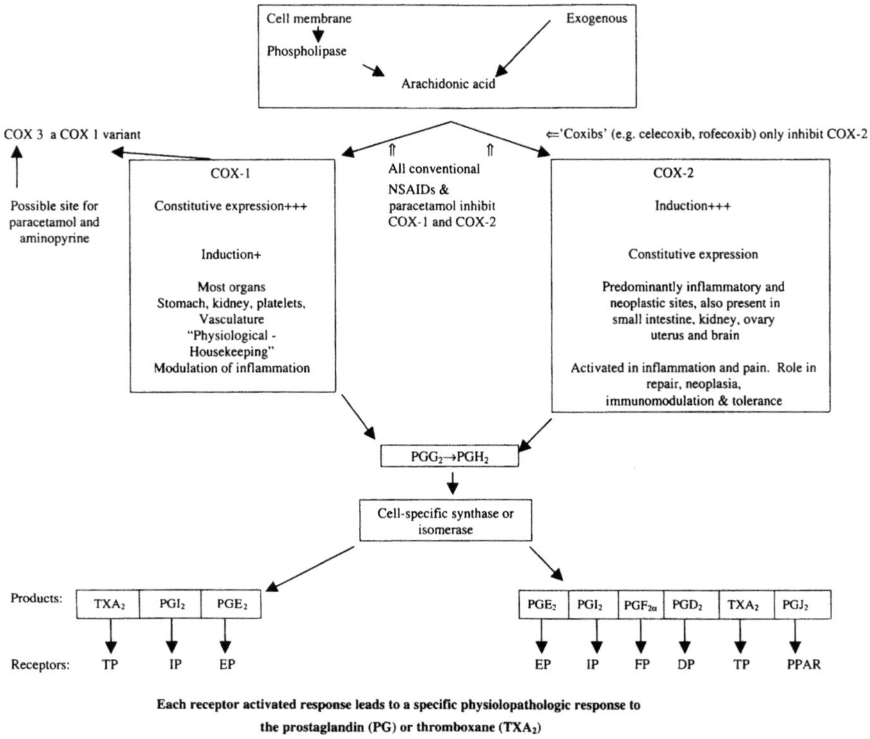

Figure 1. Roles of COX-1 and COX-2-derived prostanoids in regulating physiological (COX-1) or inflammatory (COX-2) reactions. COX-3 is a splice product of COX-1 found in the brain and some other organs and is presumed to have a role in pain. The inhibition of COX-3 by paracetamol (acetaminophen), dipyrone as well as some NSAIDs has been suggested to underlie their analgesic actions in the CNS.

14] and the importance of the nature of stimuli and the effects of the exogenous and endogenous supply of the substrate arachidonate on enzyme activity determined [15, 16]. Table 1 shows data on the *in vitro* effects of a range of NSAIDs and coxibs determined using the whole blood assay [17]. The relationship between the COX-1/COX-2 ratios for enzyme inhibition and occurrence of severe GI events is inferred from data shown in Figure 2 [18].

COX-2 selective NSAIDs

Etodolac

Etodolac ([±]-1,8-diethyl-1,3,4,9-tetrahydropyrano(3,4-b)indole-1-acetic acid Lodine®; Fig. 3) is chemically distinct from currently used NSAIDs owing to

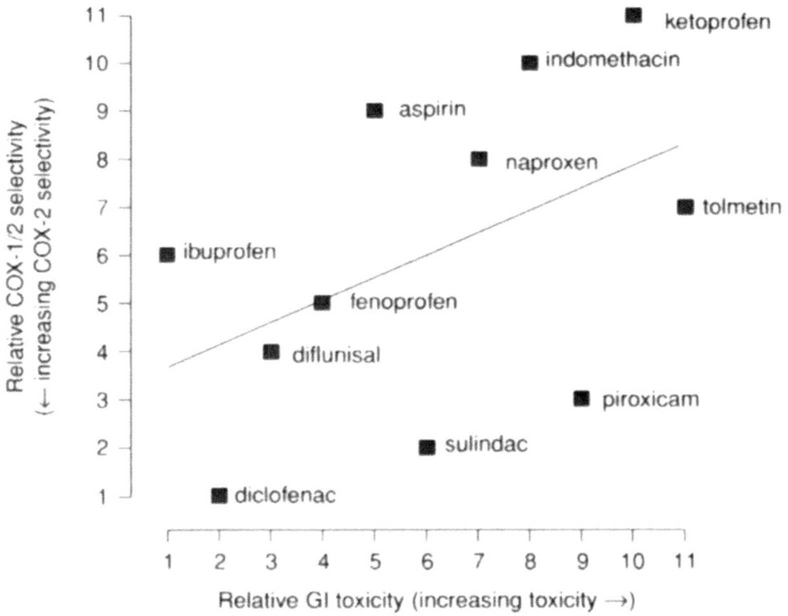

Figure 2. Relationship between COX-1/COX-2 inhibition ratios and the development of serious gastrointestinal events by unselective COX inhibitors. From Warner et al. [18] with permission of the publishers.

its tetrahydropyrano-indole nucleus. It was originally discovered in 1976 and developed by Ayerst Laboratories in Montreal (Canada) [19, 20]. Etodolac is thus among the NSAIDs whose COX-2 selectivity was established long after it was established in clinical use with a relatively favourable GI profile [20, 21]. Etodolac was originally selected in preclinical development on the basis of having the best therapeutic ratio compared with established NSAIDs and related pyranolone analogues in response to therapeutic effects in the adjuvant-induced chronic polyarthritis in rats in comparison with its gastro-ulcerogenic effects [19]. This was associated with a relative lack of acute effects of the drug on the rat gastric mucosa [19].

COX-2 selectivity

Earlier studies suggested that etodolac may spare the production of gastroprotective prostaglandins (PGs) in the gastric mucosa [22] so inferring that this may account for the low GI ulcerogenicity and bleeding of this drug in animals [19] and in human studies including those in arthritic subjects and those at high risk of serious NSAID-induced GI events [21]. In more recent studies etodolac has been shown to have COX-2 selectivity with values ranging from 10-fold in human recombinant enzymes systems, 24-fold in the modified whole blood

Table 1. Potencies of compounds as inhibitors of prostanoid formation determined in the COX-1 assay, WBA-COX-2

	COX-1 IC_{50}-μM	WBA-COX-2 IC_{50}-μM	WHMA-COX-2 IC_{50}-μM	IC_{50} ratios		Ranking at IC_{50} ratios	
				WBA COX-1	WHMA COX-1	WBA COX-1	WHMA COX-1
NSAIDs							
5-aminosalicylic acid	410	61	n.d.	0.15	n.d.	n.d.	n.d.
Aspirin	1.7	>100	7.5	>100	4.4	34	23
Diclofenac	0.075	0.038	0.020	0.5	0.3	10	9
Diflunisal	113	8.2	134	0.1	1.2	9	14
Flubiprofen	0.075	5.5	0.77	73	10	31	27
Flufenamate	3.0	9.3	n.d.	3.1	n.d.	13	n.d.
Ibuprofen	7.6	7.2	20	0.9	2.6	14	20
Indomethacin	0.013	1.0	0.13	80	10	29	24
Ketoprofen	0.047	2.9	0.24	61	5.1	31	25
Meclofenamate	0.22	0.7	0.2	3.2	0.91	22	11
Mefenamic acid	25	2.9	1.3	0.11	0.049	-	-
Naproxen	9.3	28	35	3.0	3.8	18	22
Piroxicam	2.4	7.9	0.17	3.3	0.1	17	13
Salicin	>100	>100	n.d.	-	n.d.	-	n.d.
Salicylaldehyde	>100	>100	n.d.	-	n.d.	-	n.d.
Sodium salicylate	4956	34440	482	6.9	0.10	16	15
Sulfasalazine	3242	2507	n.d.	0.8	n.d.	15	n.d.
Sulindac	>100	>100	58	-	-	-	-
Sulindac sulphide	1.9	55	1.21	29	0.64	20	10
Valeryl salicylate	42	2.3	n.d.	0.053	n.d.	-	n.d.

(Continued on next page)

Table 1. (Continued)

	COX-1 IC_{50}-μM	WBA-COX-2 IC_{50}-μM	WHMA-COX-2 IC_{50}-μM	IC_{50} ratios		Ranking at IC_{50} ratios	
				WBA COX-1	WHMA COX-1	WBA COX-1	WHMA COX-1
COX-2 Inhibitors							
Celecoxib	1.2	0.83	0.34	0.7	0.3	8	7
Etodolac	12	2.2	0.94	0.2	0.1	6	5
Meloxicam	5.7	2.1	0.23	0.37	0.040	11	6
Nimesulide	10	1.9	0.39	0.19	0.038	7	8
L745,337	>100	8.6	1.3	<0.01	<0.01	1=	1=
NS398	6.9	0.35	0.042	0.051	0.0061	5	4
Rofecoxib	63	0.84	0.31	0.013	0.0049	4	3
SC58125	>100	2.0	n.d.	>0.01	n.d.	1=	n.d.
Analgesics (non-narcotic)							
Aminopyrone	55	203	85	3.7	1.5	24	19
Paracetamol	>100	49	64	-	-	-	-

From Warner *et al.* [17].

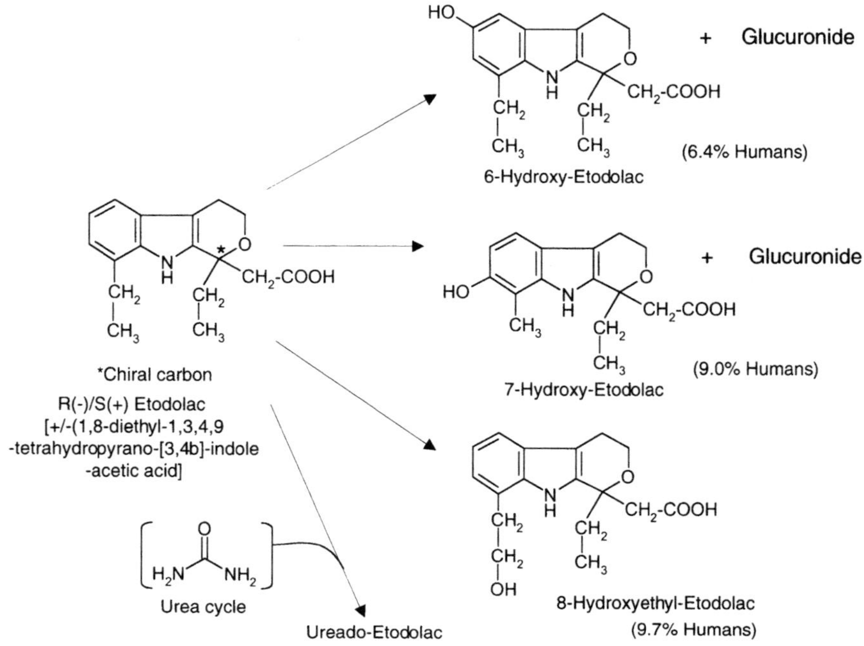

Figure 3. Pathways for the metabolism of R(–)/S(+) etodolac showing the prominent metabolites formed in humans. Based on data in [20, 26, 27].

assay up to 1000-fold in CHO cells transfected with human recombinant enzymes [21]. These data have, however, been derived from studies using the diastereoisomeric form of the drug that is commercially available. The component enantiomers of (±) etodolac (AY-24236) exhibit differing PG synthetase inhibitory activities, with the S(+) enantiomer (AY-28689) having greater potency than the R(–) (AY-28688) which is almost inactive in most assays) [23]. No attempts appear to have been made to establish if the S(+) enantiomer has the same degree of COX-2 selectivity as observed with the racemic mixture. With ibuprofen there are differing influences of the inactive R(–) enantiomers on the inhibitory effects of S(+) ibuprofen on COX-1 and COX-2 activity [24] and this may exist with the enantiomers of etodolac as well.

Despite the differences in PG synthesis inhibitory effects of the two enantiomers there is little difference in their acute gastro-irritant effects in the cholinomimetic-sensitized mouse assay [25]. However, the gastric irritancy of the racemate is lower than that of the S(+) or R(–) enantiomers [25] suggesting that there may be physical interactions between the two that in some way mask the potential irritancy of the individual enantiomers. There is no bioconversion of R(–)- to the S(+)-enantiomer of etodolac [25] in contrast to that of some propionic acids e.g., ibuprofen [24]. Thus, it is unlikely that there are any influences of bioconversion of R(–) to S(+) etodolac on gastric irritancy.

Pharmacokinetics

The pharmacokinetics of etodolac in humans, rats and dogs involves presumed cytochrome P_{450} mediated hydroxylation to the 5-, 6-, 7- and 8-hydroxymetabolites [20, 26, 27] and subsequent phenolic glucuronidation of most if not all of these hydroxyl groups (see Figure 3 for details of metabolism; Table 2 and [20, 26, 27] for details of pharmacokinetic parameters in humans and rats). There is little information available on the possible formation of the acyl glucuronides.

The main pathways of the metabolism of etodolac in humans appear similar to those in rats and dogs [19, 26–28] but there are quantitative differences in the rates and routes of elimination. Pronounced biliary excretion and subsequent enterohepatic recirculation of both unchanged drug and the glucuronides occurs in rats [26] and dogs [19] with consequent predominant excretion in the faeces and to a lesser extent the urine [19]. Thus, the expected intestinal injury in rats and dogs in long-term toxicity studies from etodolac is probably related to the entero-hepatic re-circulation of the drug [19]. However, in short-term studies this drug does not exhibit intestinal injury in these species [29–31]. There is less faecal excretion and greater urinary excretion of the drug in humans but, nonetheless, the rare occurrence of cases of colitis [31] and benign colonic strictures [32–34], which may be related to breakdown in the intestinal barrier and permeability changes that are induced by NSAIDs [35, 36]. Etodolac is rapidly absorbed after oral intake. Its pharmacokinetic profile in humans favours twice daily dosing having a plasma half-life for elimination ($t_{1/2}$) of 6–7 h after single or multiple dosages for 7 days [19]. None of the pharmacokinetic parameters are appreciably affected by age, moderate renal impairment, or in patients with stable liver cirrhosis or those with rheumatic conditions [19, 20, 37, 38]. Thus conditions in the high-risk groups of patients where drug accumulation occurs with some other NSAIDs poses less of a risk for systemic toxicity with etodolac, although obviously there would be need to have caution in these "at risk" subjects with the drug.

GI and other adverse effects

The drug also appears to have low risk for GI ulcers and haemorrhage in the elderly who are at risk for developing GI ulcers [19, 39]. Long-term clinical and epidemiological data show there is a lower risk for GI ulcers or bleeding with etodolac than other NSAIDs including ibuprofen [19, 40, 41]. It should be noted, however, that the rigor of data from such studies as compared to prospective, randomised, double blind trials with coxibs has been questioned [42]. The relative inhibitory effects of etodolac on platelet aggregation and COX-1 activity are low [43, 44] and low rates of microbleeding (from ^{51}Cr-red cell loss) suggest that even where there is evidence of mucosal injury from whatever cause there is likely to be a low risk of bleeding. Adaptation of the GI mucosa to repeated intake of etodolac 300 mg b.d. for up to 28 days has been observed and was similar to that of naproxen 500 mg b.d. but etodolac caused increased antral blood flow, which was reduced in naproxen-treated

Table 2. Principal pharmacokinetic properties of COX-2 selective drugs

Drug	C_{max}	T_{max}, hr	$T_{1/2}$, hr	V_D, litres	Renal $^{l/hr}$	Dosage mg	PKa	Comments	Refs
Etodolac	9.5–18.6 μg/ml	1–2	6.0–7.4	~10		200–600 o.d.	4.8	Metabolism probably via CYP 2C9	[17, 158]
Meloxicam	0.9–2.5 μg/ml	4.9–5.6	20–22	10–15	7.19–11.7	7.5–15 o.d.	1.09, 4.18	Metabolism via CYP 2C9	[75, 79, 82, 87, 158]
Nimesulide	2.86–9.5 μg/ml	1.22–3.17	1.22–4.95	12.6–27.3	2.11–5.18	100–200 b.i.d.	6.4–6.8	Predominant metabolism via CYP 2C9	[156, 162]
Celecoxib	705 ng/ml	2.8	11.2	429	27.7	100–200 b.i.d.	11.1	Metabolism via CYP 2C9	
Rofecoxib	207 ng/ml	2–3	17	86–91		12.5–50 o.d.	11.0	Predominant cytosolic NADH-dependent metabolism with CYP3A minor route	
Lumiracoxib	5.96–6.64 μg/ml	1.22–3.17	1.22–4.95	12.6–27.3	2.11–5.18	100–200 b.i.d.	6.4–6.8	Predominant metabolism via CYP 2C9	[358, 360]

Most values of plasma C_{max}, T_{max} and $t_{1/2}$ after multiple dosage. Values of V_D approximated to 70 kg adult.

subjects [45]. This adaptive feature appears to be reduced in elderly subjects or those who are smokers when they take etodolac or other NSAIDs [46].

Etodolac inhibits mucosal cell proliferation [47], adhesion molecule expression [48] and induction of apoptosis [49, 50], as well as inhibition of cytokine [51, 52], *Helicobacter pylori-* [53, 54] and growth factor [53]-induced production of PGE_2 or other prostanoids, similar to other COX-2 inhibitors. Prevention of colon cancer and other cancers [55–57] may also represent potential inhibitors of regeneration or repair of the GI mucosa following injury by ulcerogens. Inhibition of bone fracture healing by etodolac via inhibition of COX-2 [58] may also be considered a potential negative effect of this and other NSAIDs though the clinical significance or conditions of limitation of the use of this drug and other COX-2 inhibitors in post-orthopaedic surgical conditions have not been evaluated.

Renal impairment and other renal syndromes are rare conditions with etodolac [19] even though etodolac has been found to affect mechano-sensory neuronal function in the renal pelvic wall via a presumed COX-2 mediated mechanism [59]. There may also be other consequences of inhibiting COX-2 in the macula densa and other cells of the kidney that may promote renal impairment in at risk subjects. Etodolac has been reported to cause fulminant hepatic failure [60] though this and other hepatic events are not frequent occurrences.

Meloxicam

The discovery that this enol-carboxamide NSAID had COX-2 selectivity arose at the time when this drug had been selected for, and was well in progress, in phase III clinical trials. Thus, selection of this drug as a clinical candidate was not originally based on its COX-2 selectivity, but on the basis of conventional pharmacokinetics, pharmacodynamic, pre-clinical and phase 1 studies.

Meloxicam was first granted marketing authorization in 1995 and is now available in more than 100 countries [61]. The drug has been studied extensively in clinical trials involving more than 30,000 patients [61, 62]. It is recommended for, and has demonstrated efficacy, in osteoarthritis (OA), rheumatoid arthritis (RA) and ankylosing spondilitis, and has been used in some pain models [61, 62]. Additionally, the drug has been employed for veterinary use principally for treating of pain and inflammation in osteoarthritis in dogs where it has been found effective in at least two studies in joint inflammation [63, 64] and it has a wide range of other veterinary uses. In addition to the conventional oral formulation of the drug (Mobic®), the drug has been shown to have acceptable tissue tolerability when administered by intravenous/intramuscular or subcutaneous injection as well as in eyedrops, gels and suppositories [65], and additionally has been shown to be efficacious when given by intramuscular injection in patients with RA or OA [66].

Discovery and development of meloxicam

As indicated above the discovery of meloxicam took place in the pre-COX-2 era and thus the pharmacological profiling of the drug was based on determination of inhibitory effects of PG synthesis using bull seminal vesicles (which are now known to be comprised of COX-1) in addition to kaolin oedema, acute paw oedema in rats and screening for anti-arthritic activity using the adjuvant arthritis model [67, 68]. The initial aims were to base the drug development on the earlier oxicam class of drugs, sudoxicam and piroxicam and create a third generation of oxicam NSAID in which the pharmacokinetic properties were optimised for once daily administration (the very long half-lives of other oxicams appearing to be related to the development of GI and other serious side-effects), rapid and high distribution to the synovial fluid and improved risk to efficacy relationships [68]. In addition later studies showed that meloxicam did not have the potential to develop the so-called metabolite "C" (Fig. 4) which was associated with the development of phototoxic reactions with piroxicam [69] and as well autoradiographic studies showed that meloxicam had a very low skin penetration compared with that of other oxicams, thus reducing the potential for the development of phototoxicity reaction [69]. These earlier studies together with insight into the structure and physicochemical properties of meloxicam [70, 71] indicated that the increased lipophilicity might result in a shorter plasma half-life of the drug and that the relatively low solubility in the gastric pH range of meloxicam [70] may give a relatively low GI irritancy of this drug compared with that of established NSAIDs [61, 67, 71]. Similar arguments have been made to explain the low acute and subchronic GI irritancy of another keto-enolic NSAID, azapropazone [72]. The keto-enolic function of meloxicam produces zwitterions at high solvent polarity. Since the pKa values of meloxicam are 1.09 and 4.18 respectively the keto-enolic form is predominant in the gastric milieu whereas the enolic form at the same pH range is more likely to exist in the highly lipophilic environment [70] such as is present in mucosal membranes; the presumption here is that the enolic form of the drug is unlikely to be present following absorption of the drug and therefore

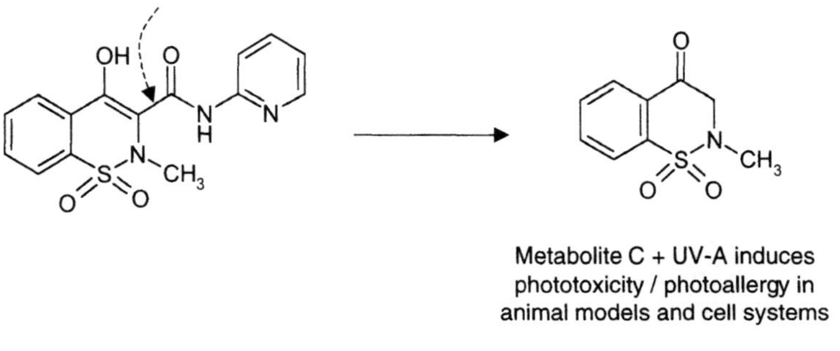

Figure 4. Formation of metabolite "C" from photo-activation of piroxicam which is presumed to underlie the development of photo-allergic reactions to this drug [69]. Inspection of the pathways of meloxicam metabolism (Fig. 4) shows that the formation of this metabolite is not possible with the latter drug.

would not, like acetylsalicylic acid, be capable of releasing a proton or protons to cause intracellular acidification and mucosal injury. Similar physicochemical explanations have been advanced to explain the differences in the gastric ulcerogenicity of low *versus* high gastrotoxic drugs [73].

Comparisons of the GI ulcerogenicity of meloxicam with standard NSAIDs as well as the therapeutic index relating ulcerogenicity to antiinflammatory activity in the adjuvant arthritis test is shown in Table 3 [74–76]. The ulcerogenicity studies were performed in normal rats and these showed a favourable therapeutic ratio of the order of about 20 compared with that of other NSAIDs which were all appreciably lower. In comparison with these data from 3 day dosage of rats the acute ulcerogenic effects of meloxicam showed high indexes (ED_{50} = 15.8 mg/kg) and in the intestine (ED_{20} = 19.7 mg/kg) and these were appreciably lower than the ulcerogenicity from indomethacin in the stomach, ED_{50} = 2.6 mg/kg and in the intestine 3.4 mg/kg [77].

In the Shay rat (in which acid secretion is measured in pyloric ligated animals), meloxicam produced a smaller increase in gastric acid production compared with diclofenac or piroxicam [77]. Furthermore, production of PGE_2 into the gastric juice of rats given meloxicam as well as gastric emptying was much lower than that of comparator NSAIDs.

Long-term toxicity studies undertaken with meloxicam compared with other similar, but not identical studies, with rofecoxib and celecoxib are shown in Table 4. These studies indicate that over a 50-week period meloxicam had a non-effect dose level of 0.8 mg/kg/day compared with that of rofecoxib 1.0 mg/kg/day while celecoxib did not achieve a non-effect dose level in a 2-year study. In comparison with the therapeutic effects in adjuvant arthritis the non-effect dose level was about 6–7 times higher than the ED_{50} dose, whereas that with rofecoxib was much closer to the therapeutic dose of 0.74 mg/kg/d for the adjuvant arthritis.

In the long-term toxicity studies the kidney was also a major target organ for the action of meloxicam, in addition to the GI mucosa, and the drug was found to delay parturition in pregnant rats. However, the dose required for

Table 3. Comparison of the gastro-intestinal effects of meloxicam with anti-arthritic effects in rats

	GI-erosions overall UD_{50} (mg/kg)	Inhibition of adjuvant arthritis ID_{50} mg/kg/d	Therapeutic ratio GI injury/anti-arthritic effects (UD_{50}/ID_{50})
Meloxicam	2.47	0.12	20
Diclofenac	2.71	1.24	2.2
Naproxen	11.1	11.8	0.9
Piroxicam`	1.07	0.76	1.4
Flubiprofen	1.4	0.96	0.3

Data from [77]. Similar ID_{50} values for inhibition of adjuvant-induced arthritis can be found in [74].

Table 4. Comparison of the GI Ulcerogenic potential of some COX-2 inhibitors in subacute and long-term toxicity studies in rats

	Meloxicam	Celecoxib	Rofecoxib
Sub-acute GI-ulcers	UD_{50}: 2.42 mg/kg[1]		No lesions at 300 mg/kg[3]
GI-bleeding ^{51}Cr-RBC	n.d.	not done	No effect at 300kg/kg[3]
Chronic GI-ulcers 52 weeks	<1 mg/kg[2] NOAEL: 0.8 mg/kg[2]	NOAEL not achieved in 2 y study[4]	<2 mg/kg[5] NOAEL: 1.0 mg/kg[5]
Therapeutic effects ID_{50} in adjuvant arthritic rats	0.12 mg/kg[1]	0.3 mg/kg[4]	0.74 mg/kg[1]

[1]Engelhardt et al. [74]
[2]Lehmann et al. [69]
[3]Chan et al. [319]
[4]Celecoxib summary basis of approval 20998, FDA [242]
[5]Rofecoxib summary basis of approval 021042/52 [318]

these toxic effects was well above those of the pharmacological effects [74]. In a 2-year carcinogenicity study in rats and mice there were no haematological/biochemical abnormalities, nor any evidence of major histopathological changes or evidence of mutagenicity.

Pharmacokinetics

The pharmacokinetic properties of meloxicam in humans (Tab. 2) are similar to that in rats [75, 78–80]. Thus, the values for plasma maximal concentration (Cp_{max}), time of maximal concentration (T_{max}), clearance (Cl), area under the plasma concentration curves (AUC's) and volumes of distribution (V_D) after single and multiple dose are similar in humans and rats ([75–80], Tab. 2). In addition the metabolic pathways (Fig. 5) are similar in those two species but for some additional minor metabolites whose identities have been established in radiolabel studies in rats [81] and where there is greater possibility experimentally to determine such minor metabolites (Fig. 5). These observations are of significance since the timing, dosing conditions and correlated pharmacological studies can be considered a good basis for relating studies in rats to those in humans – an unusual situation for NSAIDs, who often have differing patterns of metabolism and faster rates of clearances in rats compared with humans.

Oxidative metabolism of meloxicam in the human liver is by cytochromes P450 (CYP) 2C9 and 3A4 [76, 82]. There are indications from recent pharmacokinetic studies in Chinese subjects that differences in metabolism of meloxicam in "slow" *versus* "fast" metabolisers may be due to genetic polymorphisms of CYP 2C9 [83].

The synovial uptake of ^{14}C-meloxicam into the joints of arthritic rats has shown that the drug has a high rate of joint penetration [81] and this is confirmed in clinical studies [84].

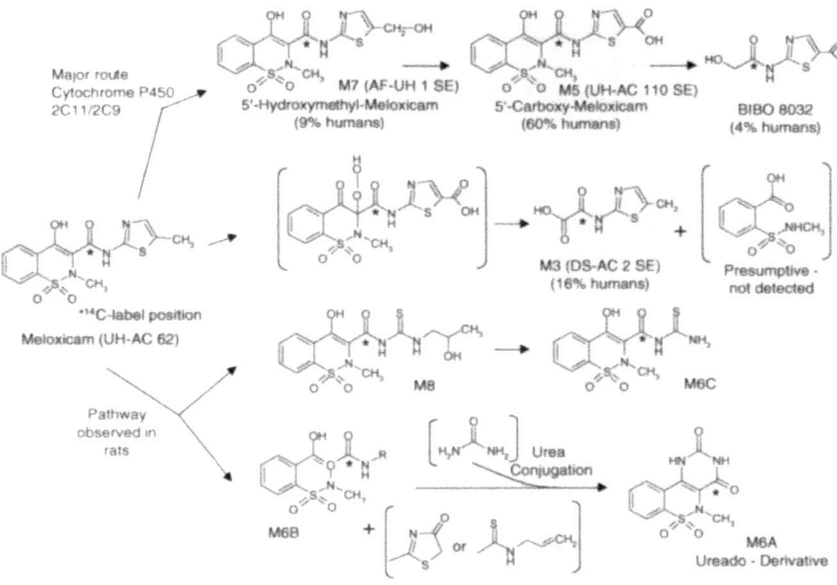

Figure 5. Pathways for the metabolism of meloxicam based on radio-label studies in rats [81] and conventional mass spectral identification of isolated in humans [75, 78–80]. The two upper pathways show the identified metabolites in humans with the compounds shown in brackets [ ] being presumed metabolites, while those in the low pathways have been additionally found in rats along with those compounds shown in the two upper pathways.

Metabolism and kinetic studies have been performed in a number of small and large animals as a basis for veterinary applications of the drug as well as being of interest in relation to comparative animal pharmacology [78, 80, 81, 85–91]. In addition to being available for use in dogs in several European countries, the oral suspension of meloxicam (Metacam®) was recently approved in the USA for use in the treatment of pain and inflammation associated with osteoarthritis [91].

Extensive clinical studies have been undertaken examining interactions with food, commonly co-administered drugs and differing routes of administration [92]. Pharmacokinetics following intramuscular injection showed that there was rapid uptake and onset of analgesia with low severity of reactions at the sites of injection [92] making this a particularly useful drug for rapid analgesia.

Pharmacological actions – COX inhibition

Meloxicam is a potent inhibitor of COX-2 *in vitro* (Tab. 1) while its metabolites do not appear to affect COX-pathways [77, 85, 93]. Confirmation of the selective inhibition of COX-2 by meloxicam has also been shown in animal models *in vivo*. Engelhardt et al. showed that orally administered meloxicam produced potent inhibition of PGE_2 production in the rat pleural cavity and mouse peritoneal cavity after carrageenan injection that is presumably due principally to COX-2 activity with IC_{50} of 0.65 mg/kg/d and 1.36 mg/kg respectively [94]. These data should be compared with data from dosing with piroxicam (IC_{50}'s 0.85 mg/kg/d, 2.62 mg/kg for rat and mouse systems) diclofenac (IC_{50}'s 5.06 mg/kg/d, 16.0 mg/kg/d) and naproxen (IC_{50}'s 14.17 mg/kg/d, 22.86 mg/kg) respectively [94]. Aspirin was virtually inactive in these systems. In the rat cotton-pellet granuloma assay meloxicam p.o. inhibited PGE_2 production with an ID_{50} of 0.88 mg/kg [77]; a lipogel formulation of meloxicam topically applied was also effective in inhibiting PGE_2 production [77].

Ogino et al. [95] showed COX-2 selective inhibition of meloxicam 3 mg/kg in the rat carageenan-inducing pleurisy model. However, they found there was a 50% reduction in gastric mucosal PGE_2 (presumably principally from COX-1) which was claimed to be not statistically significant from control.

Estimations of the COX-1 and COX-2 inhibition from plasma concentrations of meloxicam after dosing 7.5 or 15 mg/day to humans suggest that there would be 55–70% inhibition of COX-2 and 10–25% inhibition of COX-1, respectively [10, 77]. In 14 female volunteers given meloxicam 7.5 mg/day or indomethacin 25 mg t.i.d., platelet aggregation and thromboxane B_2 production were unaffected by meloxicam but almost completely inhibited by indomethacin [77]. Urinary production of PGE_2 was also not affected by meloxicam, whereas it was inhibited by 43% by indomethacin [77]. *Ex vivo* effects of meloxicam 7.5 or 15 mg given to human volunteers in the whole blood assay showed that thromboxane B_2 production was inhibited by 21 and 36% at the respective doses [77]. PGE_2 production (presumably COX-2 derived) was inhibited by 48 and 71% at 7.5 or 15.0 mg meloxicam [77].

Overall, these data from *in vivo*/*ex vivo* studies in animals and human volunteers show there is predominantly COX-2 inhibition by meloxicam, with a lesser, dose-dependent inhibition of COX-1. This could account for some inhibitory effects on gastric mucosal PGE_2 and may even be beneficial in control of COX-1 derived PGs in the analgesic pathways in the central nervous system (CNS). Enhanced uptake of meloxicam into the rat brain has been suggested to be related to binding to α_1-acid glycoprotein aiding penetration though endothelial cells of the blood-brain barrier [86]. The clinical significance of these studies is not clear but may be related to the central mechanisms of analgesia. The lack of cardiovascular events of meloxicam in clinical trials (see Chapter 7) which contrasts with the occurrence of these events in rofecoxib-treated patients, may be due some mild COX-1 inhibition and modest cardioprotection in susceptible patients and as well may contribute to the COX-1 component of central analgesia which appears of significance as well as that involving inhibition of COX-2.

A structural isomer of meloxicam, 4'-meloxicam, has effects on COX-1 as well as COX-2 in contrast to meloxicam itself [85].

Effects on oxyradical and nitric oxide (NO) production and actions
Human polymorphonuclear neutrophil (PMN) leucocytes stimulated with tumour necrosis factor-α (TNF_{α}) and formyl-methionyl-leucyl peptide (fMLP) and analysed by fluorescence activated cell sorting (FACS) for intracellular oxyradical production were found to be inhibited by 3–15 μM meloxicam in the presence of azide as an inhibitor of myeloperoxidase [96]. Inhibition of chemiluminesce from neutrophils stimulated with histone-opsonized *Streptococci* was also noted but at higher concentrations [96].

In the mouse carrageenan-induced pleurisy assay meloxicam, like that of indomethacin and nabumetone, inhibited exudate formation, PMN cell migration and myeloperoxidase at the first phase (4 h) but nitrate production was unaffected by meloxicam or indomethacin at the second peak at 48 h [97].

An *in vivo* study on adjuvant arthritic rats showed that meloxicam dose-dependently reduced the elevated levels of lipid peroxides in the plasma, glutathione peroxidase in the blood and restored to near-normal total plasma thiols and albumin, but did not alter erythrocyte superoxide dismutase [98]. It is not possible to establish from this study if there were direct effects of meloxicam on components of oxyradical production, actions as a consequence of the PG-related effects, inhibition of leucocyte recruitment or activation by the drug. No gastric lesions were observed in the stomachs of the arthritic rats with meloxicam although there was significant reduction in paw swelling [98].

Gastric lesions that develop following celiac artery clamp-induced ischaemia reperfusion in rats reach a maximum at 12 h and progress to deep ulcers at 3 days [99]. In this model, Brzozowski et al. found that there was initially reduced gastric mucosal blood flow and mucosal PGE_2 and elevated COX-2 mRNA [99]. Meloxicam, indomethacin, the COX-1 inhibitor resveratrol, rofecoxib and NS-398 increased the acute gastric lesions that develop at the early stage and delayed progression of these lesions to form ulcers [99]. The results suggested that both COX-1 and COX-2 derived PGs are important in ulcer healing [99] but there were other interpretations of these results, for example, effects on leucocyte accumulation/activation as well as effects on oxyradical production from accumulated leucocytes.

In a related study by the same group the protective effects of mild stress in rats from water immersion against subsequent severe stress were abolished by meloxicam 2 mg/kg p.o. and indomethacin 5 mg/kg i.p. as well as by NS-398 10 mg/kg p.o. coincident with reduction in protective hyperaemia, and in the case of meloxicam and indomethacin reduction in mucosal PGE_2 [100]. These results in what is a complex model of mucosal protection/ulcer formation imply blood flow/oxyradical effects as well as influence of inhibition of mucosal PG production that is elevated during the protective effects of mild stress. It is hard to discriminate the effects of meloxicam on a particular pharmacological target in these studies. Whether they have any clinical significance is obscure.

Acute gastric mucosal injury by meloxicam (3.75–30 mg/kg) in rats was found associated with increased xanthine oxidase activity but there was no increase in mucosal myeloperoxidase activity (an index of leucocyte infiltration) or the lipid peroxidase scavenging enzyme glutathione peroxidase [101].

These results suggest that mucosal injury from meloxicam may be independent of leucocytes but involve xanthine oxidase derived superoxide generation and, depending on the nature and type of stress factors, could involve inhibitory effects of the drug on mucosal PG production. Reduced mucosal glutathione concentrations may also play a role in meloxicam-induced injury [102].

Leucocyte infiltration/activation is an important target for NSAIDs including meloxicam during acute inflammation induced by agents, e.g., carrageenan, and these dugs combined with NFκB-inhibitors have been found to act synergistically on leucocytes [102, 103].

Analgesic, anti-inflammatory and antipyretic activities

Among the sites of action of meloxicam are the cytokines and leucocyte adhesion molecules. Inhibition of TNF_{α} production has been shown in lung inflammation bronchiolavage samples [104] and interleukin (IL)-8 levels in synovial fluids in antigen-induced arthritis in rabbits [105]. However, synovial production of IL-8 as well as IL-1 and IL-6 from human OA synovium in organ culture was not affected by meloxicam 0.1–10.0 μM [106]. Increased production of IL-1 which is associated with inhibition of PG production [107, 108] and which is considered a factor in promotion of cartilage damage by indomethacin in OA [109] is not evident with meloxicam.

The systemic inflammatory responses to endotoxin given i.v. were shown to be markedly reduced in heifers by prior i.v. meloxicam 0.5 mg/kg [110]. The drug produced inhibition of plasma $PGF_{2\alpha}$ metabolite, cortisol and alterations in Ca^{2+}, Fe^{2+} and Zn^{2+} induced by endotoxin but did not prevent the febrile response due to this inflammagen. Fever in guinea pigs induced by lipopolysaccharide (LPS) was reduced by meloxicam 5 mg/kg i.a. along with reduced plasma cortisol and IL-6, but TNF_{α} levels were increased [111].

Meloxicam has been found to have potent analgesic activity in conventional acute and chronic pain models in rodents [67, 75–77, 112–115]. Meloxicam has potent analgesic activity in OA or joint inflammation induced in dogs [63, 64], in chronic arthritic disease clinical trials [89, 116, 117], or in perioperative pain in dogs [118, 119] demonstrating its importance in veterinary applications in relief of pain [116–124]. Here meloxicam was comparable in analgesia with standard NSAIDs. Analgesic activity has been shown for meloxicam following surgery in rats, cats, dogs and in piglets [116–124].

It is generally considered that meloxicam has "chondroneutral" effects on cartilage in OA [125]. In organ culture of porcine, canine or human cartilage explants, meloxicam did not show any inhibitory effects on the synthesis of radiosulphate labelled proteoglycans with or without IL-1 as seen with indomethacin [64, 106], a drug which is known to accelerate cartilage destruction in OA [107]. The proportion of high molecular weight proteoglycans that were synthesized in human cartilage explants was shown to be increased by meloxicam 4 μM and substantially reduced by indomethacim 4 μM compared with controls [107]. Thus, meloxicam may be like nimesulide and aceclofenac in increasing matrix component synthesis and protecting chondrocytes against apoptosis [125]. Bovine chondrocyte production of mRNA for metalloproteinase (MMP)-1 was reduced by meloxicam, naproxen and dexamethasone but not by other NSAIDs [126]. Proteoglycanase and collagenase activities in chondrocytes were reduced by these two drugs [126]. In a rat molar model of lesion associated bone loss where IL-6, COX-2 derived PGs and MMP-1 are primarily implicated in the resorptive process, meloxicam reduced periapical bone resorption and the number of cells synthesizing IL-6 and MMP-1 [127]. These results may have relevance to the bone destruction that occurs in arthritic diseases although specific studies in animals and human models need to be undertaken to establish if these important actions of meloxicam can be confirmed along with the effects of meloxicam on oxyradicals, NO, apoptosis, COX-2, cytokines [128, 129] and uncoupling of oxidative phosphorylation [130] correlated with joint destructive changes in these models.

The mechanisms of pain control by meloxicam compared with other analgesics and NSAIDs have been studied in a number of animal models, principally in rodents. In the formalin model of nociception in rat tails which enables discrimination of central hyperalgesia from peripheral inflammation, meloxicam, lornoxicam and piroxicam produced effects on the hyperalgesia, although lornoxicam was the more effective drug of the three [131]. Single unit recordings of evoked potentials in the rat spinal dorsal horn neurons following noxious mechanical stimulation were reduced by meloxicam [131] suggesting there are direct spinal effects. In the second phase of the rat formalin test administration of the NO synthase inhibitor, N(G)-L-nitro-arginine methyl ester, (L-NAME) or the guanylate cyclase inhibitor, I*H*-(1,2,4)-oxadiazole-(4,2-a)-quinoxaline-1-one (ODQ) both blocked meloxicam induced analgesia suggesting that NO-cyclic GMP coupled reactions in the periphery contribute to the anti-nociceptic effects of meloxicam [112], this being of course in addition to its other COX-2/COX-1 effects in peripheral inflammation.

Other pharmacological effects

Like many NSAIDs, meloxicam has been found to have inhibitory effects on the growth and proliferation of rat carcinogen-induced colorectal cancer [132] and this may be related to the promotion of apoptosis [132]. Meloxicam was also found to inhibit the growth of the MAC13 tumour in mice and reduced cachexia [133] which is an important observation since not all NSAIDs exhibit this and certainly cytotoxic agents can further exacerbate weight and muscle loss in tumour bearing animals.

The relationship between inhibition of colon cancer and GI toxicity from orally-administered NSAIDs is a major issue which limits the interpretation of long-term use of NSAIDs for prevention of colon cancer. In a study in rats in which colon cancer aberrant crypt formation (ACF) was induced by 1,2-dimiethylhydrazine and a range of NSAIDs (including sulindac and its sulphone metabolite) given orally for 3 weeks it was found that meloxicam and sulindac were the most potent inhibitors of aberrant crypt formation (46%) while of the NSAIDs including sulindac sulphone and celecoxib were less potent [134]. The COX-2 inhibitors, meloxicam and celecoxib were the only drugs that were without significant gastrotoxicity [134]. Since the potency of meloxicam against ACF was twice that of celecoxib and equivalent to sulindac it would seem that meloxicam would be worth studying further in other colon cancer animal models and in humans as a potential prophylactic. Sulindac is considered by many to be among the most interesting NSAID for prevention of colon cancer because it is a pro-drug and the non-PG active sulphone metabolite is an active inhibitor of ACF. It has been suggested that sulindac itself could be a useful preventative in colon cancer. However, it is appreciably gastrotoxic and thus only meloxicam could be employed with low risk of GI toxicity. There is much debate about the mechanism of inhibition of colon cancer by NSAIDs and while COX-2 inhibition forms part of the action of some of these drugs effects on signalling pathways seem an important mechanism. This is discussed further in Chapter 11.

Neuroprotective effects of meloxicam have been shown in a few animal models. In the MPTP (1-methyl-4-phenyl-1,2,3,6-tetrahydropyridine)-induced model of Parkinson's disease in mice, a single dose of 7.5 and 50 mg/kg meloxicam i.p. significantly prevented the reduction in striatal dopamine by 36 and 40% respectively as well as almost completely preventing decline in tyrosine hydroxylase in the substantia nigra and loss of neurons in this region [135]. Aspirin 50 and 100 mg/kg i.p. had similar effects but this is the more gastrotoxic drug of the two. However, the dose of 50 mg/kg meloxicam will be well within the gastrotoxic range for this drug.

High doses of meloxicam, like that of aspirin and the TNF_{α} soluble inhibitor, etenercept, inhibited retinopathy in streptozotocin-rats by a mechanism that may involve reduction in TNF_{α}-stimulated leucocyte adhesion and COX-2 expression [136, 137]. Damage to the sciatic nerve in experimental autoimmune neuritis in rats was reduced by meloxicam, celecoxib and indomethacin but the latter was associated with significant GI and renal toxicity [138]. The authors of this study suggested that the two former drugs might be useful in the treatment of Guillain-Barre syndrome. However, the actions in this study suggest that these NSAIDs may have utility in general in those neurological conditions where there is involvement of TNF_{α} and leucocyte adhesion/activation. In conditions (including febrile reactions) where there is cytokine-mediated NO production [138, 139] as well as COX-2 derived PGs there may be neuro-protective effects, especially in conditions such as ischaemic stroke where meloxicam like other NSAIDs has been found to have protective effects in rats [140].

Cardiac ischaemia-reperfusion derived myocardial vascular injury in guinea pigs was found to reduce thromboxane B_2 production but have little other effects from meloxicam 0.3–3.0 μM whereas aspirin 100 μM had marked injurious effects [141]. Anti-fibrinogen activity has also been observed with meloxicam 0.065 mg/kg/d given for 10 days to rats exposed to various injuries following multiple laparotomies [142]. This may have some relevance to prevention of fibrin formation in cardiovascular disease.

Exercise-induced apoptosis in rats has been found to be reduced by meloxicam 11 mg/kg [143] adding further protective properties to the list of those where ischaemia may, with anti-apoptotic effects, be an important basis for protective effects of the drug.

GI and other side-effects

Meloxicam has relatively low acute GI irritancy in rats in relation to its anti-inflammatory actions when compared with that of other NSAIDs [77, 144, 145]. The low acute gastric irritancy of meloxicam in rats has been found to be related to the low potential for the drug to inhibit gastric PGE_2 production [95] although inhibition of COX-2 at the ulcer margin and increase in pro-inflammatory cytokines may contribute to promotion of the injury from meloxicam [146]. While oxyradical generation from ischaemia-reperfusion injury enhances gastric mucosal injury from meloxicam and other NSAIDs [98] with intestinal injury this may be related to drug induced effects on xanthine as well as gastric oxidase activity [101, 145]. Coadministration of an α_2-adrenergic receptor agonist (tizanidine) with meloxicam and some other NSAIDs reduced gastric lesion development in rats [115] suggesting that there may be effects of these drugs on acid secretory functions that are controlled by α_2-adrenergic receptors in the CNS.

Sub-chronic treatment with meloxicam 3.75 or 5 mg/kg/d to rats (14 or 28 days) was not associated with neutrophil infiltration into the GI mucosa or elevations in plasma transaminases indicative of liver injury or creatinine or urea reflecting renal effects as seen with piroxicam 5 mg/kg/d [147]. As with other NSAIDs, meloxicam uncouples oxidative phosphorylation [130] and the consequent reduction in mucosal or renal ATP may contribute to development of damage in these organs. Like other COX-2 selective or unselective inhibitors, meloxicam can impair the PG-regulated renal perfusion and this drug and indomethacin have been shown to reduce cortical blood flow but not mean arterial pressure in euvolemic rats [148]. However, meloxicam 1–2 mg/kg i.v had no effects on renal and cortical blood flow in furosemide Na^+-deficient hypovolemic rats whereas indomethacin 5 mg/kg i.v. compromised both renal components of blood flow [148]. These results are suggestive of some renal haemodynamic effects of meloxicam compared with other NSAIDs although their effects on renal cortical and medullary blood flow may contribute to renal effects of this drug in susceptible individuals.

In accordance with the properties of being a preferential COX-2 inhibitor, meloxicam shows a low incidence of severe adverse GI events in large-scale clinical trials and in a large-scale meta-analysis [61, 62, 149, 150].

A recent prescription-event monitoring study comparing the GI toxicity in general practice in the UK of meloxicam with rofecoxib showed that both drugs had the same incidence of gastric peptic ulcer bleeds (0.4%) and had

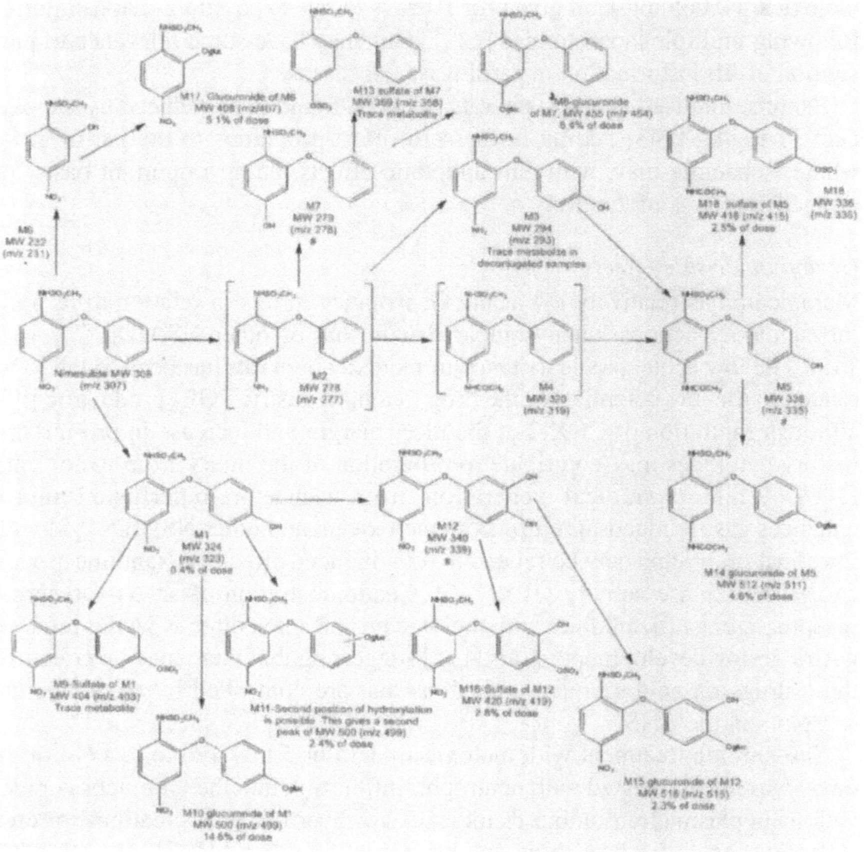

Figure 6. Pathways for the metabolism of nimesulide in humans based on studies with the radiolabelled drug [166, 176] (unpublished studies by MacPherson et al.). Taken from [166], with permission: ©2001 Prous Science. All rights reserved.

7.2% and 7.4% incidence of acid/peptic symptomatic events respectively [151]. These results are of interest in view of the substantial data on rofecoxib [8, 12, 152, 153], a low GI risk NSAID, since they show that under general practice conditions the preferential COX-2 inhibitor meloxicam has the same GI ulcer/bleeding/symptomatic profile as the highly selective COX-2 inhibitor, rofecoxib.

Nimesulide

This widely-used drug belongs to the class of sulphonamide COX-2 inhibitors [154–156] that were progenitors of some of the modern coxibs [155]. Its chemical structure (Fig. 6) is 4-nitro-2-phenoxy methane sulphonanilide and it has a pKa of 6.4–6.8 [154, 156, 157]. Nimesulide (Aulin™, Mesulid™, Nimed™ and several other trade mark names) was first developed by Riker Laboratories Inc., USA as R-805 based on their pioneering research on NSAIDs with antioxidant activities [157, 158]. In rodents and pigs this drug was found to have very low GI mucosal injury and this was a major factor in its selection for further development [158]. It was first introduced in Italy in 1985 (Helsinn Healthcare SA, Switzerland) and is now marketed successfully in over 50 countries (except in North America, Japan and UK), where in some it is a product leader being the fifth largest selling NSAID worldwide [159]. A large number of generic formulations of the drug have been developed for sale especially in Italy and India, the pharmacokinetic properties of some which are not of as high standard as the main marketed drug. A considerable number of reviews have been published on the pharmacology, pharmaceutics, pharmacokinetics, uses and safety/toxicology of nimesulide in rheumatology and other painful or inflammatory states [154, 157, 159–169]. Nimesulide is principally indicated for use in relief of pain and inflammation associated with OA and other musculo-skeletal conditions (except RA) as well as in acute pain (e.g., dental, dysmennorhea) [159, 161, 162, 166–174]. Being of relatively low cost this drug has particular advantages from a safety-pharmaco-economic viewpoint [175]. Nimesulide was shown to be the first drug to have COX-2 selectivity in humans *ex vivo* and low GI injury observed by endoscopy and in intestinal permeability studies [171].

Pharmacokinetics

The clinical pharmacokinetics in humans of nimesulide (see Tab. 2) have been comprehensively reviewed by Bernareggi [164]. The drug is rapidly absorbed with a mean C_{max} of 2.86–6.50 μg/ml being achieved 1.22–2.75 h after an oral dose of 100 mg tablets. This rapid absorption accounts for the rapid onset of analgesia of the drug [160]. The values for the volume of distribution, V_D, range from 0.1–0.2L/kg and the drug is bound to plasma proteins to 96–99% principally to albumin but also there is weak binding to α_1-acid glycoprotein and γ-globulin. The drug has a short mean elimination half-life, $t^1/_2$ = 1.8–4.73 h and low clearance with a hepatic extraction ratio Cl/F of 0.1 reflecting low impact of metabolic load in the liver despite diverse metabolic pathways (Fig. 6). The low $t^1/_2$ may be a factor accounting for the low GI adverse reactions from the drug along with lack of accumulation in the elderly [164, 173].

The principle metabolism of the drug is initially hydroxylation to 4'-hydroxynimesulide by the activity of the cytochromes P_{450} 2C9 and 2C19. This is the main metabolite detected in plasma and in a free or conjugated form in the urine [164, 166, 176]. Phenolic glucuronides and sulphates are formed [176]. Minor metabolic routes include reductive amination of the nitro-group and subsequent N-acetylation [166, 176] and it has been speculated that nitroso- and hydroxylamine metabolites may be formed during the reductive amination with subsequent covalent binding of these metabolites [166, 177, 178] but without any evidence. Further speculation that these covalent metabolites may account for liver injury from the drug [177, 178] is also without evidence. Nimesulide, like that of several other NSAIDs, has been shown to inhibit human liver phenol sulphotransferases [179].

Hepatic, but not moderate renal insufficiency or impairment, affects the kinetics of nimesulide [164]; the limited effects of mild-moderate renal impairment on the kinetics of the drug may reflect the capacity of about half of the drug to be eliminated in the faecal route [164]. Interestingly, there are no apparent or clinically significant interactions of nimesulide with commonly co-administered drugs (e.g., antacids, cimetidine, glibenclamide, furosemide, theophylline, warfarin or digoxin [164].

The kinetics and distribution of ^{14}C-nimesulide in rats is similar to that in humans [164, 180] indicating that as with meloxicam this enables comparison of effects of the drug in rats with that in humans even though the clearance rates and formation of minor hydroxylation metabolites may be greater in rats that in humans.

Pharmacology

The pharmacological effects of nimesulide have been comprehensively reviewed elsewhere [154–156, 158–160, 162, 164, 166, 173]. A comparison of acute anti-inflammatory effects in the carrageenan paw oedema model and acute gastric irritancy in rats of nimesulide with that of other established NSAIDs is shown in Table 5 [156]. It is apparent that there is a favourable therapeutic index with nimesulide compared with other NSAIDs in these acute models and this is confirmed in more recent studies [160, 170]. *In vitro* effects

Table 5. Acute anti-inflammatory and gastric irritancy effects of nimesulide on rats and calculated therapeutic indices

Drug	ED50 (mg kg^{-1})[a]	LD50 (mg kg^{-1})[b]	Therapeutic index
Nimesulide	1.25	324 (295–356)	260
Naproxen	2.1	395 (281–557)	190
Ibuprofen	13.5	923 (833–1020)	68
Difumidone	38	750 (694–811)	20
Flufenamic acid	14.7	249 (221–280)	17
Phenylbutazone	29.5	406 (375–440)	14
Acetylsalicylic acid	135	1520 (1360–1710)	11
Indomethacin	2.95	21 (19–23)	7

Data from [156, 157]. [a]Calculated in carrageenan-induced rat paw oedema from the 3-point regression of response on dose, and [b]Calculated by the method of Litchfield and Wilcoxon (95% confidence limits are given in parentheses).

of nimesulide and its metabolites have been compared in relation to the therapeutic concentrations of the drug achieved in plasma in patients given oral doses of the drug (Tab. 6) [154, 160].

Based on assessments of the actions of nimesulide at therapeutic concentrations or slightly higher it is apparent that this drug has multifactorial activities (Tab. 6) [154, 160, 164, 166, 167]. Aside from having mostly COX-2 activity with some COX-1 inhibitory effects which may be relevant for analgesia [154] nimesulide inhibits (a) the adherance and migration of neutrophils and their

Table 6. Pharmacological effects of nimesulide relevant to anti-inflammatory activity

I. Actions of nimesulide at normal doses or concentrations, or at therapeutically relevant plasma concentrations *in vitro* (up to 0.06 μg ml^{-1}; ~0.2 μM in the absence of albumin). All the effects listed are inhibitory.

Pathway	Man	Lab animals	Cells *in vitro*	Dose/conc
COX-2 activity	leucocytes *ex vivo*			100 mg b.d.
			human leucocytes	Therapeutic range
Superoxide formation	leucocytes *ex vivo*			
Histamine action	skin *in vivo*			200 mg
Histamine release		guinea-pigs		1.6 μmol kg^{-1}
Histamine release		guinea-pigs		0.1–1 mg kg^{-1} i.v.
Cytokine action		Rats		7 mg kg^{-1}
COX-2 formation			human synovocytes	0.03 μg ml^{-1}
Metalloprotease formation			human synovocytes	0.03 μg ml^{-1}
Chondrocyte apoptosis		Rat	chondrocytes	1 pM–1 μM

II. Actions of nimesulide *in vitro* at a moderately high concentration[1]

Pathway	Cells/tissue	Nimesulide	Serum %	Effect %
Urokinase synthesis	Synovial fibroblasts	0.3 μg ml^{-1}	1	50
Interleukin 6 synthesis	Synovial fibroblasts	0.3 μg ml^{-1}	1	50
Plasminogen activator inhibitor	Synovial fibroblasts	0.3 μg ml^{-1}	1	+50
Histamine action	Bronchial muscle	0.3 μg ml^{-1}	0	–20
Cyclic AMP	PMN leukocytes	1 μM	0	+40

[1] Experiments are required with lower concentrations *in vitro* or in human subjects with therapeutic doses, to investigate whether the above effects can be put into the 'therapeutically relevant' class as defined in this chapter.
From Bennett and Villa [159, 160].
Data from references in this review.

MPO enzyme and oxyradical activities, (b) release of histamine from mast cells, (c) release of leukotriene C_4 and platelet activating factor (PAF) from eosinophils, (d) production by neutrophils of the "second messenger", cyclic AMP, (e) production of cytokines (IL-1, IL-6 and possible TNF_{α}) by synovial cells, (f) production by synovial cells of NO, (g) inhibition of caspase-3 mediated apoptosis and (h) displacement of tryptophan from binding sites of albumin which, being a precursor to 5-HT (serotonin), may affect the production of increased 5-HT in the efferent neural pathways that mediate analgesia in the CNS [154, 180].

High concentrations (100 μM) of nimesulide have been found to protect against apoptosis and inhibit cytokine-mediated cartilage resorption but there are no effects on the synthesis of proteoglycans by this drug in the presence or absence of pro-inflammatory cytokines (IL-1, TNF_{α}) [181–183]. Other cartilage protective effects of nimesulide include stimulation of the synthesis of plasminogen activator inhibitor, decreased production of collagenase and other metalloproteinases, urokinase and inactivation of α_1-antitrypsin [154]. Cartilage uptake of radiolabelled nimesulide in the presence and absence of pro-inflammatory cytokines [183] is appreciable and may account for any putative protective effects of the drug on cartilage destruction.

Thus, the pharmacological actions of nimesulide that are relevant to its anti-inflammatory, analgesic and antipyretic effects involve effects on the production of a wide range of inflammatory mediators and those cells involved in the focus of inflammatory changes that occur in the joints of arthritic patients. The anti-histamine and inhibitory effects on leukotrienes and PAF are suggestive of some anti-allergic activity. While this has not been directly demonstrated it is interesting that there is evidence of some tolerance to the drug in NSAID-sensitive asthmatics [184, 185]. It is possible this is related to the effects of the drug on leukotrienes and PAF, as well as resulting in a balanced inhibition of inflammatory reactions that are manifest in allergic states. Further studies would be required to establish if there is some tolerance or antiallergic effects of nimesulide.

Molecular models of the interactions of docking of nimesulide with the active site of COX-2 have established the key roles played by valine-509 in COX-2, which is analogous to an isoleucine-523 residue in COX-1, as well as some other key amino acids [186–188]. Another interesting feature about the interactions of nimesulide with COX-isoforms was highlighted in the earlier studies by Rufer et al. [189] and work from the Riker group [190] on the interactions of nimesulide with the peroxidative (PEROX) system. Based on the original studies with phenolic compounds, e.g., MK-447 with seminal vesicle PGHS preparations, it was proposed that these act to scavenge or capture the oxyradical that is released in the conversion of PGE_2 to PGH_2 while nimesulide and some other sulphonanilides additionally act to block the initial cyclooxygenase reaction according to their respective pKa values [188]. These dual actions of nimesulide on both COX and PEROX reactions may be of significance in relation to its actions in inflammation and its low GI irritancy.

The antioxidant character of nimesulide which has been demonstrated in a number of systems [190–194] may contribute to its effects on the PEROX system. The mechanism of anti-oxidant effects of nimesulide is possibly related to the formation of its 4'-hydroxy metabolite which has been proposed to have a chain-breaking action with peroxy-radicals, as well as $OH^{\cdot}$ and $O_2^{\cdot -}$ [191].

In addition to having inhibitory effects on LTC_4 production [154], nimesulide also inhibits LPS-stimulated LTB_4 production in human leucocytes in the range of 10–100 μM as well as in human gastric explants [195]. This balanced inhibition of both 5-LOX as well as COX pathways by nimesulide may as with other dual COX-LOX inhibitors be important for its benefits as an anti-inflammatory as well as being of significance in its low GI irritancy [196].

Variable concentration-dependent effects of nimesulide have been shown on platelet aggregation [197]. At low concentrations (0.01–0.1 μM) nimesulide potentiated aggregation induced by sub-threshold concentrations of adrenaline, the effects of which were blocked by calcium antagonists and the NO-donor SNAP. At higher concentrations (1.0–100 μM) nimesulide blocked platelet aggregation induced by adrenaline and inhibited thromboxane B_2 production but the drug did not affect platelet aggregation induced by collagen, ADP, calcium ionophore or arachidonic acid. Whether or not these effects of nimesulide on adrenaline-induced platelet aggregation have any consequence for potential cardioprotective effects of the drug or in potentiating gastric bleeding in susceptible individuals is not known but should be investigated further.

Particular interest has focussed in recent years on the potential for NSAIDs to have central as well as peripheral effects in analgesia. In the formalin test in rats the injection into the formalin treated paws of the NO synthase inhibitor, L-NAME, or the soluble guanulate cyclase inhibitor, ODQ (H-1-(1,2,4)-oxadiazolo (4,2-a) quinoxalin-1) blocked the antinociceptive response to nimesulide [198]. Furthermore, the analgesic effect was potentiated by the NO-donor, SIN-1 (3-morpholino-syndnonimine). The results suggest that the peripheral analgesic effects of nimesulide and possibly also those in the CNS mediated by NO may be affected by the drug at the level of NO activity. Nociceptive effects of nimesulide have been shown from both intrathecal as well as intraperitoneal administration to mice [199] indicating this drug, like other NSAIDs, has central as well as peripheral analgesic effects [200].

The renal haemodynamic effects of noradrenaline have been shown to be potentiated by COX-2 inhibition from nimesulide and inhibition by L-NAME suggesting there are interactions between COX-2 and NO [201]. Whether these vascular actions have significance in other systems, e.g., vascular headache has yet to be determined. In a model of post-operative ileus in rats nimesulide like other COX-2 inhibitors overcame the reduction in GI transit from laparotomy but L-NAME and an iNOS inhibitor blocked the effects of the COX-2 inhibitors [202]. These results, overall, suggest that interactions between COX-2 and NO vary according to the system under investigation.

Uncoupling of oxidative phosphorylation has been shown with nimesulide, as with other NSAIDs [203–205]. Mitochondria from the livers of adjuvant arthritic rats showed greater sensitivity to the uncoupling effects of nimesulide [203]. The uncoupling effects of nimesulide evident with an IC_{50} of 5 μM were paralleled by Ca^{2+} efflux [204]. The reduction of the nitro group to an amine in nimesulide abolished the uncoupling effects of nimesulide [204]. It is not known if these uncoupling effects of nimesulide with consequent reduction in cellular ATP have any relationship to its anti-inflammatory actions. However, it has been suggested that the hepatotoxicity of nimesulide could be related to reduction in ATP production by the drug [205]. Unfortunately, the cell viability studies performed in the latter study were demonstrated at concentrations of 0.1 to 1.0 mM which exceeds concentrations that have been regarded as within the pharmacological or realistic toxicologic ranges. Other cytotoxicity studies performed at concentrations up to 100 μM did not reveal any cytotoxic effects of the drug, its major metabolites or manufacturing impurities [206].

Production of aspirin-triggered 15-epi-lipoxin A_4, which has been implicated in the anti-inflammatory actions of aspirin, and which is produced by liver cytochrome P_{450} activity, was not found to be produced by nimesulide [207].

Nimesulide has been shown to initiate apoptosis in a number of tumour cell lines [208–212] which may be related to its inhibitory effects on the growth and proliferation of chemically-induced tumours in rodents [213–216], in cell lines [217], in choline deficient F-344 male rats in whom hepatocarcinomas develop [218] and xenotransplanted melanomas in mice [219]. Paradoxically, however, nimesulide as well as ibuprofen has been shown to protect isolated chondrocytes against staurosporine-induced apoptosis [181, 220]. Thus, the state of cellular controls may dictate whether or not apoptosis is induced by this drug, or indeed possibly any other NSAID.

Membrane calcium flux is affected by nimesulide [221] and this with its inhibitory effects on peroxisomal proliferator-activated (PPAR) induction of COX-2 [222] may account for effects of the drug on PG production aside from the direct inhibition of enzyme activity. The linkage of this effect of nimesulide on PPAR induction does not appear to translate to effects on apoptosis [49]. Nimesulide is also a moderately effective inhibitor of neutrophil L-selection [49].

Cartilage protective effects of nimesulide may involve direct inhibition of cytokine-induced COX-2 [223] and the range of previously mentioned inhibitory actions on IL-1, TNF_{α}, IL-6 (but not IL-8), oxyradical and NO production, MMP's and cyclic nucleotide production [154, 159, 160, 181, 183, 190–194]. Translation of these *in vitro* effects into changes observed by magnetic resonance imaging in cartilage of patients with osteoarthritis does not appear to have revealed any direct effects of 24 weeks of treatment with nimesulide 100 mg b.i.d. compared with piroxicam 20 mg or placebo [224]. The small scale of this trial and limited duration of treatment did not allow for clinically-significant changes in cartilage structure to be identified, but there were significant changes in joint function observed with both drugs, with greater efficacy evident with the patients treated with nimesulide.

Adverse events and toxicity

A considerable number of studies have examined the pattern of adverse reactions attributed to nimesulide including regional pharmaco-epidemiological or pharmacovigilance studies [225–227] and critical evaluation of case reports to the central database of the company producing and marketing the drug [165, 173]. The relatively low incidence or frequency of cases of severe GI adverse reactions compared with those from other NSAIDs is shown in this data. The occurrence of hepatic reactions which have been reported in a number of published case reports appears no more than that of other NSAIDs [165, 173, 225–227] and is probably a class effect of the drug. Most reports are unspecific increases in the plasma levels of liver transaminases and less frequently alterations in liver function tests. Often concurrent intake of other hepatotoxic drugs (e.g., clavulanic acid, cephalosporin antibiotics, paracetamol, diclofenac) or prior or concurrent liver disease including that related to alcohol abuse has been implicated in many of the cases [165, 173, 228–230]. Skin reactions occur infrequently and are mostly minor rashes, urticaria and eruptions; very rarely have there been reports of Stevens-Johnson or Lyell's syndromes [231–234]; which are typical of those from NSAIDs [232, 234, 235], including the development of atopy being a risk factor [232]. Cases of lichenoid photodermatitis have been reported [236]. Cutaneous reaction rates to challenge test with nimesulide in patients with NSAID-related urticaria and angioedema have been observed in 21.3% of the NSAID-sensitive patients, but were 17.3% with meloxicam, 33.3% with celecoxib and 3.0% with rofecoxib [237], showing that there are appreciable differences in cross-reactivity among the COX-2 selective drugs in NSAID-sensitive patients. Using the standard local lymph node assay nimesulide has not been found to be a sensitizer [238].

Asthma and renal injury have also been observed infrequently with nimesulide [226, 239] and can be regarded as class effects of these drugs with the latter being related to effects on PGE_2-related renal haemodynamics and renal excretion [240].

The low incidence of GI ulcers and bleeding with nimesulide has been related to the relative sparing of COX-1 inhibition by the drug in the stomach and intestinal mucosa [171, 226]. This may also be related to the relatively high pKa of the drug and differences in the pattern of uptake into gastric mucosal cells compared with the more acidic NSAIDs [73].

Aside from the prevention of carcinomas, there may be potential beneficial effects of nimesulide on Alzheimer's disease [241] and other neuroprotective effects but these have yet to be extensively explored in animal models and humans with the diseases.

Coxibs

The COX-2 selective inhibitors known as coxibs were developed with the objective of reducing the risks associated with COX-1 inhibition in the GI tract, kidneys and some other systems where COX-1 inhibition was suspected as being the basis of the development of adverse reactions. This section reviews the clinical and experimental pharmacology and pharmacokinetics of this group of drugs and highlights principal toxicological and adverse reactions that have been associated with the coxibs.

Celecoxib

Celecoxib (4-[5-(4-methylphenyl)-3-(trifluoromethyl)-1*H*-pyrazol-1-yl]benzenesulphonamide; Celebrex™ (SC-58635); Figure 7) is a diaryl substituted pyrazole with a pKa of approximately 11.1 and a high degree of lipophilicity. This drug was the first of the coxibs to be introduced in 1999 by Searle/Pharmacia (now part of Pfizer Inc, USA) and was approved by the Food and Drug Administration in the USA and other countries for the treatment of signs and symptoms of RA, OA and the management of acute and chronic pain [242, 243]. The selection of celecoxib was based on an extensive structure-activity analysis of sulphonamide derivatives that had been optimised to once or twice day dosage [244]. The basis for COX-2 selectivity has been described in a number of studies [17, 242], and as with a number of other COX-2 selective and nonselective NSAIDs the enhanced supply of arachidonic acid *in vivo* can overcome some of the inhibitory effects of these drugs [245].

In vivo pharmacology was undertaken by the company that developed celecoxib (Searle) and the anti-inflammatory, analgesic, anti-pyretic and chemopreventive effects were studied in standard models principally in rodents and can be found in the NDA submission to the FDA [242].

Celecoxib has acute anti-inflammatory effects in the carrageenan paw oedema model, with an ED_{50} 7 mg/kg p.o. Chronic anti-inflammatory activity was tested in the adjuvant induced arthritis model in male Lewis rats dosed with 0.03–3 mg/kg b.i.d. p.o. with celecoxib for 10 days and a dose-dependent reduction in paw swelling with an ED_{50} of 0.3 mg/kg/d was obtained. Reduction in cartilage destruction, bone lysis, bone proliferation, soft tissue, swelling and synovial inflammation was observed in animals dosed with 1 mg/kg p.o. for 10 days. The greatest effects of the drug were evident in soft tissue oedema, cartilage destruction and bone lysis, the other effects being of lower intensity.

In vivo reduction by celecoxib of PGE_2 and 6-keto $PGF_{1\alpha}$ was observed in the carrageenan-induced air pouch in male Lewis rats with an ED_{50} of 0.2 mg/kg. Another study in the same model in fasted Lewis rats, showed dose-dependent reduction in PGE_2 in the pouch with an ED_{50} of 0.97 mg/kg, which is nearly five times less than that in the fed animals in the previous study. No explanation has been given for these differences but they perhaps highlight the role of fasting in drug bioavailabity. In the same study there was a "non-dose dependent" reduction in gastric mucosal PGE_2 production by 20–40% over the entire dose range of 0.1–200 mg/kg. Such a substantial reduction in PGE_2 production in the stomach may represent a partial inhibition of COX-1 inhibition by celecoxib.

Thus, the results obtained in the *in vivo* assays that formed part of the NDA [242] in the carrageenan air pouch and gastric mucosal assay suggest that there is some marked COX-1 inhibition although it is not as great as COX-2 inhibition by celecoxib. The inhibition of PGE_2 production in adjuvant arthritic rats, paw exudate or synovial fluid indicated that a higher dosage of 3–10 mg/kg was required to achieve between 50% and 60% inhibition of PGE_2 production [242]. Thus, differences in potency may be related to the degree of inflammation and chronicity, for in the acute carrageenan air pouch model the IC_{50} was 0.2 mg/kg in contrast to the much higher doses that were required in the chronic inflammatory model adjuvant arthritis in rats, where even doses of 10 mg/kg did not completely inhibit PGE_2 production. In comparison with published studies of NSAIDs in the standard carrageenan and adjuvant arthritis models it could be estimated that the inhibitory potency of celecoxib ranks approximately to that of naproxen.

Analgesic activity of celecoxib was investigated in the Hargreave's hyperalgesia model in Sprague-Dawley rats, which showed a dose-dependent reduction with an ED_{50} of 35 mg/kg. In the formalin paw injection model in Swiss-Webster mice, celecoxib showed an inhibition of between 67% and 88% in the dose range of approximately 30–50 mg/kg. In the phenyl benzoquinone (PBQ)-induced doxoflexion model in Swiss-Webster mice there was a 56% inhibition with 5 mg/kg p.o. There was a dose-dependent reduction in doxoflexion by 50–500 mg/kg p.o. celecoxib, in the acetic acid-induced writhing model in ICR mice.

Overall, these results show that the doses required for inhibition of PGE_2 production *in vivo* following inflammatory stimuli are much lower than those that are required for exhibiting analgesic activity. In another study by Smith et al. [246] the dose for reduction of carrageenan-induced paw oedema by celecoxib appeared to be similar to that for hyperalgesia. It is clear, therefore, that there are differences in the relation between anti-inflammatory compared with analgesic responses that depend on the model and type of stimuli used.

Anti-pyretic effects were investigated using the LPS-induced *in vivo* model in rats wherein it was found that celecoxib 30 mg/kg reduced the LPS fever by 42% but did not affect normal body temperature [247].

As evidence of chemoprotective activity celecoxib was administered to F344 rats that had been treated with azoxymethane to induce colonic aberrant cryptic foci and tumours in comparison with the effects of sulindac (which is regarded as an effective drug for the treatment of this condition) [242]. Following 52 weeks treatment with celecoxib in the diet there was a marked reduction in the numbers of adenocarcinomas and their volume in the treated rats but the numbers of crypts formed after 12 weeks treatment with celecoxib appeared unchanged; there also appeared to be no changes affected by sulindac.

Of the safety pharmacology that was undertaken as part of the NDA [242] there were some minor behavioural changes exhibited at relatively high doses of the drug (500 mg/kg), effects on increasing the hexobarbital induced sleep, reduction in electroshock and chemical induced convulsions but no effects on body temperature. There are other minor effects on the autonomic nervous system and smooth muscle involving reduction in acetylcholine and histamine induced contractions in the isolated guinea pig ileum, no effects on the cardiac or respiratory systems in dogs but there was a slight reduction in urine volume with increased urinary osmolality in rats concomitant with a reduction in renal PGE_2 production.

Single and repeated doses of celecoxib were evaluated in standard toxicology models with the GI tract and kidneys as the primary end organs of toxicity (Tab. 4) [242]. No mutagenic activity was observed *in vivo* or in the Ames' tests and there were no indications of untoward effects in special toxicology (e.g., antigenic assays).

Pharmacokinetics

The pharmacokinetics of celecoxib in humans has been reviewed by Davies et al. [248] (see Tab. 2). The essential features of drug pharmacokinetics are that it is well and completely absorbed with peak plasma concentrations being reached within 2–4 h after oral administration. The drug has a long and variable beta-elimination half-life that can range between 4.5 and 16 h. Like all NSAIDs, celecoxib is extensively protein-bound (<97%) to plasma albumin principally. However, the volume of distribution (V/F) is 455 L (5.7–7.1 L/kg) which is some five to seven times greater than most NSAIDs. This extensive volume of distribution may be related to the lipophilic nature of celecoxib as well as to its high pKa (11.1). The rate of absorption of the drug in humans is similar to that in rats and dogs.

The metabolism of celecoxib in humans and animals is principally through cytochrome P_{450} 2C9 mediated metabolism to the hydroxymethyl and subsequent carboxyl derivatives (Fig. 7) [249–253]. The Phase II metabolites include four positional isomers of the acyl glucuronide conjugates of the carboxylic acid metabolites [253]. It appears these metabolites are formed in appreciable quantities in rats [251] and rabbits [253]. The principle excretion

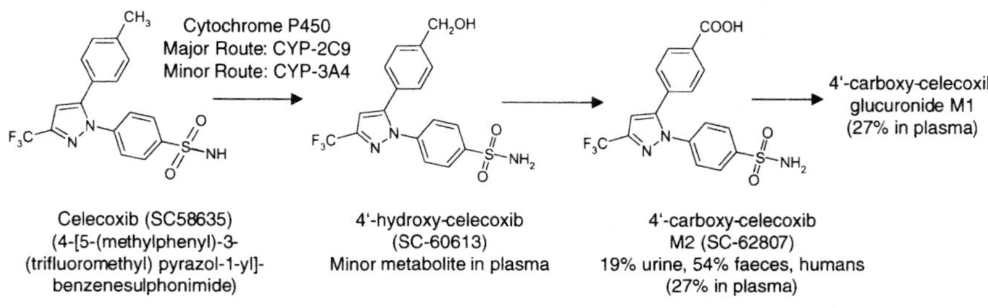

Figure 7. Pathways for the formation of the principal metabolites from celecoxib in humans [248, 249]. Oxidative metabolism via cytochromes P-450 2C9 and 3A4 is the major route of metabolism of this drug.

in the rat is in the faeces from both the intravenously administered and the orally administered radiolabelled compound [251]. In humans the carboxylic acid metabolite (SC-62807) constitutes 19% of the total dose excreted in the urine and 54% in the faeces, respectively. The hydroxymethyl and carboxylic acid metabolites constitute about one-quarter each of the total metabolites observable in plasma while unmetabolised celecoxib constitutes about one-half of the total drug present in plasma. In contrast to the situation in rats there is a much lower rate of excretion of the drug and its metabolites in the faeces in humans. However, the predominant faecal excretion of the drug, especially observed in rats, which occurs with a number of acidic NSAIDs, raises the possibility that intestinal injury may be a feature of celecoxib. To establish if this occurred Tibble et al. [254] compared celecoxib with the effects of orally administered indomethacin, which is well-known to cause intestinal mucosal changes in permeability, lesions and ulcers proceeding to peritonitis in the rat. Indomethacin also causes uncoupling of mitochondrial oxidative phosphorylation which is considered to part of its "topical" effect. Celecoxib did not cause any changes in intestinal permeability, uncoupling of oxidative phosphorylation or mucosal inflammation-injury, whereas indomethacin did [254]. These studies are important for highlighting the absence of the potential for developing intestinal injury even though some lesions were observed in long-term toxicity studies as cited in the NDA submission [242]. Since the systemic bioavailability of celecoxib in dogs when instilled into the duodenum, jejunum or colon was the same as that when given intragastrically, it appears that the absence of intestinal inflammation-injury cannot be due to poor absorption of the drug from the intestinal tract [255]. The absence of intestinal injury which is an important safety feature of both celecoxib as well as rofecoxib appeared to be related to the sparing of COX-1 in the intestine since there were no effects on PGE_2 content of the intestinal mucosa 24 h after oral administration of the drug or of rofecoxib [256]. It is interesting that these studies showed that the selective COX-1 inhibitor SC560 also failed to cause mucosal injury but when SC560 and rofecoxib were given simultaneously intestinal injury occurred [256]. It appeared from these studies that the administration of SC560 alone caused upregulation of COX-2 in the small intestine and this may have had a partial protective effect even though COX-1 induced injury may have been evident with this drug [256].

The predominance of celecoxib metabolism through cytochrome P_{450} 2C9 raises issues about the potential for drug interactions and the possibility of unequal metabolism through different genetic variants of this cytochrome [248, 250]. The occurrence of polymorphisms of cytochrome P_{450} 2C9 may account for the variable hepatotoxicity that has been reported with this drug. Interestingly slow and fast metabolisers of celecoxib in beagle dogs have been related to novel variants of the cytochromes P_{450} 2D15, 3A12 and variance of the 2D15 predominantly accounted for the polymorphism in canine celecoxib metabolism [250]. The relationship of this to hepatic injury which has been reported with this drug [257] has not been investigated, at least in dogs.

Pharmacology and toxicology

Some of the aspects concerning the pharmacology and toxicology of celecoxib have been discussed in the earlier part of this section but here it is useful to illustrate aspects relating to the mechanisms of celecoxib in inflammation, algesia and the toxicity of the drug.

An interesting study comparing the effects of celecoxib or rofecoxib with those of indomethacin, dexamethasone and the COX-1 inhibitor SC560 was performed in rats in which acute inflammation was induced by injection of carrageenan into the hind paws or pleural cavity [258]. There was differentiation of the effects of COX-1 and COX-2 inhibition inasmuch as the paw volumes, though reduced by all the drugs except the COX-1 inhibitor SC560, were accompanied by a reduction in PGE_2 content in the pleural exudates, and a reduction in NO by dexamethasone and indomethacin but not by the COX-2 inhibitors. Also, in the rat hind paw dexamethasone, indomethacin and SC560, but not the two COX-2 inhibitors inhibited myeloperoxidase activity, an index of leucocyte infiltration in the hind paw. Of the two pro-inflammatory cytokines investigated, indomethacin increased IL-1 production while dexamethasone decreased IL-1 and TNF_{α} whereas the two COX-2 inhibitors had no effect on the concentration of these cytokines in the carrageenan-induced pleural exudate. In contrast, in the paws indomethacin, dexamethasone and celecoxib, but not rofecoxib, decreased the concentration of IL-1 whereas indomethacin and rofecoxib increased TNF_{α} production, while dexamethasone decreased its production. These somewhat complex results reflect potentially differences in the actions of these drugs on different components of the acute inflammatory process. It must be said, however, that the lack of full dose-response studies makes it difficult to provide an overview of what relative effects these drugs have on the different components of acute inflammation. In addition, the selective nature of the experiments in rat paws as compared with those in the carrageenan-induced pleural inflammation has undoubtedly contributed to the complexity of the results that have been obtained. Furthermore, the lack of time-dependent studies also mitigates against clear interpretation of the effects of the drugs. Nonetheless there are indications from these studies that there may be weaker effects of celecoxib and rofecoxib on acute inflammation compared with that of potent established drugs such as indomethacin and dexamethasone but this obviously requires further investigation.

In an acute model of lung injury induced by carrageenan administration in rats, celecoxib administration reduced the infiltration of PMN leucocytes and subsequent lipid peroxidation, reduced PGE_2, TNF_{α} and IL-1β production and the upregulation of adhesion molecules ICAM and P-selectin and the production of nitro-tyrosine [259]. This is an interesting study for highlighting the multiple actions of celecoxib which may be related in part to its properties as a sulphonamide rather like that of nimesulide which is also a sulphonamide derivative.

An interesting differentiation between the anti-oedemic effects of COX-2 inhibitors and other NSAIDs and the effects on hyperalgesia or hypoalgesia were shown in the study by Francischi et al. [260] using a model in which carrageenan was injected into the hind paw and the hyperalgesic response was measured 24 h later concomitant with anti-oedemic effects being measured by plethysmography. It was shown that celecoxib abolished the inflammatory hyperalgesia and induced hypoalgesia without reducing the oedema at the doses given. Similar differentiation was established with the other COX-2 inhibitors, rofecoxib and SC236, but the results were less marked than for celecoxib. In the formalin footpad nociceptive model, peripheral administration of celecoxib did not produce anti-nociceptive effects whereas diclofenac and the COX-1 inhibitor, resveratrol, did produce dose-dependent anti-nociceptive effects during the second phase of the formalin-induced nociception [261]. Again, as with the studies by Francischi et al. [260] it appears that celecoxib has less effect on the inflammatory as compared with the direct neural effects of pain. This may be due to more significant effects of the drug in inhibiting COX-2, and possibly even to a small extent COX-1, in the CNS pathways of pain and may be linked to the high lipophilicity of the drug and its extensive volume of distribution. Unfortunately, studies linking the actions of celecoxib and other drugs with localised uptake into the CNS have not been established to verify this point. The possibility that there may be direct effects of celcoxib like that of other NSAIDs on ion currents mediated by glutamate or glycine sensitive neurones has been indicated from the studies of Shin et al. [262] using the niastatin perforated patch clamp method for measurement of ion currents in the periacqueductal gray (PAG) neurones. It was found that application of celecoxib resulted in decreased glycine induced ion currents and increased glutamate induced ion currents. The latter results are puzzling in view of the established effects of NSAIDs on glutamate sensitive neurons and clearly further investigation is required.

Other potentially important pharmacological effects of celecoxib have been extensively investigated including anti-cancer activity, demonstrated in models of colon and other cancer [263–276, 278, 279]. These anti-tumour effects are probably multifactorial but in a large number of these studies appear to be related to anti-angiogenic effects, the consequences of inhibition of COX-2 in general inflammatory responses and possibly (though not convincingly) due to NFκB and other intracellular transductional events [277, 278]. These extensive studies in a variety of tumour models supported by the studies showing the role of COX-2 inhibition in head and neck xenograph tumours [278] are impressive for showing consistently the relatively potent effects of celecoxib like that of many other NSAIDs in the inhibition of carcinogenesis and tumour growth and proliferation which with ongoing clinical studies especially in colorectal cancer may provide proof of principle of the clinical efficacy of celecoxib [279].

Neuroprotective effects have been indicated in a number of model systems including the age-related degenerative changes and accompanying production in the hippocampus of IL-1, TNF_{α} and PGE_2 [280]. Repeated oral treatment with celecoxib on these degenerative neural markers and the accompanying behavioural changes were effective, providing the treatment was started early on in the development of the aging process [280]. In the model of MMP-induced neurodegeneration with accompanying development of degenerative changes in dopamine sensitive neurones, celecoxib 2.5–50 mg/kg was ineffective [281]. This was shown to be related to the lack of effect on hydroxylamine formation; in contrast salicylate 50 and 100 mg/kg was very effective. Diclofenac 5–10 mg/kg, which also was ineffective in controlling oxyradical production, likewise was ineffective in this model [281]. Thus, celecoxib would be unlikely to be effective in the adjuvant therapy in Parkinson's disease as has been proposed for aspirin or salicylate.

In a transgenic mouse model of amyotrophic lateral sclerosis (ALS) which accompanies glutamate mediated excitatory activity, oxyradical damage and apoptosis with accompanying elevations of COX-2-derived PGE_2, it was found that celecoxib markedly inhibited the production of PGE_2 in the spinal cords of the ALS mice and significantly delayed the onset of weakness and weight loss and loss of survival which develops in these mice [282]. It appeared that the spinal cords of these mice showed significant preservation of spinal neuronal integrity and the prevention of astrogliosis and microglial activation by this COX-2 inhibitor. While these results are at variance with the lack of effects on oxyradical production [281] it is clear that in the spinal cord there are protective effects of celecoxib that may be related to its multiple actions as well as COX-2 inhibition [282].

Renal toxicity with selective COX-2 inhibitors has attracted considerable interest since these effects appear to be mediated by COX-2 inhibition in the same way as experienced with non selective COX inhibitors [283–287]. The consequent effects of actions of COX-2 inhibitors on alterations in glomerular filtration rate, sodium retention, the development of peripheral oedema, hypertension and long-term rare effects in causing nephrotic syndrome, acute interstitial nephritis and chronic interstitial nephritis, hypercalaemia, are of particular concern especially since some of these can be related to the development of hypertension [286]. Given that these are probably class effects common with all NSAIDs there do appear to be some differences between the relative effects of celecoxib compared with that of rofecoxib. In the study by Welton et al. (as part of the Success VI Study Group), which was a double-blind trial for 6 weeks in patients with OA greater than 65 years of age and taking antihypertensive agents, there were half as many patients that had received celecoxib who experienced oedema, systolic and diastolic blood pressure changes and mean arterial blood pressure alterations compared with those on rofecoxib during the same period [282]. Thus, there may be significant differences between the relative cardio-renal effects of different coxibs but this requires independent investigation and detailed experimental studies. While renal prostacyclin and thromboxane production is known to be altered by celecoxib and other NSAIDs, alterations in the expression of iNOS in the cortex do not seem to be implicated in the development of renal damage [288]. Alterations in the production of renin in the activity of the renin-angiotensin system have been known to be affected by aspirin and some other NSAIDs but apparently in WKY hypertensive and normotensive rats COX-2 does not appear to be related to the stimulation of renin production [289]. There are indications that COX-2 inhibition by celecoxib is not related to alterations in renin response in this model. In another study in which hypertension was induced by the chronic administration of *N*-omega-nitro-L-arginine methylester it was found that celecoxib significantly elevated blood pressure in both normal and hypertensive rats and this was related to elevation of plasma arginine-vasopressin levels, plasma urea and reduction in plasma renin [290]. Increased leucocyte adherence to the endothelium of both normal and hypertensive rats were also observed [290]. In ureteric obstructive rats significant reduction in creatinine levels occurred with celecoxib and accompanying urine volume and urine sodium values were increased together with certain morphological alterations [291]. In normotensive salt-depleted subjects where there is a compensatory change in sodium and potassium alterations, it was found that celecoxib, while having no effect on systemic blood pressure, caused a short-term transient decrease in renal blood flow and glomerular filtration rate with evident sodium retention within the first 3 days of treatment [292]. It appears, therefore, that there are functional alterations in sodium and water retention that can be achieved even in normotensive individuals [292]. Furthermore, in rats with experimental membranous glomerular nephritis there were alterations in albumin urea, mesangiolysis and refraction of glomerular endothelial cells whereas proliferation and activation of mesangial cells was unaffected [293]. Thus, selective COX-2 inhibitors including rofecoxib may impair glomerular capillary repair after mesangiolysis and thus these inhibitors should be used in caution with patients exhibiting endocapillary glomerular disorders [293].

The hepatic effects of celecoxib have already been discussed and related to the cytochrome P_{450} metabolism of the drug and the potential interactions highlighted with other drugs that affect the cytochromes or other hepatotoxic drugs. Of particular interest however, since NSAIDs are frequently taken with paracetamol (acetaminophen) is the potential for interactions between COX-2 inhibitors and paracetamol since both are potentially hepatotoxic. In COX-2 "knock-out" (–/–) and –/+ mice, paracetamol induced hepatotoxicity and lethality were markedly increased compared with COX-2 (+/+) mice and this appeared to be unrelated to alterations in drug protein adduct formation [294]. The COX-2 deficiency in COX knock-out mice appeared to be related to profoundly impaired induction of heat shock proteins and this suggested that PGs may be important endogenous stress signals regulating heat shock protein production. The increased liver toxicity in mice treated with celecoxib also appeared to be related to these drug effects [294] suggesting that there may be important molecular and regulatory alterations that take place during COX-2 inhibition linking COX-2 regulated PG production to that of liver injury.

One of the major issues which confounds the identification of NSAIDs with greater GI safety are the kinds of experimental models which are employed to discover new drugs and screen them for GI toxicity as well as the use of long-term toxicity studies [295, 296]. The unfortunate fact is that most of the drug screening takes place in essentially normal animals and since drugs are being used for arthritic disease it would be more appropriate if the drugs were initially screened in animal models that replicate the chronic inflammatory diseases that are experienced in patients [295]. Thus, we have advocated the use of a range of disease and physical stress models in order to quickly screen for the GI toxicity of NSAIDs and these models have been used extensively and are well-defined [295, 296]. It is not surprising, therefore, that in studies in arthritic rats the acute gastric toxicity of both celecoxib and rofecoxib has been shown to be markedly enhanced compared with that in normal animals [297, 298]. It is also not surprising that the expression of COX-2 mRNA and COX-2 immunoreactivity was enhanced in arthritic but not all normal animals [297]. The inference from the study of Kato et al. [297, 298] was that the inhibition of elevated expression of COX-2 by celecoxib and rofecoxib would somehow be responsible for the enhanced gastric injury.

This is a naïve interpretation since it is most logical that the arthritic process has undoubtedly compromised mucosal protective functions and integrity. The extensive effects that are known to occur with adjuvant arthritic rats on liver and other metabolism and detoxification [295] were not considered in this study. Delayed ulcer healing and impaired angiogenesis, both influenced by COX-2, also contribute to the ulcerogenic effects of COX-2 inhibitors including celecoxib [299, 300].

The beneficial GI effects of celecoxib (5 mg/kg/d) and other COX-2 inhibitors have been explored in the inhibition of the development of colitis induced by dinitrobenzene sulphonic acid in rats [301]. In addition, celecoxib reduced haemorrhagic diarrhoea and weight loss [301]. Celecoxib also reduced the myeloperoxidase activity in the mucosa suggesting that there was a reduction in infiltration of PMN leucocytes. The malondialdehyde activity, which is a measure of oxyradicals, was reduced as well as levels of nitrotyrosine which is formed from peroxynitrite reaction with tyrosine [301]. Despite these potentially important effects on GI inflammation models it is, however, evident that celecoxib like that of other NSAIDs can decrease epithelial proliferation, and this may have consequences in treating inflammatory bowel diseases.

Increasing concerns about the relative safety of NSAIDs for female reproduction have been of ongoing concern [302]. COX-1 as well as COX-2 inhibition appears to underlie deficient abnormalities and ovulation, fertilisation, implantation and decidualisation in mice [303]. While non-selective as well as selective inhibitors of COX-2 have been employed as potent tocolytic agents they unfortunately have potential risks for constriction of the foetal ductus arteriosus [304]. In studies in sheep the closure of the lamb ductus arteriosus and ductal blood flow was more profoundly affected by indomethacin than celecoxib [304], implying that the use of celecoxib as a tocolytic agent may have some safety advantages. However, studies in rabbits indicate that when the newborn animals are exposed to hyperoxia there was a lower thromboxane B_2 to 6 keto $PGF_{1\alpha}$ which was linked to the effect of celecoxib on reducing

8-epi-$PGF_{2\alpha}$ and this may have indirect influences on thromboxane production [305]. Clearly more detailed investigations are required and this is addressed further in Chapter 10.

Valdecoxib and parecoxib

In the search for other diarylisosoxisole COX-2 inhibitors having sulphonamide moieties, Searle identified valdecoxib (Bextra™) (4-[5methyl-3-phenyloxisol-4-yl]-benzene sulphonamide) as an orally effective COX-2 inhibitor [306]. Parecoxib (*N*-2[[(5-methyl-3-phenyloxisol-4yl)-phenyl]sulphonyl] propane amide) is the pro-drug precursor of valdecoxib for use as an injectable form in acute pain relief [307]. These two second generation coxibs are particularly interesting therapeutic developments. In the human *ex vivo* whole blood assay valdecoxib (SC65872) appeared to have approximately 26-fold selectivity for COX-2 and the hydroxymethyl metabolite was virtually devoid of any COX-1 activity [306]. The overall potency of COX-1/COX-2 inhibition of valdecoxib was lower than that of the progenitor celecoxib in the whole blood assay but it appeared to have significantly greater selectivity in the assay involving human recombinant isoenzymes. The potency of valdecoxib in rat adjuvant arthritis was some 10-fold greater than that of celecoxib; the IC_{50} being 0.032 mg/kg/d compared with that of celecoxib 0.37 mg/kg/d. Valdecoxib was slightly less potent in the carrageenan oedema assay.

Parecoxib was shown to form valdecoxib and the half-life of valdecoxib is approximately 7–8 h while that of the pro-drug precursor is approximately 30 min to 1 h [307]. The drug was shown to reverse hyperalgesia [307] and is the first injectable COX-2 selective inhibitor that is being employed clinically.

Valdecoxib has been recently introduced for the treatment of OA and RA, pain and treatment of primary dysmenorrhea and has all the conventional contraindications and special considerations as with any other NSAID or coxib [308]. It appears to undergo cytochrome P_{450} 2C9 and 3A4 mediated metabolism and thus not surprisingly any inducers or inhibitors of these isoforms will inevitably influence the plasma concentrations and relative safety of valdecoxib. In the treatment of RA and OA, valdecoxib 10, 20 or 40 mg once daily was as effective as naproxen 50 mg twice daily and superior to placebo [309, 310]. In a randomised 12-week double-blind study in patients with osteoarthritis of the hip, valdecoxib 5 and 10 mg once daily was as effective as 500 mg twice daily naproxen but had fewer GI adverse events [310]. In an upper GI endoscopy study performed in 1052 osteoarthritis patients, valdecoxib 10 mg once daily or 20 mg once daily was compared with ibuprofen 800 mg t.i.d. or diclofenac 75 mg t.i.d. over a 12-week period. The incidence of gastroduodenal ulcers over the 12-week period was 5% in the patients receiving 10 mg once daily valdecoxib, 4% in those receiving 20 mg once daily, 7% in those receiving placebo, 16% in those with ibuprofen and 17% in those with diclofenac [311]. With there being serious concerns about the potential for upper GI mucosal injury in the elderly, a study was recently performed in subjects between 65–76 years old, who received valdecoxib at the supratherapeutic dosage of 40 mg twice daily compared with naproxen 500 mg twice daily. [312]. Upper GI mucosal injury was evaluated by oesophago-gastroduodenal endoscopy at day 7. The patients who received valdecoxib had no evident gastroduodenal ulcers while those that received placebo had an incidence of 3% and those that received naproxen 18%. These studies are indicative of there being a favourable GI safety with valdecoxib in elderly healthy subjects [312].

Increasing concerns about the risk of cardiovascular thrombotic events from coxibs [313] has led to studies being undertaken with valdecoxib to compare the incidence of myocardial infarction, myocardial ischaemia and unstable angina, cerebrovascular disorders or peripheral arterial ischaemia in patients receiving valdecoxib 10–18 mg once daily compared with that of diclofenac 75 mg b.i.d. ibuprofen 800 mg t.i.d. or naproxen 500 mg b.i.d. [314]. These studies have shown that in 4531 patients that received valdecoxib with a total exposure of 1340 patient years there was no difference in the total number of serious thrombotic events or any of the other individual events as compared with traditional NSAIDs but these did, however, show a slightly increased risk of any event above placebo [314].

Thus, valdecoxib may represent a new second generation sulphonamide coxib that can be given once daily (this compares with the twice daily dosage of celecoxib) and this may be associated with a lower potential risk of cardiovascular thrombotic events than possibly experienced with other coxibs, but this aspect has yet to be clarified.

Rofecoxib

Rofecoxib (MK996; VIOXX™; Merck Inc; 4-(4'-methylsulphonylphenyl)-3-phenyl-2,5*H*)-furanone is probably among the most potent of the coxib inhibitors that has been employed clinically so far. This drug being a furanone is of a different chemical class than celecoxib, valdecoxib and parecoxib. The initial discovery and the overall clinical pharmacokinetics and pharmacodynamic properties of rofecoxib have been previously reviewed [315–317] and the details of the NDA and clinical studies can be found at the FDA website [318]. The initial development of rofecoxib at Merck Frosst Laboratories, Montreal [318–320], followed an extensive and fundamental programme of other eicosanoid inhibitors and clearly benefited from an enormous effort that was made in those laboratories to understand the molecular, biological and cellular aspects of inflammatory processes. The database from which rofecoxib was developed was probably amongst the largest in the pharmaceutical industry devoted to the discovery of new NSAIDs. In the standard benchmark whole blood COX-1 and COX-2 assays, rofecoxib has amongst the highest ranking of all COX-2 selective drugs for selectivity; with COX-1-COX-2 selectivity ratios of 0.013 (Tab. 1).

In the rat carrageenan paw oedema model, the drug exhibited an ED_{50} of 1.5 mg/kg p.o. in the carrageenan-induced paw hyperalgesic model this anti-oedemic effect overlaps that of the hyperalgesia where the ED_{50} is 1 mg/kg [319]. In the adjuvant arthritis model rofecoxib had an ED_{50} of 0.74 mg/kg/d and showed significant protective effects against cartilage and bone destruction measured radiographically 14 or 21 days after induction of arthritis in Lewis rats. The inhibition of secondary paw swelling paralleled that of the reduction in radiographical evidence of joint damage. The antipyretic effects of rofecoxib were investigated in the LPS-induced pyresis in rats with ED_{50} of 0.24 mg/kg being obtained [319].

Using the radiochromium labelled red cell loss technique for GI bleeding in rats and in squirrel monkeys rofecoxib was shown to have no effects up to 200 mg/kg per day for 5 days [319]. Regrettably in the radiochromium GI blood loss studies there was no information provided about the integrity of the gastrointestinal mucosa. It is well known that radiochromium blood loss is an ineffective measure of GI injury and only adequately reflects the actions of aspirin and other drugs that inhibit platelet aggregation and so may potentiate blood loss resulting from mucosal injury [321]. Since rofecoxib does not affect platelet function [321], it is not surprising that there is little evident blood loss from the GI tract.

A considerable number of GI safety studies were performed with rofecoxib [152, 153, 318, 322–324] including GI endoscopy in patients with RA, which was the initial primary indication for this drug [324–325]. All these studies showed that rofecoxib had somewhat lower gastric ulcer incidence and incidence of peptic ulcer bleeds than the lower ulcerogenic group of NSAIDs but, as highlighted earlier, the relative reductions in these severe ulcer events are probably relatively small when reviewed from long-term data [326]. Considerable concern has, however, been expressed about the potential for rofecoxib to be a serious cause of renal injury; this aspect having been reviewed in the previous section [285–287]. The conclusion overall is that renal injury is probably a COX-2 mediated event but is probably potentiated in long half-life coxibs that are very potent because of the high binding to COX-2 in the macula densa which is an underlying feature of all inhibitors of cyclooxygenases [327, 328].

Rofecoxib has had a reasonably favourable safety profile in general practice with predictable adverse reactions [329]. In a long-term randomised control trial in 784 patients with OA of the hip or knee randomised to receive rofecoxib 12.5 mg or 25 mg once daily or diclofenac 15 mg t.i.d., it was shown that rofecoxib was effective in inhibiting the symptoms of pain and WOMAC scores [330]. An interesting feature about this study was that the liver transaminases present in the plasma of patients that received diclofenac were all consistently elevated from 2 weeks to the end of the study whereas there were no changes in transaminases evident with rofecoxib or placebo over this period [330]. While the elevation of liver transaminases is considered to be an unspecific event associated with intake of all NSAIDs interestingly there were no indications of any liver changes evident with rofecoxib.

Pharmacokinetics

The pharmacokinetics of rofecoxib in humans (see Tab. 2) are essentially quite complex and may be variable depending on patients and other conditions. The drug has a variable GI absorption with the maximum concentration occurring between 2 and 9 h. The mean oral bioavailability is 93% following acute administration [317] and it has a terminal half-life of elimination of approximately 17 h at steady-state. Like celecoxib, the volume of distribution of rofecoxib is between 86–91 L and much greater than normally expected with NSAIDs (their average being 7–10 L). This may be related to the high lipophilicity of this drug and like celecoxib, it has a high pKa value. Rofecoxib has a lower percentage of binding to human plasma proteins, this being 87%. The drug also crosses the blood brain barrier in rats and has been shown to cross the placenta in rats and rabbits thus raising the possibility of foetotoxic events. The metabolic routes of rofecoxib are shown in Figure 8 [317]. An interesting feature about rofecoxib, which is not evident with celecoxib (which should be recalled has a predominant cytochrome P_{450}-based metabolism), is that the drug is primarily metabolised by cytosolic reductases in the liver with only a minor role for the cytochromes P_{450} [317]. There is apparently a reversible metabolism of rofecoxib to the 5-hydroxy derivative at least in rats, but in rats the 5-hydroxymetabolite is the major metabolite. The complex

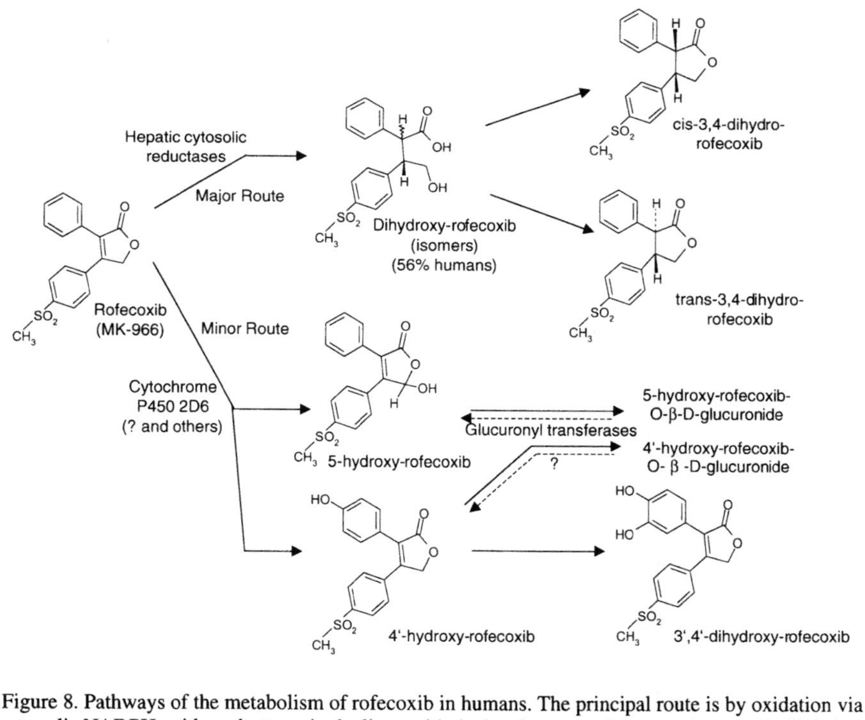

Figure 8. Pathways of the metabolism of rofecoxib in humans. The principal route is by oxidation via cytosolic NADPH oxido-reductases in the liver, with the involvement of the cytochromes P-450 being relatively minor [317, 318].

metabolism of rofecoxib progresses to the formation of the 3-4-dihydrodiol and 4-hydroxy derivatives that can then progress to form glucuronides. Relatively little unchanged drug occurs in the urine as it is primarily metabolised in the liver. The potential drug interactions at the level of the minor cytochrome P_{450} system probably involves intestinal cytochrome P_{450} 3A but not that of the liver; the intestinal enzymes being inhibited by concurrent administration of midazolam [317]. Cimetidine, which is an inhibitor of cytochrome P_{450} 2D6, slightly increases the maximum plasma concentration of rofecoxib but this does not lead to the necessity to reduce dosage. However, potent inducers of cytochrome P_{450} among them rifampin, carbamazepine, phenobarbitone and phenytoin may decrease levels of rofecoxib, thus implying cytochrome metabolism even by the minor route can have small but nonetheless clinically significant effects on the pharmacokinetics of the drug.

The ADME and general pharmacokinetics of rofecoxib have been investigated in rats and dogs [318]. The oral absorption of rofecoxib was relatively rapid in both species with peak levels achieved at 0.5 h in rats and 1.5 h in dogs which is appreciably shorter than that in humans [317, 318]. The bioavailability was poor in dogs being 26% and the drug was primarily excreted by the biliary route in both species [318]. The metabolism of rofecoxib is quite extensive in both species forming the major metabolites as identified in Figure 8. The distribution of radiolabelled drug following intravenous administration was high in most tissues at 5–30 mins with predominant amounts being in the liver, skin, fat, prostate and bladder [318]. Using synthetic metabolites of rofecoxib it was found using the whole blood assay *in vitro* that none of the metabolites had any effects on COX-2 or COX-1 activity [320].

The mechanisms underlying the biliary excretion of rofecoxib and its 5-hydroxymetabolite were investigated in intact and biliary-canulated rats following administration of ^{14}C- or ^{18}O-labelled compounds. The biliary excretion of both rofecoxib and its 5-hydroxy metabolite were dependent on uninterrupted bowel flow since this was blocked by ligation of the bowel duct. The secondary peak of drug which occurs in the plasma at approximately 10 h after administration of rofecoxib or its 5-hydroxymetabolite was abolished by bile duct canulation. The reversible reaction of 5-hydroxy rofecoxib to the parent drug was also demonstrated. These results suggested that a metabolic sequence involving 5-hydroxylation of rofecoxib followed by biliary elimination of the corresponding glucuronide and de-conjugation of the glucuronide in the lower intestinal tract presumably by β-glucuronidases from enteric bacteria yielded the corresponding hydroxy-acid. This spontaneously cyclised to regenerate parent rofecoxib and was reabsorbed into the systemic circulation [331]. The biliary excretion in rats following oral administration is about 58–76% and that following intravenous administration 37–94%, showing there is a considerable degree of overlap when the radiolabelled drug is administered by these two routes [318, 331]. The biliary excretion in dogs is appreciably lower than that from orally administered drug being 26.5% and following intravenous administration 28.2% [318, 331]. The urinary excretion correspondingly is increased and 16.8% following oral dosage and 32% following intravenous administration. This contrasts with the markedly lower urinary excretion of the drug in rats which is about 1–2% [331, 332].

In vitro studies on the cytochromes P_{450} show that there was no impairment of testosterone metabolism, thus indicating that there was no likely effect of rofecoxib on cytochrome P_{450} 3A4 even though it appeared that rofecoxib could be metabolised by this isoform and cytochrome P_{450} 1A2 to a weak extent. Overall, it appeared that rofecoxib only had a very weak ability to inhibit cytochromes P_{450}.

As indicated earlier, rofecoxib is primarily metabolised by cytosolic enzymes and investigations indicated that this was an NADPH-dependent system. It appears that the oxidative pathway of metabolism of rofecoxib predominates in the rat whereas the reductive pathway (involving NADPH) predominates in humans as well as in monkeys and is most marked in dogs [317, 332]. These important metabolic differences may have considerable significance for the interpretation of the effects of rofecoxib in rats as compared with dogs. It appears that the metabolism in dogs is closer to that of humans.

A full dose ranging pharmacokinetic study in which the drug was administered at oral doses in healthy men, covering the range of therapeutic and supratherapeutic doses (25 to 375 mg/d) was undertaken over a 14-day period and these data were compared with the effects on COX-1 and COX-2 activity using the whole blood assay [321]. The kinetics of rofecoxib were found to be complex and non-linear. The elimination half-life ranged from 9.9–17.5 h after multiple dosages with an accumulation ratio of approximately two overall doses. The C_{max} after 14 days with the standard 25 mg dose of the drug was 382 ng/ml and this ranged up to 3,169 ng/ml following the last dose of 375 mg per day. There appeared to be no differences in the area under the curve (AUC_{0-24}) at the 250 and 375 mg dosage levels suggesting that there is a limitation to absorption at the high doses. Furthermore, the geometric mean of the C_{max} values was not different for the doses of 100–150 and the 375 mg dose but they were all greater than the 25 mg dose following single oral administration of the drug. For each of the doses the C_{max} values were significantly greater after multiple doses than single dose administration of rofecoxib. While there were no effects of any of the dosages on COX-1 derived serum thromboxane production there were markedly different results obtained with the dosages on LPS stimulated PGE_2, which is presumed to be from COX-2. The 25 mg dose, given as a single dose or after multiple doses, did not cause complete inhibition of COX-2 activity. There was an ~40–50% inhibition over the time course of COX-2 derived PGE_2 following single dosage and 50–70% inhibition following multiple dosage of 25 mg rofecoxib. The higher doses, however, caused almost complete inhibition of COX-2 activity. These results are of interest in view of the wide-spread usage of the 25 mg dose for control of pain and inflammation, where it is clear that the effects on COX-2 activity in the circulation are not complete.

Pharmacology and toxicology

The pre-clinical anti-inflammatory, analgesic and antipyretic activity as described in the NDA and in the earlier section are indicative of a classical NSAID in terms of the properties in standard model systems. Toxicokinetic studies formed in rats, mice, rabbits and dogs and the long-term (53 weeks) studies at the maximum performed in dogs and rats have been described in the NDA [318] and all revealed a classical pattern of NSAID related toxicity with primary toxic effects occurring in the GI tract and renal systems. There was no evident genotoxicity [318]. Aspects of safety pharmacology were investigated and there were no influences on cardiovascular function, autonomic responses, respiratory function or gastric secretion in dogs following administration of 25 mg/kg of rofecoxib. GI motility, motor function, autonomic responses and spinal reflexes were not affected in mice when given 1–100 mg/kg p.o. of the drug.

In female dogs administration of 5 or 25 mg/kg rofecoxib caused a significant decrease in urine volume, sodium excretion and urinary PGE_2 levels. Glomerular filtration rate, renal plasma flow, plasma renin activity and anti-diuretic hormone levels were unaltered. The results suggest there may be some effect of rofecoxib at the level of tubular functions but not at the glomerulus in dogs. The changes in urinary volume and sodium excretion appear to be secondary to the reduction in renal PGE_2.

In volume depleted normotensive male Wistar rats given 10 mg/kg of rofecoxib p.o., there was a reduction in glomerular filtration rate, renal blood flow, sodium excretion and urine volume. Overall, the study suggested that rofecoxib had similar effects to that of many NSAIDs on renal functions and that these were primarily at the level of tubular functions.

The renal effects of rofecoxib as well as those of other COX-2 inhibitors have received much attention since there have been a considerable number of reports of renal toxicity associated with these drugs in humans. In a study undertaken in healthy elderly adults on fixed sodium intake, rofecoxib 50 mg once daily, indomethacin 50 mg t.i.d. or placebo was compared over a 2-week period [333]. Glomerular filtration rate was unaffected by rofecoxib whereas it was inhibited by indomethacin thus the conclusions of a lack of a glomerular effect suggested from the preclinical toxicology studies were confirmed in humans. Rofecoxib caused a clinically insignificant and transient retention of sodium (a feature seen with most NSAIDs) and this appeared to be related to the selective inhibition of COX-2. Since glomerular filtration was impaired by indomethacin it was suggested that this was a COX-1 related effect, though it must be said that no studies with direct acting COX-1 inhibitors have been undertaken to support this view in humans.

There have been extensive investigations of the safety profile of rofecoxib in controlled clinical trials sponsored by the company marketing the drug (Merck Inc., Review and Evaluation of Pharmacology and Toxicology; [334]). These studies showed that in elderly, osteoarthritic patients having varying degrees of alterations in creatinine clearance, rofecoxib 50 mg but not 25 mg or the lower dose of 12.5 mg significantly altered creatinine clearance, but this was within the range of effects seen by ibuprofen 2.4 g daily and diclofenac 150 mg t.i.d. [327]. Some of the OA patients reported peripheral oedema and mean changes in systolic and diastolic blood pressure occurred with rofecoxib, with systolic blood pressure increasing ~2 mm over all the dosage groups whereas there was lower degree of increase in systolic blood pressure with ibuprofen and diclofenac. These results have suggested that under controlled clinical trial conditions there may be effects of rofecoxib, like that of other COX-2 inhibitors, on components of renal function and cardiovascular activity.

In the prescription event monitoring studies performed in 15,268 patients that received rofecoxib primarily for OA (24% total patients) under general practice conditions in the UK there were three reports of acute renal failure [329]. It is notable that despite claims that the drug rofecoxib was primarily developed to have appreciably lower GI toxicity in humans that there were 110 reports of serious GI events including 76 upper GI bleeds and peptic ulcers and one perforated colon. There were also 111 thromboembolic events reported, one case of Stephen Johnson's syndrome, severe anaphylaxis and angio-oedema [329]. Cardiovascular complications have been a well-known problem with the COX-2 inhibitors [335] and it has been suggested that this may relate in part to the inhibitory effects of the COX-2 inhibitors on the vascular production of COX-2 derived prostacyclin [336].

Investigating the mechanistic aspects of the actions of rofecoxib in relationship to anti-inflammatory analgesic and anticancer effects has largely implicated the actions of the drug on the COX-2 system. However, there are indications that there may be some actions of the drug on signal transduction pathways that may be of considerable significance. Thus, the binding of the transcription factor NFκB to DNA was inhibited at high concentrations of 10–100 μmol/l whereas the binding of AP1 was markedly increased at 100 μmol/l concentrations of the drug [337, 338]. These concentrations are relatively high and thus caution should be noted about the interpretation of these effects in relationship to known pharmacologically active concentrations of the drug.

Reduction in polyps in APCΔ716 mice were shown with both rofecoxib and sulindac so that in relation to colon cancer there may be inhibitory effects of rofecoxib like that of sulindac on vascular angiogenesis and transcription factor activation linked to the development of human adenomatous polyposis [339]. In another study 10 μmol/l rofecoxib attenuated NFκB activation and the expression of NFκB dependent genes in mouse peritoneal macrophages [338]. The drug was also found to protect against LPS-induced apoptosis by the impairment of the inflammatory response presumably mediated by the NFκB system as well as that involving mitogen activated protein kinase p38 and the extracellular regulated kinase p44 as well as the cJUN-JUN N-terminal kinase [338].

The relationship of the signal transduction effects of rofecoxib in causing inhibition of NFκB binding to DNA and the paradoxical increase in activating protein 1 (AP1) binding high concentrations of the drug [337] have been related to the effects of rofecoxib *in vivo* in the spinal cord of rats in whom oedema had been induced by injection of zymosan [340]. At doses of 1 and 10 mg/kg the spinal cord induction of iNOS was increased but COX-2 protein expression was decreased. This suggested that the opposite effects of rofecoxib in activating AP1 and inhibiting NFκB may explain the paradoxical changes in the spinal cord on the nitric oxide and COX systems.

In the uric acid induced pain model in the rat prior treatment with rofecoxib or the selective COX-1 inhibitor SC560 showed similar analgesic effects although the effects of the latter were somewhat lower than the effects of rofecoxib [340]. Paradoxically, however, when SC560 was given after the induction of the pain response by uric acid, it produced greater inhibition than that with rofecoxib [340]. Thus there may be differential effects on COX-1 and COX-2 responses that explain some of the analgesic effects of NSAIDs [338, 340].

In the hot plate test in rats the prior treatment with p-chlorophenylalanine inhibited the analgesic activity of rofecoxib [341]. It was found that rofecoxib significantly increased 5HT concentrations as well as $5HT_2$ receptors in the frontal cortex suggesting that, as with many other NSAIDs and analgesics, rofecoxib may influence the serotonergic pathway and that this may contribute to the central analgesic effects of this drug [341].

Interestingly the anti-nociceptive and anti-inflammatory effects of melatonin which have been shown to be potentiated by indomethacin are only slightly affected by rofecoxib [342]. This suggests that the major effect of the interaction may be due to COX-1 inhibition or perhaps the linked inhibition of COX-1 or COX-2 by indomethacin. Rofecoxib was also found to attenuate the neurogenic component of adjuvant induced temporomandibular chronic inflammation in Freund's adjuvant injected rats [343]. As with other NSAIDs and celecoxib, rofecoxib has been shown to inhibit a number of tumour models [344, 345] and potentiation of cytokine-related angiogenic responses in combination with radiation exposure [339].

As noted previously, coxibs have been shown to be more ulcerogenic in the stomach and intestine in adjuvant arthritic rats [298]. It was suggested by Kato and Takeuchi [298] that the increased ulcerogenic response in arthritic rats was due to upregulation of iNOS and COX-2 but it is also noted that these authors did not consider the other extensive effects of arthritic processes in this model.

Impairment of ulcer healing has been implicated with rofecoxib administration in rats, possibly by impairment of growth factors such as hepatocyte growth factor and gastrin [346]. Laudanno and co-workers observed that the ulcerogenic effects of rofecoxib were revealed to be markedly increased in rats that had been exposed to stress, duodenal ulceration by injection of cysteamine or gastric ulcers induced by acetic acid [347]. An interesting phenomenon observed by these authors was that in unstressed animals there was an increased neutrophilia but not leucocyte accumulation in the mucosa in rats treated with rofecoxib or celecoxib [347]. Thus there may be some systemic effects that are caused by these coxibs that predispose to the development of mucosal injury when the animals are exposed to stress or other ulcerogens. There may also be some indirect effects on "ischaemic preconditioning" relating to the actions of PGs, NO, adenosine and activation of sensory nerves during the development of gastric mucosal injury that is induced by alterations in blood flow [348].

The potential for COX-2 rebound to be an important factor relating to wound healing as well as the ulcerogenic effects following chronic administration of coxibs has been highlighted by studies in rats where prolonged administration of coxibs has been investigated [349]. Colville-Nash and Gilroy have speculated that COX-2 may help resolve pathology, e.g., GI ulceration, inflammation, wound healing, etc. possibly by generation of an alternative series of PGs such as the cyclopentenone PGs [349]. The exact influence of these and the relationship to resolve resolution of injury has yet to be established but this is an interesting concept.

In terms of untoward effects of COX-2 inhibition it has been suggested that COX-2 function is important for bone fraction healing [350] using implanted tibiae into the ear chambers then given 4 weeks oral administration of rofecoxib, 12 mg/kg/d or naproxen 110 mg/kg. It was found that both drugs decreased bone growth as determined by antibody staining for vitronectin receptor reflecting osteoclast-like cells [351]. Whether these studies in animal model systems implicating inhibition of bone growth by coxibs as well as NSAIDs has any clinical significance has yet to be determined.

Etoricoxib

Etoricoxib (MK-O663) (5-chloro-2-(6-methylpyridine-3-yl)-3-(4-methylsulphonylphenyl) pyridine is another methyl sulphonylphenyl derivative that was developed by Merck and recently introduced as a second generation coxib. The pharmacokinetic and pharmacological properties of etoricoxib have been reviewed [352] and the main claim for unique pharmacology with etoricoxib is that it has a greater selectivity for COX-2 in relationship to COX-1 than either that of its progenitor rofecoxib, celecoxib or valdecoxib [352]. Indeed the selectivity in the whole blood assay is indicated by the ratio of COX-2 to COX-1 being 106 for etoricoxib compared with 35 for rofecoxib, 30 for valdecoxib, 7.6 for celecoxib, 7.3 for nimesulide, 2 for meloxicam, 2.4 for etodolac and various other ranges of 0.2–3.0 for other NSAIDs [352]. It is claimed that etoricoxib is least likely to interfere with the antiplatelet effects of aspirin via the irreversible inactivation of COX-1, an issue which has been raised with some other NSAIDs. The GI blood loss from etoricoxib 120 mg once daily was compared with ibuprofen 800 mg t.i.d. and placebo for 28 days and it was found that etoricoxib did not cause any significant blood loss measured by radiochromium technique [352]. Like that of rofecoxib, etoricoxib is recommended for once a day administration due to its long elimination half-life of ~22 h [352]. Like that of the other coxibs, etoricoxib has a large volume of distribution, approximately 120 L but is more extensively bound to plasma proteins (92%) in comparison with rofecoxib [352]. In contrast to rofecoxib, etoricoxib undergoes hydroxylation by cytochromes P_{450} 3A4 and 2D6, 2C9, 1A2 and 2C19 may also play a role [352]. Thus any drugs interacting with these cytochromes will inevitably lead to some alterations in pharmacokinetic or pharmacodymaic properties and safety.

The effects of etoricoxib in various cyclooxygenase systems were extensively investigated and the selectivity confirmed for COX-2 [353]. In the carrageenan paw oedema in rats, etoricoxib was more potent having an ED_{50} of 0.6 mg/kg compared with that of rofecoxib of 1.5 mg/kg and indomethacin 2 mg/kg [353]. In the carrageenan induced paw hyperalgesia, etoricoxib showed a three-fold greater potency compared with rofecoxib whereas in the adjuvant arthritis model rofecoxib (ED_{50} 0.6 mg/kg/d) was almost identical to that of rofecoxib (0.7 mg/kg/d) [17]. Antipyretic activity administration of etoricoxib in rats was approximately three- to four-fold less than with rofecoxib but within the range of that observed with indomethacin [353].

The combined pharmacokinetics and influences on circulating COX-1 and COX-2 enzymatic activity were investigated in healthy volunteers who received doses of 5, 25, 125 and 500 mg of etoricoxib or 10, 50 and 250 mg once daily of etoricoxib [354]. The drug was found to be rapidly absorbed with a maximum concentration at 1–1.5 h and following multiple oral dosage the accumulation ratio averaged approximately 1.6 [354]. There was no evident inhibition of thromboxane production using the whole blood assay *ex vivo* but the percent inhibition of LPS induced PGE_2 in the whole blood assay correlated with the plasma concentrations of the drug. The lower doses of 25 and 50 mg etoricoxib only produced partial inhibition of LPS induced PGE_2 production from COX-2 whereas the higher doses of 100 mg or greater did achieve at least 80% inhibition after single or repeated dosage of the drug [354]. In dose ranging studies performed in patients with OA doses of 30 to 90 mg etoricoxib were approximately equally effective to diclofenac 50 mg t.i.d. over a 6 week treatment period [355] and the 19 mg dose of etoricoxib employed in two studies in RA [356, 357] appeared to be similar to that in efficacy to naproxen 500 mg b.i.d. It would appear, therefore, in relationship to COX-2 inhibition *ex vivo* [298] that although there is about 70–80% inhibition of COX-2 derived PGE_2 at the dose range for effects in OA and RA [355–357] that it may not be that there is complete inhibition.

Lumiracoxib

Lumiracoxib is a structurally distinct COX-2 selective inhibitor that does not have the sulphur containing moiety but instead possesses a carboxylic acid, giving it weakly acidic properties (pKa 4.7) [358, 359] and is structurally similar to that of diclofenac. The drug is currently under development by Novartis (Switzerland, USA) but relatively little basic information is available on this drug to date. The pharmacokinetics of the drug have been reported recently at a number of international meetings [359–366]. The drug is rapidly absorbed with peak drug values appearing in the plasma at 2 h and with a relatively short elimination half-life of 5.8 h [359]. In patients with RA there is a marked uptake of the drug and the 4-hydroxy or 5-carboxy metabolite into the synovial fluid following lumiracoxib 400 mg for 7 days. The values of this metabolite in synovial fluid were approximately that of the plasma concentrations, lumiracoxib peaked at about 6 h in synovial fluid at about half the concentration seen in plasma [359]. It is clear that there is a relatively high uptake of the drug into the synovial fluid and this may have therapeutic advantages for the control of pain and inflammation in joints. There does not appear to be any accumulation of the drug in elderly OA patients as the clearance rates are identical after 28 days treatment with lumiracoxib 200 or 400 mg once daily [360]. Antacids or the proton pump inhibitor omeprazole do not affect the pharmacokinetics of lumiracoxib nor does the administration of warfarin or an oral contraceptive [361]. Lumiracoxib also does not influence the pharmacokinetic parameters of methotrexate in patients with RA. The GI absorption of lumiracoxib was investigated in healthy male volunteers and the regions of uptake and bioavailability were investigated [363]. It was shown that lumiracoxib was rapidly and efficiently absorbed from all regions of the GI tract with no particular region being favoured [363]. The longer median maximum time of absorption in the stomach relative to that of the small bowel and colon probably reflects dissolution effects in the local pH environment. Daily doses of 100 to 400 mg lumiracoxib once daily were shown to be effective in controlling the symptoms of pain and inflammation in patients with OA of the hand, hip or knee [364–366]. Further studies including pre-clinical investigations on lumiracoxib are awaited with much interest. This acidic coxib may present with interesting properties that separate it from the established coxibs.

Conclusions

In this chapter a review of established and newly available COX-2 selective drugs has revealed that there are important pharmacokinetic and pharmacodynamic differences among these drugs. The eventual safety and clinical significance of many of these studies especially with the newer COX-2 selective inhibitors has yet to be fully evaluated as further extensive long-term pharmacoepidemiological and clinical studies are awaited with interest especially to establish if the preclinical investigations of these drugs have actually translated into favourable therapeutic profiles.

Acknowledgements
My thanks to Mrs Veronica Rainsford-Koechli and Mrs Marguerite Lyons for assistance with the typing of this manuscript and to Drs Guenther Trummlitz and Joanne van Ryn for the preparation of Figures 3–5, 7 and 8.

References

1 Smith WM, Meade EA, De Witt DL (1994) Interactions of PGH synthase isozymes-1 and –2 with NSAIDs. *Ann NY Acad Sci* 755: 50–57
2 Vane JR, Botting, RM (1995) A better understanding of anti-inflammatory drugs based on isoforms of cyclooxygenase (COX-1 and COX-2). In: B Samuelsson et al (eds): *Advances in Prostaglandin, Thromboxane and Leukotriene Research, Volume 23.* Raven Press Ltd., New York, 41–48
3 Vane JR, Botting RM (1995) New insights into the mode of action of anti-inflammatory drugs. *Inflamm Res* 44: 1–10
4 Dubois RN, Abramson SB, Crofford L, Gupta RA, Simon LS, Van de Putte LBA, Lipsky PE (1998) Cyclooxygenase in biology and disease. *FASEB J* 12: 1063–1073
5 Kam PCA, See AU-L (2000) Cyclo-oxygenase isoenzymes: physiological and pharmacological role. *Anaesthesia* 55: 442–449
6 Jouzeau J-Y, Terlain B, Abid A, Nédélec E, Netter P (1997) Cyclo-oxygenase isoenzymes. How recent findings affect thinking about nonsteroidal anti-inflammatory drugs. *Drugs* 53: 563–582
7 Kulkarni SK, Jain NK, Singh A (2000) Cyclooxygenase isoenzymes and newer therapeutic potential for selective COX-2 inhibitors. *Methods Findings Exp Clin Pharmacol* 22: 291–298
8 Vane J, Botting J, Botting R (eds) (1995) *Improved non-steroid anti-inflammatory drugs COX-2 enzyme inhibitors.* Kluwer Academic Publishers, Dordrecht and William Harvey Press, London
9 Vane J, Botting RM (eds): (2001) *Therapeutic roles of selective COX-2 inhibitors.* William Harvey Press, London
10 Fenner H (1997) Differentiating among nonsteroidal anti-inflammatory drugs by pharmacokinetic and pharmacodynamic profiles. *Semin Arthritis Rheum* 26: 28–33
11 Brooks P, Emery P, Evans JF, Fenner H, Hawkey CJ, Patrono C, Smolen J, Breedveld F, Day R, Dougados M et al (1999) Interpreting the clinical significance of the differential inhibition of cyclooxygenase-2 and cyclooxygenase-2. *Rheumatology* 38: 779–788
12 Fitzgerald GA, Patrono C (2001) The coxibs, selective inhibitors of cyclooxygenase-2. *N Engl J Med* 345: 433–442
13 Herbette LF, Vecchiarelli, Trummlitz G (1996) NSAID mechanism of action: the role of intracellular pharmacokinetics. In: J Vane, J Botting, R Botting (eds): *Improved Non-Steroid Anti-Inflammatory Drugs COX-2 Enzyme Inhibitors.* Kluwer Academic Publishers, Dordrecht, William Harvey Press London, 85–102
14 Trummlitz G, van Ryn J (2002) Designing selective COX-2 inhibitors: molecular modelling approaches. *Curr Opin Drug Discovery Development* 5: 550–561
15 Chulada PC, Langenbach R (1997) Differential inhibition of murine prostaglandin synthase-1 and –2 by nonsteroidal anti-inflammatory drugs suing exogenous and endogenous sources of arachidonic acid. *J Pharmacol Exp Ther* 280: 606–613
16 Hulkower KI, Otis ER, Wernimont AK, Bell RL (1997) Stimulus dependence of non-steroidal anti-inflammatory drug potency in a cellular assay of prostaglandin H synthase-2. *Eur J Pharmacol* 331: 79–85
17 Warner TD, Giuliano F, Vojnovic I, Bukasa A, Mitchell JA, Vane JR (1999) Nonsteroid drug selectivities for cyclo-oxygenase-1 rather than cyclo-oxygenase-2 are associated with human gastrointestinal toxicity: a full *in vitro* analysis. *Proc Natl Acad Sci USA* 96: 7563–7568
18 Warner TD, Pairet M, Van Ryn J (2001) Test systems for inhibitors of cyclooxygenase-1 and cyclooxygenase-2. In: JR Vane, RM Botting (eds): *Therapeutic roles of selective COX-2 inhibitors.* William Harvey Press, London, 76–94
19 Sanda M, Jacob GB, Fliedner Jt L, Kennedy J, Gotz M (1987) Etodolac. In: AJ Lewis, DE Furst DE (eds): *Nonsteroidal Anti-inflammatory Drugs, Mechanisms and Use.* Marcel Dekker Inc, New York, 349–370
20 Jones R (2001) Etodolac: Clinical profile of an established cyclooxygenase-2 inhibitor. In: JR Vane, RM Botting (eds): *Therapeutic Roles of Selective COX-2 inhibitors.* William Harvey Press, London, 482–497
21 Glaser KB (1995) Cyclooxygenase selectivity and NSAIDs: cyclooxygenase-2 selectivity of etodolac (Lodine). *Inflammopharmacology* 3: 335–345
22 Dvornik D, Lee DKH (1989) Theoretical mechanism for the gastrointestinal safety of etorolac: selective sparing of cytoprotective prostaglandins. *Clin Rheumatol* 8 (Suppl. 1): 16–24
23 Denurson CA, Humber LG, Abraham NA, Schilling G, Martel RR, Pace-Asciak C (1983)

Resolution of etodolac and anti-inflammatory and prostaglandin synthetase inhibiting properties of the enantiomers. *J Med Chem* 26: 1778–1780
24 Rainsford KD (1999) Pharmacology and toxicology of ibuprofen. In: KD Rainsford (ed.): *Ibuprofen: A Critical Bibliographic Review*. Taylor and Francis, London, 143–275
25 Rainsford KD (1995) Gastric ulcerogenicity in mice of enantiomers of non-steroidal anti-inflammatory drugs having differing potency as prostaglandin synthesis inhibitors. *Pharmaceutical Sci* 1: 169–171
26 Ogiso T, Kitagawa T, Iwaki M, Tanino T (1997) Pharmacokinetic analysis of enterohepatic circulation of etodolac and effect of hepatic and renal injury on the pharmacokinetics. *Biol Pharm Bull* 20: 405–410
27 Strickmann DB, Blaschke G (2001) Isolation of an unknown metabolite of the non-steroidal anti-inflammatory drug etodolac and its identification as 5-hydroxy etodolac. *J Pharmaceutical Biomedical Analysis* 25: 977–984
28 Brocks DR, Jamali F (1990) The pharmacokinetics of etodolac enantiomers in the rat. Lack of pharmacokinetic interaction between enantiomers. *Drug Metabolism Disp* 18: 471–475
29 Laudanno OM, Cesolari JA, Esnarriaga J, Flaherty P, Vada J, Guastalli G, San Miguel P, Bedini OA (1998) *In vivo* selectivity of nonsteroidal anti-inflammatory drugs on COX-1-COX-2 and gastrointestinal ulcers, in rats. *Acta Gastroenterol Latinoam* 28: 249–255
30 Reimer ME, Johnston SA, Leib MS, Duncan RB Jr, Reimer DC, Marini M, Gimber TK (1999) The gastroduodenal effects of buffered aspirin, carprofen, and etodolac in healthy dogs. *J Vet Intern Med* 13: 472–477
31 Laudanno OM, Cesolarij JA, Esnarriaga J, San Miguel P, Bedini OA (2000) *In vivo* selectivity of nonsteroidal anti-inflammatory drugs and gastrointestinal ulcers in rats. *Dig Dis Sci* 45: 1359–1365
32 Wilcox GM, Porensky RS (1997) Acute colitis associated with etodolac. *J Clin Gastroenterol* 25: 367–368
33 Eis MJ, Watkins BM, Philip A, Welling RE (1998) Nonsteroidal-induced benign strictures of the colon: a case report and review of the literature. *Am J Gastroenterol* 93: 120–121
34 Gopal DV, Katon RM (1999) Endoscopic balloon dilation of multiple NSAID-induced colonic strictures: case report and review of literature on NSAID-related colopathy. *Gastrointest Endosc* 50: 120–123
35 Bjarnason I, Zanelli G, Smith T, Prouse P, Williams P, Smethurst P, Delacey G, Gumpel MJ, Levi AJ (1987) Nonsteroidal anti-inflammatory drug-induced intestinal inflammation in humans. *Gastroenterology* 93: 480–489
36 Bjarnason I, Hayllar J, MacPherson AJ, Russell AS (1993) Side-effects of nonsteroidal anti-inflammatory drugs on the small and large intestine in humans. *Gastroenterology* 104: 1832–1847
37 Svendsen KB, Bech JN, Sorensen TB, Pedersen EB (2000) A comparison of the effects of etodolac and ibuprofen on renal haemodynamics, tubular function, renin, vasopressin and urinary excretion of albumin and alpha-glutathione-S-transferase in healthy subjects: a placebo-controlled cross-over study. *Eur J Clin Pharmacol* 56: 383–388
38 Boni JP, Korth-Bradley JM, Martin P, Simcoe DK, Richards LS, Rennebohm R, Walson PD (1999) Pharmacokinetics of etodolac in patients with stable juvenile rheumatoid arthritis. *Clin Ther* 21: 1715–1724
39 Rothstein R (1998) Safety profiles of leading nonsteroidal anti-inflammatory drugs. *Am J Med* 105 (5A): 39S–43S
40 Russell RI (1999) Defining patients at risk of non-steroidal anti-inflamamtory drug gastropathy. *Ital J Gastroenterol Hepatol* 31:S14–S18
41 Neustadt DH (1997) The influence of age, gender, Helicobacter pylori and smoking on gastric mucosal adaptation to non-steroidal anti-inflammatory drugs. *Aliment Pharmacol Ther* 11: 907–912
42 Micklewright R, Lane S, Linley W, McQuade C, Thompson F, Maskrey N (2003) NSAIDs, gastroprotection and cyclo-oxygenase-II-selective inhibitors. *Aliment Pharmacol Ther* 17: 321–332
43 Wilcox GM, Porensky RS (1997) Acute colitis associated with etodolac. *J Clin Gastroenterol* 25: 367–368
44 Kawai S, Nishida S, Kato M, Furumaya Y, Okamoto R, Koshino T, Mizushima Y (1998) Comparison of cyclooxygenase-1 and –2 inhibitory activities of various nonsteroidal anti-infalmamtory drugs using human platelets and synovial cells. *Eur J Pharmacol* 347: 87–94
45 Lipscomb GR, Wallis N, Armstrong G, Goodman MJ, Rees WD (1995) Gastric mucosal adapta-

tion to etodolac and naproxen. *Aliment Pharmacol Ther* 9: 379–385
46 Lipscomb GR, Campbell F, Rees WD (1997) The influence of age, gender, Helicobacter pylori and smoking on gastric mucosal adaptation to non-steroidal anti-inflammatory drugs. *Aliment Pharmacol Ther* 11: 907–912
47 Kusuhara H, Komatsu H, Sumichika H, Sugahara K (1999) Reactive oxygen species are involved in the apoptosis induced by nonsteroidal anti-inflammatory drugs in cultured gastric cells. *Eur J Pharmacol* 383: 331–337
48 Noda M, Tasumi Y, Tomizawa M, Takama T, Mitsufuji S, Sugihara H, Kashima K, Hattori T (2002) Effects of etodolac, a selective cyclooxygenase-2 inhibitor, on the expression of E-cadherin-catenin complexes in gastrointestinal cell lines. *J Gasdtroenterol* 37: 896–904
49 Kusunoki N, Yamazaki R, Kawai S (2002) Induction of apoptosis in rheumatoid synovial fibroblasts by celecoxib, but not by other selective cyclooxygenase 2 inhbitors. *Arthritis Rheum* 46: 3159–3167
50 Yamazaki R, Kusunoki N, Matsuzaki T, Hashimoto S, Kawai S (2002) Selective cycloxygenase-2 inhibitors show a differential ability to inhibit proliferation and induce apoptosis of colon adenocarcinoma cells. *FEBS Lett* 531: 278–284
51 Sato TA, Keelan JA, Blumenstein M, Mitchell MD (2002) Efficacy and specificity of non-steroidal anti-inflammatory drugs for the inhibition of cytokine-stimulated prostaglandin E(2) secretion by amnion-derived WISH cells. *Prostaglandins Leukotrienes Essential Fatty Acids* 66: 525–527
52 Sato TA, Keelan JA, Mitchell MD (2002) Regulation of prostaglandin production in an immortalized human myometrial cell line by cytokines and non-steroidal anti-inflammatory drugs. *Eur J Obstet Gynecol Reprod Biol* 100: 158–162
53 Takahashi M, Katayama Y, Takada H, Kuwayama H, Terano A (2000) The effect of NSAIDs and a COX-2 specific inhibitor on Helicobacter pylori-induced PGE2 and HGF in human gastric fibroblasts. *Aliment Pharmacol Ther* 14: 44–49
54 Laine L, Cominelli F, Sloane R, Casini-Raggi V, Marin-Sorensen M, Weinstein WM (1995) Interaction of NSAIDs and Helicobacter pylori on gastrointestinal injury and prostaglandin production: a controlled double-blind trial. *Aliment Pharmacol Ther* 9: 127–135
55 Kishimoto Y, Takata N, Jinnai T, Morisawa T, Shiota G, Kawasaki H, Hasegawa J (2000) Sulindac and a cyclooxygenase-2 inhibitor, etodolac, increase APC mRNA in the colon of rats treated with azoxymethane. *Gut* 47: 812–819
56 Chen WS, Wei SJ, Liu JM, Hsiao M, Kou-Lin J, Yang WK (2001) Tumor invasiveness and liver metastasis of colon cancer cells correlated with cyclooxygenase-2 (COX-2) expression and inhibited by a COX-2 selective inhibitor, etodolac. *In J Cancer* 91: 894–899
57 Kamijo T, Sato T, Nagatomi Y, Kitamura T (2001) Induction of apoptosis by cyclooxygenase-2 inhibitors in prostate cancer cell lines. *Int J Urol* 8: S35–S39
58 Endo K, Sairyo K, Komatsubara S, Sasa T, Egawa H, Yonekura D, Adachi K, Ogawa T, Murakami R, Yasui N (2002) Cyclooxygenase-2 inhibitor inhibits the fracture healing. *J Physiol Anthropol Appl Human Sci* 21: 235–238
59 Kopp UC, Cicha MZ, Smith LA, Haeggstrom JZ, Samuelsson B, Hokfelt T (2000) Cyclooxygenase-2 involved in stimulation of renal mechanosensitive neurons. *Hypertension* 35: 373–378
60 Mabee CL, Mabee SW, Baker PB, Kirkpatrick RB, Levine EJ (1995) Fulminant hepatic failure associated with etodolac use. *Am J Gastroenterol* 90: 659–661
61 Degner F, Richardson B (2001) Review of gastrointestinal tolerability and safety of meloxicam. *Inflammopharmacology* 9: 71–80
62 Degner F, Lanes S, van Ryn J, Sigmund R (2001) Pharmacological and clinical profile of meloxicam. In: JR Vane, RM Botting (eds): *Therapeutic roles of selective COX-2 inhibitors.* William Harvey Press, London, 498–523
63 Van Bree H, Justus C, Quirke JF (1994) Preliminary observations on the effects of meloxicam in a new model for acute intra-articular inflammation in dogs. *Vet Res Commun* 18: 217–224
64 Rainsford KD, Skerry TM, Chindemi P, Delaney K (1999) Effects of the NSAIDs meloxicam and indomethacin on cartilage proteoglycan synthesis and joint responses to calcium pyrophosphate crystals in dogs. *Vet Res Commun* 23: 101–113
65 Stei P, Kruss B, Wiegleb J, Trach V (1996) Local tissue tolerability of meloxicam, a new NSAID: indications for parenteral, dermal and mucosal administration. *Brit J Rheumatol* 35: 44–50
66 Ghozlan PR, Bernhardt M, Vélicitat P, Bluhmki E (1996) Tolerability of multiple administration of intramuscular meloxicam: a comparison with intramuscular piroxicam in patients with rheuma-

toid arthritis or osteoarthritis. *Brit J Rheumatol* 35: 51–55
67 Engelhardt G (1996) Pharmacology of meloxicam, a new non-steroidal anti-inflammatory drug with an improved safety profile through preferential inhibition of COX-2. *Brit J Rheumatol* 35: 4–12
68 Trummlitz G, Engelhardt G (1987) Meloxicam: Structural aspects of a new NSAID with improved therapeutic properties. Poster presented at the XIth European Congress of Rheumatology, Athens, 28 June to 4 July 1987
69 Lehmann HA, Baumeister M, Lützen L, Wiegleb J (1996) Meloxicam: a toxicology overview. *Inflammopharmacology* 4: 105–123
70 Luger P, Daneck K, Engel W, Trummlitz G, Wagner (1996) Structure and physicochemical properties of meloxicam, a new NSAID. *Eur J Pharm Sci* 4: 175–187
71 Schoenfeld P (1999) Gastrointestinal safety profile of meloxicam: a meta-analysis and systematic review of randomised controlled trials. *Am J Med* 107 (6A): 48S–54S
72 McCormack K (1982) Mathematical model for assessing risk of gastrointestinal reactions to NSAIDs. In: KD Rainsford (ed.): *Azapropazone – 20 years of clinical use.* Kluwer Academic Publishers, London, 81–93
73 Rainsford KD (1999) Profile and mechanisms of gastrointestinal and other side-effects of nonsteroidal anti-inflammatory drugs (NSAIDs). *Am J Med* 107 (6A): 27S–36S
74 Engelhardt G, Homma D, Schnitzler C (1995) Meloxicam: a potent inhibitor of adjuvant arthritis in the Lewis rat. *Inflamm Res* 44: 548–555
75 Schmid J, Busch U, Heinzel G, Bozler G, Kaschke S, Kummer M (1995) Pharmacokinetics and metabolic pattern after intravenous infusion and oral administration to healthy subjects. *Drug Metab Dispos* 23: 1206–1213
76 Chesne C, Guyomard C, Guillouzo A, Schmid J, Ludwig E, Sauter T (1998) Metabolism of meloxicam in human liver involves cytochromes P4502C9 and 3A4. *Xenobiotica* 28: 1–13
77 US Food and Drug Administration (2002) MELOXICAM, MOBIC™ Tablets, 7.5 mg. Boehringer-Ingelheim Pharmaceuticals Inc (Sponsor). New Drug Application – NDA 20–938
78 Schmid J, Busch U, Trummlitz G, Prox A, Kaschke S, Wachsmuth H (1995) Meloxicam: Metabolic profile and biotransformation products in the rat. *Xenobiotica* 25: 1219–1236
79 Turck D, Roth W, Busch U (1998) A review of the clinical pharmacokinetics of meloxicam. *Br J Rheumatol* 35 (Suppl 1): 13–16
80 Busch U, Schmid J, Heinzel G, Schmaus H, Baierl J, Huber C, Roth W (1998) Pharmacokinetics of meloxicam in animals and the relevance to humans. *Drug Metab Dispos* 26: 576–584
81 Busch U, Engelhardt G (1990) Distribution of (^{14}C) meloxicam in joints of rats with adjuvant arthritis. *Drugs Exp Clin Res* 16: 49–52
82 Ludwig E, Schmid J, Beschke K, Ebner T (1999) Activation of human cytochrome P-450 3A4-catalyzed meloxicam 5'-methylhydroxylation by quinidine and hydroquinidine *in vitro*. *J Pharmacol Exp Ther* 290: 1–8
83 Xu HY, Zhong DF, Zhao LM, Zhang YF, Zhang BJ (2001) Pharmacokinetics of meloxicam in healthy Chinese volunteers. *Yao Xue Xue Bao* 36: 71–73
84 Lapicque F, Vergne P, Jouzeau JY, Loeuille D, Gillet P, Vignon E, Thomas P, Velicitat P, Turck D, Guillaume C et al (2000) Articular diffusion of meloxicam after a single oral dose: relationship to cyclo-oxygenase inhibition in synovial cells. *Clin Pharmacokinet* 39: 369–382
85 Pairet M, van Ryn J, Schierok H, Mauz A, Trummlitz G, Engelhardt G (1998) Differential inhibition of cyclooxygenases-1 and -2 by meloxicam and its 4'-isomer. *Inflamm Res* 47: 270–276
86 Jolliet P, Simon N, Bree F, Urien S, Pagliara A, Carrupt PA, Testa B, Tillement JP (1997) Blood-to-brain transfer of various oxicams: effects of plasma binding on their brain delivery. *Pharm Res* 14: 650–656
87 Poulsen Nautrup B, Horstermann D (1999) Pharmacodynamic and pharmacokinetic aspects of the non-inflammatory non-steroidal agent meloxicam in dogs. *Dtsch Tierarztl Wochenschr* 106: 94–100
88 Lees P, Sedgwick AD, Higgins AJ, Pugh KE, Busch U (1991) Pharmacodynamics and pharmacokinetics of meloxicam in the horse. *Brit Vet J* 147: 97–108
89 Yabe T, Honma M, Katsuki S, Wiegleb J, Luetzen L, Pueschner H, Lehmann H (1997) Single and repeated dose toxicity studies of meloxicam by oral administration in minipigs. *Pharmacometrics* 53: 197–212
90 Baert K, De Backer P (2003) Comparative pharmacokinetics of three non-steroidal anti-inflammatory drugs in five bird species. *Comp Biochem Physiol C Toxicol Pharmacol* 134: 25–33

91 US Food and Drug Administration (2000) New Animal Drug Application NADA 141-213. Metacam® (meloxicam) 0.5 mg/mL and 1.5 mg/mL Oral Suspension "indicated for the control of pain and inflammation associated with osteoarthritis in dogs. Boehringer Ingelheim Vetmedica, Inc. (Sponsor). Approval date April 15 2003

92 Euller-Ziegler L, Velicitat P, Bluhmki E, Türck D, Scheuerer S, Combe B (2001) Meloxicam: a review of its pharmacokinetics, efficacy and tolerability following intramuscular administration. *Inflamm Res* 50 (Suppl 1): S5–S9

93 Engelhardt G, Boegel R, Schnitzer C, Utzmann R (1996) Meloxicam: influence on arachidonic acid metabolism, part I. *In vitro* findings. *Biochem Pharmacol* 51: 21–28

94 Engelhardt G, Boegel R, Schnitzer C, Utzmann R (1996) Meloxicam: influence on arachidonic acid metabolism, part II. *In vivo* findings. *Biochem Pharmacol* 51: 29–38

95 Ogino K, Hatanaka K, Kawamura M, Ohno T, Harada Y (2000) Meloxicam inhibits prostaglandin E(2) generation via cyclooxygenase 2 in the inflammatory site but not that via cyclooxygenase 1 in the stomach. *Pharmacology* 61: 244–250

96 Rainsford KD, Ginsburg I, Gadd SJ (1997) A comparison between the effects of meloxicam and other NSAIDs on the production of oxyradicals by human polymorphonuclear leucocytes. *Immunopharmacology* 5: 9–19

97 Saleh TS, Calixto JB, Medeiros YS (1999) Effects of anti-inflammatory drugs upon nitrate and myeloperoxidase levels in the mouse pleurisy induced by carrageenan. *Peptides* 20: 949–956

98 Agha AM, El-Khatib AS, Al-Zuhair H (1999) Modulation of oxidant status by meloxicam in experimentally induced arthritis. *Pharmacol Res* 40: 385–392

99 Brzozowski T, Konturek PC, Konturek SJ, Sliwowski Z, Drozdowicz D, Stachura J, Pajdo R, Hahn EG (1999) Role of prostaglandins generated by cyclooxygenase-1 and cyclooxygenase-2 in healing of ischemia-reperfusion-induced gastric lesions. *Eur J Pharmacol* 385: 47–61

100 Brzozowski T, Konturek PC, Konturek SJ, Drozdowicz D, Pajdo R, Pawlik M (2000) Expression of cyclooxygenase (COX-1 and COX-2) in adaptive cytoprotection induced by mild stress. *J Physiol Paris* 94: 83–91

101 Villegas I, Martin MJ, La Casa C, Motilva V, Alarcon de la Lastra C (2000) Effects of meloxicam on oxygen radical generation in rat gastric mucosa. *Inflamm Res* 49: 361–366

102 Villegas I, Martin MJ, La Casa C, Motilva V, De La Lastra CA (2002) Effects of oxicam inhibitors of cyclooxygenase on oxidative stress generation in rat gastric mucosa. A comparative study. *Free Radic Res* 36: 769–777

103 Frode-Saleh TS, Calixto JB (2000) Synergistic anti-inflammatory effect of NfkappaB inhibitors and steroidal or non-steroidal anti-inflammatory drugs in the pleural inflammation induced by carrageenan in mice. *Inflamm Res* 49: 330–337

104 Bednarek D, Szuster-Ciesielska A, Zdzisinska B, Kondracki M, Paduch R, Kandefer-Szerszen M (1999) The effect of steroidal and non-steroidal anti-inflammatory drugs on the cellular immunity of calves with experimentally-induced local lung inflammation. *Vet Immunol Immunopathol* 71: 1–15

105 Lopez-Armada MJ, Sanchez-Pernaute O, Largo R, Diez-Ortego I, Palacios I, Egido J, Herero-Beaumont G (2002) Modulation of cell recruitment by anti-inflammatory agents in antigen-induced arthritis. *Ann Rheum Dis* 61: 1027–1030

106 Rainsford KD, Ying C, Smith FC (1997) Effects of meloxicam, compared with other NSAIDs, on cartilage proteoglycan metabolism, synovial prostaglandin E_2, and production of interleukins 1, 6 and 8, in human and porcine explants in organ culture. *J Pharm Pharmacol* 49: 991–998

107 Bonta IL, Elliott GR (1992) Non-steroidal anti-inflammtory drugs and the augmented lipoxygenase pathway: conceivable impact on joint conditions. In: KD Rainsford, GP Velo (eds): *Side-effects of Non-inflammatory Drugs 3*. Kluwer Academic Publishers, Dordrecht, 269–274

108 Rainsford KD (2003) Cytokines and eicosanoids in arthritis. In: P.B. Curtis-Prior (ed.): *Eicosanoids*. Wiley, Chichester, 347–358

109 Bayliss MT, Roughley PJ (1985) The properties of proteoglycan prepared from human articlular cartilage by using associative cesium chloride gradients of high and low starting densities. *Biochem J* 232: 111–117

110 Konigsson K, Odensvik K, Kindahl H (2002) Endocrine, metabolic and clinical effects of intravenous endotoxin injection after pre-treatment with meloxicam in heifers. *J Vet Med A Physiol Pathol Clin Med* 49: 408–414

111 Roth J, Hubschle T, Pehl U, Ross G, Gerstberger R (2002) Influence of systemic treatment with cyclooxygenase inhibitors on lipopolysaccharide-induced fever and circulating levels of

cytokines and cortisol in guinea-pigs. *Pflugers Arch* 443: 411–417
112 Aguirre-Banuelos P, Granados-Soto V (2000) Evidence for the participation of the nitric oxide-cyclic GMP pathway in the antinociceptive action of meloxicam in the formalin test. *Eur J Pharmacol* 395: 9–13
113 Laird JM, Carter AJ, Grauert M, Cervero F (2001) Analgesic activity of a novel use-dependent sodium channel blocker, crobenetine, in mono-arthritic rats. *Br J Pharmacol* 134: 1742–1748
114 Chidiac JJ, Rifai K, Hawwa NN, Massaad CA, Jurjus AR, Jabbur SJ, Saade NE (2002) Nociceptive behaviour induced by dental application of irritants to rat incisors: a new model for tooth inflammatory pain. *Eur J Pain* 6: 55–67
115 Jain NK, Kulkarni SK, Singh A (2002) Modulation of NSAID-induced antinociceptive and ant-inflammatory effects by alpha2-adrenoceptor agonists with gastroprotective effects. *Life Sci* 70: 2857–2869
116 Doig PA, Purbrick KA, Hare JE, McKeown DB (2000) Clinical efficacy and tolerance of meloxicam in dogs with chronic osteoarthritis. *Can Vet J* 41: 296–300
117 Moreau M, Dupuis J, Bonneau NH, Desnoyers M (2003) Clinical evaluation of a nutraceutical, carprofen and meloxicam for the treatment of dogs with osteoarthritis. *Vet Rec* 152: 323–329
118 Mathews KA, Pettifer G, Foster R, McDonell W (2001) Safety and efficacy of preoperative administration of meloxicam, compared with that of ketoprofen and butorphanol in dogs undergoing abdominal surgery. *Am J Vet Res* 62: 882–888
119 Lascelles BD, Henderson AJ, Hackett IJ (2001) Evaluation of the clinical efficacy of meloxicam in cats with painful locomotor disorders. *J Small Anim Pract* 42: 587–593
120 Nell T, Bergman J, Hoeijmakers M, Van Laar P, Horspool LJ (2002) Comparison of vedaprofen and meloxicam in dogs with musculoskeletal pain and inflammation. *J Small Anim Pract* 43: 208–212
121 Slingsby LS, Waterman-Earson AE (2002) Comparison between meloxicam and carprofen for postoperative analgesia after feline ovariohysterectomy. *J Small Anim Pract* 43: 286–289
122 Reyes L, Tinworth KD, Li KM, Yau DF, Waters KA (2002) Observer-blinded comparison of two nonopioid analgesics for postoperative pain in piglets. *Pharmacol Biochem Behav* 73: 521–528
123 Budsberg SC, Cross AR, Quandt JE, Pablo LS, Runk AR (2002) Evaluation of intravenous administration of meloxicam for perioperative pain management following stifle joint surgery in dogs. *Am J Vet Res* 63: 1557–1563
124 Singsby LS, Waterman-Pearson AE (2000) Postoperative analgesia in the cat after ovariohysterectomy by use of carprofen, ketoprofen, meloxicam or tolfenamic acid. *J Small Anim Pract* 41: 447–450
125 Din C (2002) Do NSAIDs affect the progression of osteoarthritis? *Inflammation* 26: 139–142
126 Sadowski T, Steinmeyer J (2001) Effects of non-steroidal anti-inflammatory drugs and dexamethasone on the activity and expression of matrix metalloproteinase-1, matrix metalloproteinase-3 and tissue inhibitor of metalloproteinases-1 by bovine articular chondrocytes. *Osteoarthritis Cartilage* 9: 407–415
127 Lin SK, Kok SH, Kuo MY, Wang TJ, Wang JT, Yeh FT, Hsiao M, Lan WH, Hong CY (2002) Sequential expressions of MMP-1, TIMP-1, IL-6, and COX-2 genes in induced periapical lesions in rats. *Eur J Oral Sci* 110: 246–253
128 Bezerra MM, de Lima V, Alencar VB (2000) *In vitro* effects of nonsteroidal anti-inflammatory drugs on cyclooxygenase activity in dogs. *Am J Vet Res* 61: 802–810
129 Gomez-Gaviro MV, Gonzalez-Alvaro I, Dominguez-Jimenez C, Peschon J, Black RA, Sanchez-Madrid F, Diaz-Gonzalez F (2002) Structure-function relationship and role of tumor necrosis factor-alpha-converting enzyme in the down-regulation of L-selectin by non-steroidal anti-inflammatory drugs. *J Biol Chem* 277: 38212–38221
130 Moreno-Sanchez R, Bravo C, Vasquez C, Ayala G, Silveira LH, Martinez-Lavin M (1999) Inhibition and uncoupling of oxidative phosphorylation by nonsteroidal anti-inflammatory drugs: study in mitochondria, submitochondrial particles, cells, and whole heart. *Biochem Pharmacol* 57: 743–752
131 Pitcher GM, Henry JL (2001) Meloxicam selectively depresses the after discharge of rat spinal dorsal horn neurones in response to noxious stimulation. *Neurosci Lett* 305: 45–48
132 Brown WA, Skinner SA, Malcontenti-Wilson C, Vogiagis D, O'Brien PE (2001) Non-steroidal anti-inflammatory drugs with activity against either cyclooxygenase 1 or cyclooxygenase 2 inhibit colorectal cancer in a DMH rodent. *Gut* 48: 660–666
133 Hussey HJ, Tisdale MJ (2000) Effect of the specific cyclooxygenase-2 inhibitor meloxicam on

tumour growth and cachexia in a murine model. *Int J Cancer* 87: 95–100
134 Brown WA, Skinner SA, Malcontenti-Wilson C, Misajon A, DeJong T, Vogiagis D, O'Brien PE (2000) Non-steroidal anti-inflammatory drugs with different cyclooxygenase inhibitory profiles that prevent aberrant crypt foci formatin but vary in acute gastrotoxicity in a rat model. *J Gastroenterol Hepatol* 15: 1386–1392
135 Teisman P, Ferger B (2001) Inhibition of the cyclooxygenase isoenzymes COX-1 and COX-2 provide neuroprotection in the MPTP-mouse model of Parkinson's disease. *Synapse* 39: 167–174
136 Joussen AM, Poulaki V, Mitsiades N, Kirchhof B, Koizumi K, Dohmen S, Adamis AP (2002) Nonsteroidal anti-inflammatory drugs prevent early diabetic retinopathy via TNF-alpha suppression. *FASEB J* 16: 438–440
137 Pop-Busui R, Marinescu V, Van Huysen C, Li F, Sullivan K, Greene DA, Larkin D, Stevens MJ (2002) Dissection of metabolic, vascular, and nerve conduction interrelationships in experimental diabetic neuropathy by cyclooxygenase inhibition and acetyl-L-carnitine administration. *Diabetes* 51: 2619–2628
138 Miyamoto K, Oka N, Kawasaki T, Miyake S, Yamamura T, Akiguchi I (2002) New cyclooxygenase-2 inhibitors for treatment of experimental autoimmune neuritis. *Muscle Nerve* 25: 280–282
139 Mohn C, Lomniczi A, Faletti A, Scorticati C, Elverdin JC, McCann SM, Rettori V (2001) Effects of aminoguanidine and meloxicam on nitric oxide and prostaglandin E production induced by lipopolysaccharide in the hypothalamus and anterior pituitary of the rat. *Neuroimmunomodulation* 9: 276–285
140 Gupta YK, Chaudhary G, Sinha K (2002) Enhanced protection by melatonin and meloxicam combination in a middle cerebral artery occlusion model of acute ischemic stroke in rat. *Can J Physiol Pharmacol* 80: 210–217
141 Heindl B, Becker BF (2001) Aspirin, but not the more selective cyclooxygenase (COX)-2 inhibitors meloxicam and SC 58125, aggravates postischaemic cardiac dysfunction, independent of COX function. *Naunyn Schmiedebergs Arch Pharmacol* 363: 233–240
142 Moya M, Campana V, Gavotto A, Spitale L, Simes J, Palma J (2002) Hyperfibrinogenemia in rats treated with meloxicam. *Jpn Heart J* 43: 559–566
143 Arslan S, Erdem S, Sivri A, Hascelik Z, Tan E (2002) Exercise-induced apoptosis of rat skeletal muscle and the effect of meloxicam. *Rheumatology International* 21: 133–136
144 Laudanno OM, Cesolari JA, Esnarriaga J, San Miguel P, Bedini OA (2000) *In vivo* selectivity of nonsteroidal antiinflamamtory drugs and gastrointestinal ulcers in rats. *Dig Dis Sci* 45: 1359–1365
145 Villegas I, Alarcon de la Lastra C, La Casa C, Motilva V, Martin MJ (2001) Effects of food intake and oxidative stress on intestinal lesions caused by meloxicam and prioxicam in rats. *Eur J Pharmacol* 414: 79–86
146 Brzozowski T, Konturek PC, Konturek SJ, Sliwowski Z, Pajdo R, Drozdowicz D, Ptak A, Hahn EG (2001) Classic NSAID and selective cyclooxygenase (COX)-1 and COX-2 inhibitors in healing of chronic gastric ulcers. *Microscopy Research and Technique* 53: 343–353
147 Villegas I, Alarcon de la Lastra C, Martin MJ, Motilva V, La Casa Garcia C (2002) Gastric damage induced by subchronic administration of preferential cyclooxgyenase-1 and cyclooxygenase-2 inhibitors in rats. *Pharmacology* 66: 68–75
148 Birck R, Krzossok S, Knoll T, Braun C, van Der Woude FJ, Rohmeiss P (2000) Preferential COX-2 inhibitor, meloxicam, compromises renal perfusion in euvolemic and hypovolemic rats. *Exp Nephrol* 8: 173–180
149 Hawkey C, Kahan A, Steinbrück K, Alegre C, Baumelou E, Bégaud B, Dequeker J, Isomäki H, Littlejohn G, Mau J et al on behalf of the Melissa Study Group (1998) Gastrointestinal tolerability of meloxicam compared to diclofenac in osteoarthritis patients. *Brit J Rheumatol* 37: 937–945
150 Dequeker J, Hawkey C, Kahan A, Steinbrück K, Alegre C, Baumelou E, Bégaud B, Isomäki H, Littlejohn G, Mau J et al on behalf of the Select Study Group (1998) Improvement in gastrointestinal tolerability of the selective cyclooxygenase (COX)-2 inhibitor, meloxicam, compared with piroxicam: results of the safety and efficacy large-scale evalution of COX-inhibiting therapies (select) trial in osteoarthritis. *Brit J Rheumatol* 37: 946–951
151 Layton D, Heeley E, Hughes K, Shakir SAW (2003) Comparison of the incidence rates of selected gastrointestinal events reported for patients prescribed rofecoxib and meloxicam in general practice in England using prescription-event monitoring data. *Rheumatology* 42: 622–631
152 Hawkey CJ (2001) NSAIDs and COX-2 inhibitors: what can we learn from large outcomes trials? The gastroenterologist's perspective. *Clin Exp Rheumatol* 19 (Suppl 25): S23–S30

153 Hawkey CJ (1999) COX-2 inhibitors. *Lancet* 353: 307–314
154 Ward A, Brogden RN (1988) Nimesulide. A preliminary review of its pharmacological properties and therapeutic efficacy in inflammation and pain states. *Drugs* 36: 732–753
155 Kalgutkar AS (1999) Selective cyclooxygenase-2 inhibitors as non-ulcerogenic anti-inflammatory agents. *Exp Opin Ther Patents* 9: 831–849
156 Singla AK, Chawla M, Singh A (2000) Nimesulide: some pharmaceutical and pharmacological aspects – an update. *J Pharm Pharmacol* 52: 467–486
157 Swingle KF, Moore GGI, Grant TJ (1976) 4-nitro-2-phenoxymethanesulfonanilide (R 805): a chemically novel anti-inflammatory agent. *Arch Int Pharmacodyn* 221: 132–139
158 Rainsford KD (1996) Mode of action, uses, and side-effects of anti-inflammatory drugs. In: KD Rainsford (ed.): *Advances in Anti-Rheumatic Therapy*. CRC Press, Boca Raton. 59–111
159 Bennett A (2001) Nimesulide: A well-established cyclooxygenase-2 inhibitor with many other pharmacological properties relevant to inflammatory diseases. In: JR Vane, RM Botting (eds): *Therapeutic roles of selective COX-2 inhibitors*. William Harvey Press, London, 524–540
160 Bennett A, Villa G (2000) Nimesulide: an NSAID that preferentially inhibits COX-2, and has various unique pharmacological activities. *Expert Opinion on Pharmacotherapy* 1: 277–286
161 Bennett A, Berti F, Ferreira SH (eds) (1993) Nimesulide: A Multifactorial Therapeutic Approach to the Inflammatory Process? A 7-year clinical experience. *Drugs* 46: 1–283
162 Davis R, Brogden RN (1994) Nimesulide. An update of its pharmacodynamic and pharmacokinetic properties, and therapeutic efficacy. *Drugs* 48: 431–454
163 Rabasseda X (1997) Safety profile of nimesulide: ten years of clinical experience. *Drugs of Today* 33: 1–10
164 Bernareggi A (1998) Clinical pharmacokinetics of nimesulide. *Clin Pharmacokinet* 35: 247–274
165 Rainsford KD (1998) An analysis from clinico-epidemiological data of the principal adverse events from the COX-2 selective NSAID, nimesulide, with particular reference to hepatic injury. *Inflammopharmacology* 6: 203–221
166 Rainsford KD (2001) Nimesulide: Overview of properties and applications. *Drugs of Today* 37 (Supp B): 3–7
167 Bennett A (2001) Clinical importance of the multifactorial actions of nimesulide. *Drugs of Today* 37 (Supp B): 9–14
168 Fitzgerald D, McCrory C (2001) Pharmacology of COX-2 inhibition in man: anti-inflammatory and analgesic effects of nimesulide. *Drugs of Today*, 37 (Supp B): 15–20
169 Huskisson EC (2001) Nimesulide, a balanced drug for the treatment of osteoarthritis. *Clin Exp Rheumatol* 19 (I): S21–S25
170 Tavares IA, Borrelli F, Welsh NJ (2001) Inhibition of gastric acid secretion by nimesulide: a possible factor in its gastric tolerability. *Clin Exp Rheumatol* 19: S13–S15
171 Shah AA, Thjodleifsson B, Murray FE, Kay E, Barry M, Sigthorsson G, Gudjonsson H, Oddsson E, Price AB, Fitzgerald DJ et al (2001) Selective inhibition of COX-2 in humans is associated with less gastrointestinal injury: a comparison of nimesulide and naproxen. *Gut* 48: 339–346
172 Huskisson EC (2001) Nimesulide, a balanced drug for the treatment of osteoarthritis. *Clin Exp Rheumatol* 19 (Suppl 22): S21–S25
173 Rainsford KD (1999) Relationship of nimesulide safety to its pharmacokinetics: assessment of adverse reactions. *Rheumatology* 38: 4–10
174 Wober W (1999) Comparative efficacy and safety of nimesulide and diclofenac in patients with acute shoulder, and a meta-analysis of controlled studies with nimesulide. *Rheumatology* 38: 33–38
175 Liaropoulos L (1999) Economic comparison of nimesulide and diclofenac, and the incidence of adverse events in the treatment of rheumatic disease in Greece. *Rheumatology* 38: 39–46
176 Carini M, Aldini G, Stefani R, Marinello C, Facino RM (1998) Mass spectrometric characterization and HPLC determination of the main urinary metabolites of nimesulide in man. *J Pharm Biomed Anal* 18: 201–211
177 Boelsterli UA (2002) Mechanisms of NSAID-induced hepatotoxicity: focus on nimesulide. *Drug Safety* 25: 633–648
178 Boelsterli UA (2002) Nimesulide and hepatic adverse effects: roles of reactive metabolites and host factors. *Int J Clin Pract Suppl* 128: 30–36
179 Vietri M, De Santi C, Pietrabissa A, Mosca F, Pacifici GM (2000) Inhibition of human liver phenol sulfotransferase by nonsteroidal anti-inflammatory drugs. *Eur J Clin Pharmacol* 56: 81–87
180 Carrasco Portugal MC, Granados Soto V, Camacho Vieyra GA, Parez Urizar J, Flores Murrieta FJ

(2000) A simple method for determination of nimesulide in rat blood samples by high performance liquid chromatography. *J Liquid Chromatography and Related Technologies* 23: 2237–2244
181 Mukherjee P, Rachita C, Aisen PS, Pasinetti GM (2001) Non-steroidal anti-inflammatory drugs protect against chondrocyte apoptotic death. *Clin Exp Rheumatol* 19: S7–S11
182 Sanchez C, Mateus MM, Defresne MP, Crielaard JM, Reginster JY, Henrotin YE (2002) Metabolism of human articular chondrocytes cultured in alginate beads. Longterm effects of interleukin 1beta and nonsteroidal antiinflamamtory drugs. *J Rheumatol* 29: 772–782
183 Rainsford KD, Omar H, Ashraf A, Hewson AT, Bunning RAD, Rishiraj R, Shepherd P, Seabrook RW (2002) Recent pharmacodynamic and pharmacokinetic findings on oxaprozin. *Inflammopharmacology* 10: 185–239
184 Bavbek S, Celik G, Ediger D, Mungan D, Demirel YS, Misirligil Z (1999) The use of nimesulide in patients with acetylsalicylic acid and nonsteroidal anti-inflammtory drug intolerance. *J Asthma* 36: 657–663
185 Sanchez-Borges M, Capriles-Hulett A (2000) Atopy is a risk factor for non-steroidal anti-inflammatory drug sensitivity. *Ann Allergy Asthma Immunol* 84: 101–106
186 Rowlinson SW, Kiefer JR, Prusakiewicz J, Pawlilz JL, Kozak KR, Kalgutkar AS, Stallings WC, Kurumbail RG, Marnett LJ (2003) A novel mechanism of cyclooxygenase-2 inhibition involving interactions with Ser-530 and Tyr-385. *J Biol Chem* Aug 18, 2003, in press
187 García-Nieto R, Pérez C, Checa A, Gago F (1999) Molecular model of the interaction between nimesulide and human cyclooxygenase-2. *Rheumatology* 38: 14–18
188 Julemont F, de Leval X, Michaux C, Damas J, Charlier C, Durant F, Pirotte B, Dogne JM (2002) Spectral and crystallographic study of pyridinic analogues of nimesulide: determination of the active form of methanesulfonamides as COX-2 selective inhibitors. *J Med Chem* 45: 5182–5185
189 Rufer C, Schillinger E, Böttcher I, Repenthin W, Herrmann Ch (1982) Non-steroidal anti-inflammatories – XII. *Biochem Pharmacol* 22: 3591–3596
190 Swingle KF, Bell RL, Moore GGI (1985) Anti-inflammatory activity of antioxidants. In: KD Rainsford (ed.): *Anti-Inflammatory and Anti-Rheumatic Drugs. Volume III: Anti-Rheumatic Drugs, Experimental Agents and Clinical Aspects of Drug Use.* CRC Press, Boca Raton, 105–126
191 Orhan H, Dogruer DS, Cakir B, Sahin G, Sahin MF (1999) The *in vitro* effects of new non-steroidal anti-inflammatory compounds on antioxidant system of human erythrocytes. *Exp Toxicol Pathol* 51: 397–402
192 Maffei Facino R, Carini M, Aldini G, Saibene L, Morelli R (1995) Differential inhibition of superoxide, hydroxyl and peroxyl radicals by nimesulide and its main metabolite. *Drug Res* 45: 10–17
193 Zheng SX, Mouithys-Mickalad A, Deby-Dupont GP, Deby CM, Maroulis AP, Labasse AH, Lamy ML, Crielaard JM, Reginster JY, Henrotin YE (2000) *In vitro* study of the antioxidant properties of nimesulide and 4-OH nimesulide: effects of HRP- and luminal-dependent chemiluminescence produced by human chondrocytes. *Osteoarthritis and Cartilage* 8: 419–425
194 Mouithys-Mickalad AM, Zheng SX, Deby-Dupont GP, Deby CM, Lamy MM, Reginster JY, Henrotin YE (2000) *In vitro* study of the antioxidant properties of non steroidal anti-inflammatory drugs by chemiluminescence and electron spin resonance (ESR). *Free Radic Res* 33: 607–621
195 Tavares IA, Bishai PM, Bennett A (1995) Activity of nimesulide on constitutive and inducible cyclooxygenases. *Drug Res* 45: 1–3
196 Shah AA, Murray FE, Fitzgerald DJ (1999) The *in vivo* assessment of nimesulide cyclooxygenase-2 selectivity. *Rheumatology* 38: 19–23
197 Saeed SA, Afzal MN, Shah BH (1998) Dual effects of nimesulide, a COX-2 inhibitor, in human platelets. *Life Sci* 63: 1835–1841
198 Islas Cadena M, Aguirre Banuelos P, Granados Soto V (1999) Evidence for the participation of the nitric oxide-cyclic GMP pathway in the antinociceptive effect of nimesulide. *J Pharmacol Toxicol Methods* 42: 87–92
199 Miranda HF, Lopez J, Sierralta F, Correa A, Pinardi G (2001) NSAID antinociception measured in a chemical and thermal assay in mice. *Pain Res Manag* 6: 190–196
200 Sandrini G, Tassorelli C, Cecchini AP, Alfonsi E, Nappi G (2002) Effects of nimesulide on nitric oxide-induced hyperalgesia in humans – a neurophysiological study. *Eur J Pharmacol* 450: 259–262
201 Lopez R, Llinas MT, Roig F, Salazar FJ (2003) Role of nitric oxide and cyclooxygenase-2 in regulating the renal hemodynamic response to norepinephrine. *Am J Physiol Regul Integr Comp Physiol* 284: R488–R493

202 Korolkiewicz RP, Ujda M, Dabkowski J, Ruczynski J, Rekowski P, Petrusewicz J (2003) Differential salutary effects of nonselective and selective COX-2 inhibitors in postoperative ileus in rats. *J Surg Res* 109: 161–169
203 Caparroz-Assef SM, Salgueiro-Pagadigorria CL, Bersani-Amado CA, Bracht A, Kelmer-Bracht AM, Ishii-Iwamoto EL (2001) The uncoupling effect of the nonsteroidal anti-inflammatory drug nimesulide in liver mitochondria from adjuvant-induced arthritic rats. *Cell Biochem Funct* 19: 117–124
204 Mingatto FE, dos Santos AC, Rodrigues T, Pigoso AA, Uyemura SA, Curti C (2000) Effects of nimesulide and its reduced metabolite on mitochondria. *Brit J Pharmacol* 131: 1154–1160
205 Mingatto FE, Ridrigues T, Pigoso AA, Uyemura SA, Curti C, Santos AC (2002) The critical role of mitochondrial energetic impairment in the toxicity of nimesulide to hepatocytes. *J Pharmacol Exp Ther* 303: 601–607
206 Rainsford KD, Seabrook RW, Spencer S, Hewson AT (2001) Effects of nimesulide and its metabolites or manufacturing intermediates on the viability and growth of the human hepatoma HepG2 cell line. *Life Sci* 69: 2965–2973
207 Titos E, Chiang N, Serhan CN, Romano M, Gaya J, Pueyo G, Claria J (1999) Hepatocytes are a rich source of novel aspirin-triggered 15-epi-lipoxin A(4). *Am J Physiol* 277: C870–C877
208 Tardieu D, Jaeg JP, Deloly A, Corpet DE, Cadet J, Petit CR (2000) The COX-2 inhibitor nimesulide suppresses superoxide and 8-hydroxy-deoxyguanosine formation, and stimulates apoptosis in mucosa during early colonic inflammation in rats. *Carcinogenesis* 21: 973–976
209 Hida T, Kozaki K, Muramatsu H, Masuda A, Shimizu S, Mitsudomi T, Sugiura T, Ogawa M, Takahashi T (2000) Cyclooxygenase-2 inhibitor induces apoptosis and enhances cytotoxicity of various anticancer agents in non-small cell lung cancer cell lines. *Clin Cancer Res* 6: 2006–2011
210 Zhang Z, DuBois RN (2000) Par-4, a proapoptotic gene, is regulated by NSAIDs in human colon carcinoma cells. *Gastroenterology* 118: 1012–1017
211 Tian G, Yu JP, Luo HS, Yu BP, Yue H, Li JY, Mei Q (2002) Effect of nimesulide on proliferation and apoptosis of human hepatoma SMMC-7721 cells. *World J Gastroenterol* 8: 483–487
212 Maria DA, Ribeiro OG, Pizzocaro KF, De Franco M, Cabrera WK, Starobinas N, Gallois V, Siqueira M, Seman M, Ibanez OM (2000) Resistance to melanoma metastases in mice selected for high acute inflammatory response. *Carcinogenesis* 22: 337–342
213 Nakatsugi S, Ohta T, Kawamori T, Mutoh M, Tanigawa T, Watanabe K, Sugie S, Sugimura T, Wakabayashi K (2000) Chemoprevention by nimesulide, a selective cyclooxygenase-2 inhibitor, of 2-amino-1-methyl-6-phenylimidazo[4,5-b]pyridine (PhIP)-induced mammary gland carcinogenesis in rats. *Jpn J Cancer Res* 91: 886–892
214 Kawamori T, Wakabayashi K (2002) COX-2 and prostanoid receptors: good targets for chemoprevention. *J Environ Pathol Toxicol Oncol* 21: 149–153
215 Kubatka P, Kalicka K, Chamilova M, Ahlersova E, Ahlers I, Bojkova B, Adamekova E (2002) Nimesulide and melatonin in mammary carcinogenesis prevention in female Sprague-Dawley rats. *Neoplasma* 49: 255–259
216 Furukawa F, Nishikawa A, Lee IS, Kanki K, Umemura T, Okazaki K, Kawamori T, Wakabayashi K, Hirose M (2003) A cyclooxygenase-2 inhibitor, nimesulide, inhibits post-initiation phase of N-nitrosobis (2-oxopropyl)amine-induced pancreatic carcinogenesis in hamsters. *Int J Cancer* 104: 269–273
217 Vardar S, Baylas H, Huseyinov A (2003) Effects of selective cyclooxygenase-2 inhibition on gingival tissue levels of prostaglandin E2 and prostaglandin F2alpha and clinical parameters of chronic periodontitis. *J Periodontol* 74: 57–63
218 Yamazaki R, Kusunoki N, Matsuzaki T, Hashimoto S, Kawai S (2002) Selective cyclooxygenase-2 inhibitors show a differential ability to inhibit proliferation and induce apoptosis of colon adenocarcinoma cells. *FEBS Lett* 531: 278–284
219 Denda A, Kitayama W, Murata A, Kishida H, Sasaki Y, Kusuoka O, Tsujiuchi T, Tsutsumi M, Nakae D, Takagi H et al (2002) Increased expression of cyclooxygenase-2 protein during rat hepatocarcinogenesis caused by a choline-deficient, L-amino acid-defined diet and chemopreventive efficacy of a specific inhibitor, nimesulide. *Carcinogenesis* 23: 245–256
220 Mukherjee P, Pasinetti GM (2002) Altered gene expression during nimesulide-mediated inhibition of apoptotic death in human chondrocytes. *Int J Clin Pract Suppl* 128: 20–23
221 Wang D, An SJ, Wang WH, McGiff JC, Ferrari NR (2001) CaR-mediated COX-2 expression in primary cultured in TAL cells. *Am J Physiol Renal Physiol* 281: F658–F654
222 Kalajdzic T, Faour WH, He QW, Fahmi H, Martel-Pelletier J, Pelletier JP, Di Battista JA (2002)

Nimesulide, a preferential cyclooxygenase 2 inhibitor, suppresses peroxisome proliferators-activated receptor induction of cyclooxygenase 2 gene expression in human synovial fibroblasts: evidence for receptor antagonism. *Arthritis Rheum* 46: 494–506
223 Fahmi H, He Y, Zhang M, Martel-Pelletier J, Pelletier JP, Di Battista JA (2001) Nimesulide reduces interleukin-1beta-induced cyclooxygenase-2 gene expression in human synovial fibroblasts. *Osteoarthritis Cartilage* 9: 332–340
224 Sharma S, Rastogi S, Gupta V, Rohtagi D, Gulati P (1999) Comparative efficacy and safety of nimesulide *versus* piroxicam in osteoarthritis with special reference to chondroprotection. *Am J Ther* 6: 191–197
225 Garcia Rodriguez LA, Cattaruzzi C, Tronçon MG, Agostinis L (1998) Risk of hospitalization for upper gastrointestinal tract bleeding associated with ketorolac, other nonsteroidal anti-inflammtory drugs, calcium antagonists and other antihypertensive drugs. *Arch Intern Med* 158: 33–39
226 Conforti A, Leone R, Moretti U, Mozzo F, Velo G (2001) Adverse drug reactions related to the use of NSAIDs with a focus on nimesulide: results of spontaneous reporting fgrom a Northern Italian area. *Drug Safety* 24: 1081–1090
227 Macia MA, Carvajal A, del Pozo JG, Vera E, del Pino A (2002) Hepatotoxicity associated with nimesulide: data from the Spanish Pharmacovigilance System. *Clin Pharmacol Ther* 72: 596–597
228 Villa G (1999) NSAIDs and hepatic reactions. *Lancet* 353: 846
229 Chitturi S, Farrell GC (2001) Drug-induced cholestasis. *Sem Gastrointest Dis* 12: 113–124
230 Elmalem E (2000) Nimesulide, clavulanic acid and hepatitis. *J Intern Med* 248: 168–169
231 Kanwar AJ, Kaur S, Thami GP (2000) Nimesulide-induced purpura. *Dermatology* 201: 376
232 Sanchez-Borges M, Capriles-Hulett A (2000) Atopy is a risk factor for non-steroidal anti-inflammatory drug sensitivity. *Ann Allergy Asthma Immunol* 84: 101–106
233 Nettis E, Marcandrea M, Ferrannini A, Tursi A (2001) Tolerability of nimesulide and paracetamol in patients with NSAID-induced urticaria/angioedema. *Immunopharmacol Immunotoxicol* 23: 343–354
234 Naldi L, Conforti A, Venegoni M, Troncon MG, Caputi A, Ghiotto E, Cocci A, Moretti U, Velo G, Leone R (1999) Cutaneous reactions to drugs. An analysis of spontaneous reports in four Italian regions. *Br J Clin Pharmacol* 48: 839–846
235 Anonymous (1993) Cutaneous reactions to analgesic-antipyretics and nonsteroidal anti-inflammatory drugs. Analysis of reports to the spontaneous reporting system of the Gruppo Italiano Studi Epidemiologici in Dermatologia. *Dermatology* 186: 164–169
236 Tursen U, Kaya TI, Kokturk A, Dusmez D (2001) Lichenoid photodermatitis associated with nimesulide. *Int J Dermatol* 40: 767–768
237 Sanchez Borges M, Capriles-Hulett A, Caballero-Foneseca F, Perez CR (2001) Tolerability to new COX-2 inhibitors in NSAID-sensitive patients with cutaneous reactions. *Ann Allergy Asthma Immunol* 87: 201–204
238 Kanikkannan N, Jackson T, Shaik MS, Singh M (2001) Evaluation of skin sensitization potential of melatonin and nimesulide by murine local lymph node assay. *Eur J Pharm Sci* 14: 217–220
239 Mangalvedhekar SS, Gogtay NJ, Phadke AV, Gore S, Shah JM, Shah SM, Kshirsagar NA (2000) Adverse drug reactions postal survey-bronchial asthma and angioedema with nimesulide. *J Assoc Physicians India* 48: 548
240 Llinas MT, Rodriguez F, Moreno C, Salazar FJ (2000) Role of cyclooxygenase-2-derived metabolites and nitric oxide in regulating renal function. *Am J Physiol Regul Integr Comp Physiol* 279: R1641-R1646
241 Aisen PS, Schmeidler J, Pasinetti GM (2002) Randomized pilot study of nimesulide treatment of Alzheimer's disease. *Neurology* 58: 1050–1054
242 Food and Drug Administration (USA) Celebrex® (celecoxib capsules). Centre for Drug Evaluation. (CDER, New and Generic Drug Approvals: 1998–2003), NDA Application # 20-998, approval date 12/31/98. Pharmacological Review, part 1-7, posted 6/9/99. *www.fda.gov/cder/approval/index.htm* (Accessed Dec 2003)
243 Clemett D, Goa KL (2000) Celecoxib: a review of its use in osteoarthritis, rheumatoid arthritis and acute pain. *Drugs* 59: 957–980
244 Penning TD, Talley JJ, Bertenshaw SR, Carter JS, Collins PW, Docter S, Graneto MJ, Lee LF, Malecha JW, Miyashiro JM et al (1997) Synthesis and biological evaluation of the 1,5-diarylpyrazole class of cyclooxygenase-2 inhibitors: identification of 4-[5-(4-methylphenyl)-3-(trifluoromethyl)-*1H*-pyrazol-1-yl]benzene sulfonamide (SC-58635, cele-

coxib). *J Med Chem* 40: 1347–1365
245 Hamilton LC, Mitchell JA, Tomlinson AM, Warner TD (1999) Synergy between cyclo-oxygenase-2 induction and arachidonic acid supply *in vivo*: consequences for nonsteroidal anti-inflammatory drug efficacy. *FASEB J* 13: 245–251
246 Smith CJ, Zhang Y, Koboldt CM, Muhammad J, Zweifel BS, Shaffer A, Talley JJ, Masferrer JL, Seibert K, Isakson (1998) Pharmacological analysis of cycloxygenase-1 in inflammation. *Proc Natl Acad Sci USA* 95: 13313–13318
247 Gierse JK, Koboldt CM, Walker MC, Seibert K, Isakson PC (1999) Kinetic basis for selective inhibition of cyclo-oxygenases. *Biochem J* 339: 607–614
248 Davies NM, McLachlan AJ, Day RO, Williams KM (2000) Clinical pharmacokinetics and pharmacodynamics of celecoxib. A selective cyclo-oxygenase-2 inhibitor. *Clin Pharmacokinet* 38: 225–242
249 Tang C, Shou M, Mei Q, Rushmore TH, Rodrigues AD (2000) Major role of human liver microsomal cytochrome P450 2C9 (CYP2C9) in the oxidative metabolism of celecoxib, a novel cyclooxgyenase-II inhibitor. *J Pharmacol Exp Ther* 293: 453–459
250 Paulson SK, Engel L, Reitz B, Bolten S, Burton EG, Maziasz TJ, Yan B, Schoenhard GL (1999) Evidence for polymorphism in the canine metabolism of the cyclooxygenase 2 inhibitor, celecoxib. *Drug Metabolism and Disposition* 27: 1133–1142
251 Paulson SK, Zhang JY, Breau AP, Hribar JD, Liu NW, Jessen SM, Lawal YM, Cogburn JN, Gresk CJ, Markos CS et al (2000) Pharmacokinetics, tissue distribution, metabolism, and excretion of celecoxib in rats. *Drug Metabolism and Disposition* 28: 514–521
252 Paulson SK, Zhang JY, Jessen SM, Lawal Y, Liu NW, Dudkowski CM, Wang YF, Change M, Yang D, Findlay JW et al (2000) Comparison of celecoxib metabolism and excretion in mouse, rabbit, dog, cynomolgus monkey and rhesus monkey. *Xenobiotica* 30: 731–744
253 Zhang JY, Wang Y, Dudkowski C, Yang D, Chang M, Yuan J, Paulson SK, Breau AP (2000) Characterization of metabolites of celecoxib in rabbits by liquid chromatography/tandem mass spectrometry. *J Mass Spectrom* 35: 1259–1270
254 Tibble JA, Sigthorsson G, Forster R, Bjarnason I (2000) Comparison of the intestinal toxicity of celecoxib, a selective COX-2 inhibitor, and indomethacin in the experimental rat. *Scand J Gastroenterol* 35: 802–807
255 Paulson SK, Vaughn MB, Jessen SM, Lawal Y, Gresk CJ, Yan B, Maziasz TJ, Cook CS, Karim A (2001) Pharmacokinetics of celecoxib after oral administration in dogs and humans: effect of food and site of absorption. *J Pharmacol Exp Ther* 297: 638–645
256 Tanaka A, Hase S, Miyazawa T, Takeuchi K (2002) Up-regulation of cyclooxygenase-2 by inhibition of cyclooxygenase-1: a key to nonsteroidal anti-inflammatory drug-induced intestinal damage. *J Pharmacol Exp Ther* 300: 754–761
257 O'Beirne JP, Cairns SR (2001) Drug points: Cholestatic hepatitis in association with celecoxib. *BMJ* 323: 23
258 Pinheiro RM, Calixto JB (2002) Effect of the selective COX-2 inhibitors, celecoxib and rofecoxib in rat acute models of inflammation. *Inflamm Res* 51: 603–610
259 Cuzzocrea S, Mazzon E, Sautebin L, Dugo L, Serraino I, De Sarro A, Caputi AP (2002) Protective effects of celecoxib on lung injury and red blood cells modification induced by carrageenan in the rat. *Biochem Pharmacol* 63: 785–795
260 Francischi JN, Chaves CT, Moura AC, Lima AS, Rocha OA, Ferreira-Alves DL, Bakhle YS (2002) Selective inhibitors of cyclo-oxygenase-2 (COX-2) induce hypoalgesia in a rat paw model of inflammation. *Br J Pharmacol* 137: 837–844
261 Torres-Lopez JE, Ortiz MI, Castaneda-Hernandez G, Alonso-Lopez R, Asomoza-Espinosa R, Granados-Soto V (2002) Comparison of the antinociceptive effect of celecoxib, diclofenac and resveratrol in the formalin test. *Life Sci* 70: 1669–1676
262 Shin MC, Jang MH, Chang HK, Kim YJ, Kim EH, Kim CJ (2003) Modulation of cyclooxygenase-2 on glycine- and glutamate-induced ion currents in rat periaqueductal gray neurons. *Brain Res Bull* 59: 251–256
263 Kawamori T, Rao CV, Seibert K, Reddy BS (1998) Chemoprotective activity of celecoxib, a specific cyclooxygenase-2 inhibitor, against colon carcinogenesis. *Cancer Res* 58: 409–412
264 Jacoby RF, Seibert K, Cole CE, Kelloff G, Lubet RA (2000) The cyclooxygenase-2 inhibitor celecoxib is a potent preventative and therapeutic agent in the min mouse model of adenomatous polyposis. *Cancer Res* 60: 5040–5044
265 Alshafie GA, Abou-Issa HM, Seibert K, Harris RE (2000) Chemotherapeutic evaluation of cele-

coxib, a cyclooxygenase-2 inhibitor, in a rat mammary tumor model. *Oncol Rep* 7: 1377–1381
266 Grubbs CJ, Lubet RA, Koki AT, Leahy KM, Masferrer JL, Steele VE, Kelloff GJ, Hill DL, Seibert K (2000) Celecoxib inhibits N-butyl-N-(4-hydroxybutyl)-nitrosamine-induced urinary bladder cancers in male B6D2F1 mice and female Fischer-344 rats. *Cancer Res* 60: 5599–5602
267 Grosch S, Tegeder I, Niederberger E, Brautigam L, Geisslinger G (2001) COX-2 independent induction of cell cycle arrest and apoptosis in colon cancer cells by the selective COX-2 inhibitor celecoxib. *FASEB J* 15: 2742–2744
268 Rao CV, Indranie C, Simi B, Manning PT, Connor JR, Reddy BS (2002) Chemopreventive properties of a selective inducible nitric oxide synthase inhibitor in colon carcinogenesis, administered alone or in combination with celecoxib, a selective cyclooxygenase-2 inhibitor. *Cancer Res* 62: 165–170
269 Leahy KM, Ornberg RL, Wang Y, Zweifel BS, Koki AT, Masferrer JL (2002) Cyclooxygenase-2 inhibition by celecoxib reduces proliferation and incudes apoptosis in anagiogenic endothelial cells *in vivo*. *Cancer Res* 62: 625–631
270 Medhurst SJ, Walker K, Bowes M, Kidd BL, Glatt M, Muller M, Hattenberger M, Vaxelaire J, O'Reilly T, Wotherspoon G, Winter J, Green J, Urban L (2002) A rat model of bone cancer pain. *Pain* 96: 129–140
271 Orengo IF, Gerguis J, Phillips R, Guevara A, Lewis AT, Black HS (2002) Celecoxib, a cyclooxygenase 2 inhibitor as a potential chemopreventive to UV-induced skin cancer: a study in the hairless mouse model. *Arch Dermatol* 138: 751–755
272 Lu S, Zhang X, Badawi AF, El-Sohemy A, Archer MC (2002) Cyclooxygenase-2 inhibitor celecoxib inhibits promotion of mammary tumorigenesis in rats fed a high fat diet rich in n-6 polyunsaturated fatty acids. *Cancer Letters* 184: 7–12
273 Wenger FA, Kilian M, Bisevac M, Khodadayan C, von Seebach M, Schimke I, Guski H, Muller JM (2002) Effects of celebrex and zyflo on liver metastasis and lipid peroxidation in pancreatic cancer in Syrian hamsters. *Clin Exp Metastasis* 19: 681–687
274 Howe LR, Subbaramaiah K, Patel J, Masferrer JL, Deora A, Hudis C, Thaler HT, Muller WJ, Du B, Brown AM, Dannenberg AJ (2002) Celecoxib, a selective cyclooxygenase 2 inhibitor, protects against human epidermal growth factor receptor 2 (HER-2)/neu-induced breast cancer. *Cancer Res* 62: 5405–5407
275 Trifan OC, Durham WF, Salazar VS, Horton J, Levine BD, Zweifel BS, Davis TW, Masferrer JL (2002) Cyclooxygenase-2 inhibition with celecoxib enhances antitumor efficacy and reduces diarrhea side-effect of CPT-11. *Cancer Res* 62: 5778–5784
276 Blomme EA, Chinn KS, Hardy MM, Casler JJ, Kim SH, Opsahl AC, Hall WA, Trajkovic D, Khan KN, Tripp CS (2003) Selective cycloxygenase-2 inhibition does not affect the healing of cutaneous full-thickness incisional wounds in SKH-1 mice. *Br J Dermatol* 148: 211–223
277 Neiderberger E, Tegeder I, Vetter G, Schmidtko A, Schmidt H, Euchenhofer C, Brautigam L, Grosch S, Geisslinger G (2001) Celecoxib loses its anti-inflammatory efficacy at high doses through activation of NFκB. *FASEB J* 15: 1622–1624
278 Zweifel BS, Davis TW, Ornberg RL, Masferrer JL (2002) Direct evidence for a role of cyclooxygenase 2-derived prostaglandin E2 in human head and neck xenograft tumors. *Cancer Res* 62: 6706–6711
279 Subbaramaiah K, Dannenberg AJ (2003) Cyclooxygenase 2: a molecular target for cancer prevention and treatment. *Trends Pharmacol Sci* 24: 96–102
280 Casolini P, Catalani A, Zuena AR, Angelucci L (2002) Inhibition of COX-2 reduces the age-dependent increase of hippocampal inflammatory markers, corticosterone secretion, and behavioural impairments in the rat. *J Neurosci Res* 68: 337–343
281 Sairam K, Saravanan KS, Banerjee R, Mohanakumar KP (2003) Non-steroidal anti-inflammatory drug sodium salicylate, but not diclofenac or celecoxib, protects against 1-methyl-4-phenyl pyridinium-induced dopaminergic neurotoxicity in rats. *Brain Res* 966: 245–252
282 Drachman DB, Frank K, Dykes-Hoberg M, Teismann P, Almer G, Przedborski S, Rothstein JD (2002) Cyclooxygenase 2 inhibition protects motor neurons and prolongs survival in a transgenic mouse model of ALS. *Ann Neurol* 52: 771–778
283 Appel GB (2001) COX-2 inhibitors and the kidney. *Clin Exp Rheumatol* 19 (Suppl 25): S37–S40
284 Boyd IW, Mathew TH, Thomas MC (2000) COX-2 inhibitors and renal failure: the triple whammy revisited. *Medical Journal of Australia* 173: 274
285 Brater DC, Harris C, Redfern JS, Bertz BJ (2001) Renal effects of COX-2 selective inhibitors. *Am J Nephrol* 21: 1–15

286 Swan SK, Rudy DW, Lasseter KC, Ryan CF, Buechel KL, Lambrecht LJ, Pinto MB, Dilzer SC, Obrda O, Sundblad KJ et al (2000) Effect of cyclooxygenase-2 inhibition on renal function in elderly persons receiving a low-salt diet. *Ann Int Med* 133: 1–9
287 Whelton A, Fort JG, Puma JA, Normandin D, Bello AE, Verburg KM (2001) Cyclooxygenase-2-specific inhibitors and cardiorenal function: a randomized, controlled trial of celecoxib and rofecoxib in older hypertensive osteoarthritis patients. *Amer J Ther* 8: 85–95
288 Kiyataka M, Rich KA, Ingram M, Yamamoto T, Bing RJ (2002) Nitric oxide, anti-inflammatory drugs on renal prostaglandins and cyclooxygenase-2. *Hypertension* 39: 785–789
289 Hocherl K, Wolf K, Castrop H, Ittner KP, Bucher M, Kees F, Grobecker HF, Kurtz A (2001) Renocortical expression of renin and of cyclooxygenase-2 in response to angiotensin II AT1 receptor blockade is closely coordinated but not causally linked. *Pflugers Arch* 442: 821–827
290 Muscara MN, Vergnolle N, Lobren F, Triggle CR, Elliott SN, Asfaha S, Wallace JL (2000) Selective cyclo-oxygenase-2 inhibition with celecoxib elevates blood pressure and promotes leukocyte adherence. *Br J Pharmacol* 129: 1423–1430
291 Ozturk H, Ozdemir E, Otcu S, Buyukbayram H, Ihsan Dokucu A (2002) Renal effects on a solitary kidney of specific inhibition of cyclooxygenase-2 after 24 h of complete ureteric obstruction in rats. *Urol Res* 30: 223–226
292 Rossat J, Maillard M, Nussberger J, Brunner HR, Burnier M (1999) Renal effects of selective cyclooxygenase-2 inhibition in normotensive salt-depleted subjects. *Clin Pharmacol Ther* 66: 76–84
293 Kitahara M, Eitner F, Ostendorf T, Kunter U, Janssen U, Westenfeld R, Matsui K, Kerjaschki D, Floege J (2002) Selective cyclooxygenase-2 inhibition impairs glomerular capillary healing in experimental glomerulonephritis. *J Am Soc Nephrol* 13: 1261–1270
294 Reilly TP, Brady JN, Marchick MR, Bourdi M, George JW, Radonovich MF, Pise-Masison CA, Pohl LR (2001) A protective role for cyclooxygenase-2 in drug-induced liver injury in mice. *Chem Res Toxicol* 14: 1620–1628
295 Rainsford KD (1991) Uncoupling the toxicological morass in the development of new antirheumatic drugs—is there any hope? *Br J Rheumatol* 30: 161–163
296 Rainsford KD (1989) Gastrointestinal side-effects. In: CY Chang, AJ Lewis (eds): *Pharmacological Methods in the Control of Inflammation.* Alan R Liss, New York, 343–362
297 Kato S, Ogawa Y, Kanatsu K, Okayama M, Watanabe T, Arakawa T, Takeuchi K (2002) Ulcerogenic influence of selective cyclooxygenase-2 inhibitors in the rat stomach with adjuvant-induced arthritis. *J Pharmacol Exp Ther* 303: 503–509
298 Kato S, Takeuchi K (2002) Alteration of gastric ulcerogenic and healing responses in rats with adjuvant-induced arthritis. *Jpn J Pharmacol* 89: 1–6
299 Ma L, del Soldato P, Wallace JL (2002) Divergent effects of new cyclooxygenase inhibitors on gastric ulcer healing: shifting the angiogenic balance. *Proc Natl Acad Sci USA* 99: 13243–13247
300 Tibble J, Sigthorsson G, Caldwell C, Palmer RH, Bjarnason I (2001) Effects of NSAIDs on cryoprobe-induced gastric ulcer healing in rats. *Aliment Pharmacol Ther* 15: 2001–2008
301 Cuzzocrea S, Mazzon E, Serraino I, Dugo L, Centorrino T, Ciccolo A, Sautebin L, Caputi AP (2001) Celecoxib, a selective cyclo-oxygenase-2 inhibitor reduces the severity of experimental colitis induced by dinitrobenzene sulfonic acid in rats. *Eur J Pharmacol* 431: 91–102
302 Ostensen M (2001) Drugs in pregnancy. Rheumatological disorders. *Best Pract Res Clin Obstet Gynecol* 15: 953–969
303 Reese J, Zhao X, Ma WG, Brown N, Maziasz TJ, Dey SK (2001) Comparative analysis of pharmacologic and/or genetic disruption of cyclooxygenase-1 and cyclooxygenase-2 function in female reproduction in mice. *Endocrinology*, 142: 3198–3206
304 Takahashi Y, Roman C, Chemtob S, Tse MM, Lin E, Heymann MA, Clyman RI (2000) Cyclooxygenase-2 inhibitors constrict the fetal lamb ductus arteriosus both *in vitro* and *in vivo*. *Am J Physiol Regul Integr Comp Physiol* 278: R1496-R1505
305 Ozaki N, Beharry K, Nishihara KC, Akmal Y, Ang JG, Modanlou HD (2002) Differential regulation of prostacyclin and thromboxane by dexamethasone and celecoxib during oxidative stress in newborn rabbits. *Prostaglandins Other Lipid Mediat* 70: 61–78
306 Talley JJ, Brown DL, Carter JS, Graneto MJ, Koboldt CM, Masferrer JL, Perkins WE, Rogers RS, Shaffer AF, Zhang YY et al (2000) 4-[5-methyl-3-phenylisoxazol-4-yl]-benzenesulfonamide, valdecoxib: a potent and selective inhibitor of COX-2. *J Med Chem* 43: 775–777
307 Talley JJ, Bertenshaw SR, Brown DL, Carter JS, Graneto MJ, Kellogg MS, Koboldt CM, Yuan J, Zhang YY, Seibert K (2000) N-[[(5-methyl-3-phenylisoxazol-4-yl)-phenyl]sulfonyl]propan-

amide, sodium salt, paracoxib sodium: a potent and selective inhibitor of COX-2 for parenteral administration. *J Med Chem* 43: 1661–1663

308 US Food and Drug Adminstration, Centre for Drug Evaluation and Research (2002) Bextra (valdecoxib) tablets. Company: G. D. Searle and Co. Application No.: 21-341. Approval Date: 11/16/01 *www.fda.gov/cder/foi/nda/2001/21-341_Bextra.htm* (Accessed 25/10/2002)

309 Benson W, Weaver A, Espinoza L, Zhao WW, Riley W, Paperiello B, Recker DP (2002) Efficacy and safety of valdecoxib in treating the signs and symptoms of rheumatoid arthritis: a randomized, controlled comparison with placebo and naproxen. *Rheumatology* 41: 1008–1016

310 Makarowski W, Zhao WW, Bevirt T, Recker DP (2002) Efficacy and safety of the COX-2 specific inhibitor valdecoxib in the management of osteoarthritis of the hip: a randomized, double-blind, placebo-controlled comparison with naproxen. *Osteoarthritis and Cartilage* 10: 290–296

311 Sikes DH, Agrawal NM, Zhao WW, Kent JD, Recker DP, Verburg KM (2002) Incidence of gastroduodenal ulcers associated with valdecoxib compared with that of ibuprofen and diclofenac in patients with osteoarthritis. *Eur J Gastroenterol Hepatol* 14: 1101–1111

312 Goldstein JL, Kivitz AJ, Verburg KM, Recker DP, Kent JD (2003) A comparison of the upper gastrointestinal mucosal effects of valdecoxib, naproxen and placebo in healthy elderly subjects. *Aliment Pharmacol Ther* 18: 125–132

313 Strand V, Hochberg MC (2002) The risk of cardiovascular thrombotic events with selective cyclooxygenase-2 inhibitors. *Arthritis and Rheumatism* 47: 349–355

314 White WB, Verburg KM, Whelton A (2003) The COX-2 specific inhibitor, valdecoxib, is not associated with an increased risk of cardiovascular thrombotic events in arthritis patients

315 Prasit P, Wang Z, Brideau C, Chan C-C, Charleson S, Cromlish W, Ethier D, Evans JF, Ford-Hutchinson AW, Gauthier JY et al (1999) The discovery of rofecoxib, (MK 966, VIOXX$^{\forall}$, 4-(4'-methylsulfonylphenyl)-3-phenyl-2(5*H*)-furanone], an orally active cyclooxygenase-2 inhibitor. *Bioorganic and Medicinal Chemistry Letters* 9: 1773–1778

316 Scott LJ, Lamb HM (1999) Rofecoxib: new drug profile. *Drugs* 58: 499–506

317 Ahuja N, Singla AK, Singh A, Singh B (2003) Rofecoxib: an update on physicochemical, pharmaceutical, pharmacodynamic and pharmacokinetic aspects. *J Pharm Pharmacol* 55: 859–894

318 Food and Drug Adminstration (USA): Vioxx® (rofecoxib). Centre for Drug Evaluation (CDER, New and Generic Drug Approvals: 1998–2003), NDA Application # 21-042, approval date 5/20/99. Pharmacological Review posted 4/17/00. *www.fda.gov/cder/approval/index.htm* (Accessed Dec 2003)

319 Chan CC, Boyce S, Brideau C, Charleson S, Cromlish W, Ether D, Evans J, Ford-Hutchinson AW, Forrest MJ, Gauthier JY et al (1999) Rofecoxib (Vioxx, MK-0966); 4-(4'-methylsulfonylphenyl)-3-phenyl-2-(5*H*)-furanone]: a potent and orally active cyclooxygenase-2 inhibitor. Pharmacological and biochemical profiles. *J Pharmacol Exp Ther* 290: 551–560

320 Nicoll-Griffith DA, Yergey JA, Trimble LA, Silva JM, Li C, Chauret N, Gauthier JY, Grimm E, Leger S, Roy P et al (2000) Synthesis, characterization, and activity of metabolites derived from the cyclooxygenase-2 inhibitor reofecoxib (MK-0966, Vioxx). *Bioorg Med Chem Lett* 10: 2683–2686

321 Greenberg HE, Gottesdiener K, Huntington M, Wong P, Larson P, Wildonger L, Gillen L, Dorval E, Waldman SA (2000) A new cyclooxygenase-2 inhibitor, rofecoxib (VIOXX®), did not alter the antiplatelet effects of low-dose aspirin in healthy volunteers. *J Clin Pharmacol* 40: 1509–1515

322 Feldman M, McMahon AT (2000) Do cyclooxygenase-2 inhibitors provide benefits similar to those of traditional nonsteroidal anti-inflammatory drugs, with less gastrointestinal toxicity? *Ann Intern Med* 132: 134–143

323 Lanza FL, Rack MF, Simon TJ, Quan H, Bolognese JA, Hoover ME, Wilson FR, Harper SE (1999) Specific inhibition of cyclooxygenase-2 with MK-0966 is associated with less gastroduodenal damage than either aspirin or ibuprofen. *Aliment Pharmacol Ther* 13: 761–767

324 Bombardier C, Laine L, Reicin A, Shapiro D, Burgos-Vargas R, Davis B, Day R, Ferraz MB, Hawkey CJ, Hochberg MC et al (2000) Comparison of upper gastrointestinal toxicity of rofecoxib and naproxen in patients with rheumatoid arthritis. *N Engl J Med* 343: 1520–1528

325 Depré M, Ehrich E, Van Hecken A, De Lepeleire I, Dallob A, Wong P, Porras A, Gertz BJ, De Schepper PJ (2000) Pharmacokinetics, COX-2 specificity, and tolerability of supratherapeutic doses of rofecoxib in humans. *Eur J Clin Pharmacol* 56: 167–174

326 Schoenfeld P (2001) An evidence-based approach to the gastrointestinal safety profile of COX-2 selective ani-inflammatories. *Gastroenterology Clinics of North America* 30: 1027–1044

327 Daniels B, Gertz B, Morrison B, Seidenberg B (2003) Renal safety profile of rofecoxib, a spe-

cific inhibitor of COX-2, in controlled clinical trials. European League of Rheumatology Congress (EULAR), Lisbon, Portugal
328 Nantel F, Meadows E, Denis D, Connolly B, Metters KM, Giaid A (1999) Immunolocalization of cyclooxygenase-2 in the macula densa of human elderly. *FEBS Letters* 457: 475–477
329 Layton D, Riley J, Wilton LV, Shakir SAW (2003) Safety profile of Rofecoxib as used in general practice in England: results of a prescription-event monitoring study. *Br J Clin Pharmacol* 55: 166–174
330 Cannon GW, Caldwell JR, Holt P, McLean B, Seidenberg B, Bolognese J, Ehrich E, Mukhopadhyay S, Daniels B (2000) Rofecoxib, a specific inhibitor of cyclooxygenase 2, with clinical efficacy comparable with that of diclofenac sodium. *Arthritis and Rheumatism* 43: 978–987
331 Baillie TA, Halpin RA, Matuszewski BK, Geer LA, Chavez-Eng CM, Dean D, Braun M, Doss G, Jones A, Marks T et al (2001) Mechanistic studies on the reversible metabolism of rofecoxib to 5-hydroxyrofecoxib in the rat: evidence for transient ring opening of a substituted 2-furanone derivative using stble isotope-labeling techniques. *Drug Metab Dispos* 29: 1614–1628
332 Halpin RA, Geer LA, Zhang Kem Marks TM, Dean DC, Jones AN, Melillo D, Doss G, Vyas KP (2000) The absorption, distribution, metabolism and excretion of rofecoxib, a potent and selective cyclooxygenase-2 inhibitor, in rats and dogs. *Drug Metab Dispos* 28: 1244–1254
333 Catella-Lawson F, McAdam B, Morrison BW, Kapoor S, Kujubu D, Antes L, Lasseter KC, Quan H, Gertz BJ, Fitzgerald GA (1999) Effects of specific inhibition of cyclooxygenase-2 on sodium balance, hemodynamics, and vasoactive eicosanoids. *J Pharmacol Exp Ther* 289: 735–741
334 Langman MJ, Jensen DM, Watson DJ, Harper SE, Zhao P, Quan H, Bolognese JA, Simon TJ (1999) Adverse upper gastrointestinal effects of rofecoxib compared with NSAIDs. *J Am Med Assn* 282: 1929–1933
335 Strand V, Hochberg MCC (2002) The risk of cardiovascular thrombotic events with selective cyclooxygenase-2 inhibitors. *Arthritis and Rheumatism* 47: 349–355
336 McAdam BF, Catella-Lawson F, Maradini IA, Kapoor S, Lawson JA, FitzGerald GA (1999) Systemic biosynthesis of prostacyclin by cyclooxygenase (COX)-2. The human pharmacology of a selective inhibitor of COX-2. *Proc Natal Acad Sci USA* 96: 272–277
337 Niederberger E, Tegeder I, Schafer C, Seegel M, Grosch S, Geisslinger G (2003) Opposite effects of rofecoxib on nuclear factor kappaB and activating protein-1 activation. *J Pharmacol Exp Ther* 304: 1153–1160
338 Callejas NA, Fernandez-Martinez A, Castrillo A, Bosca L, Martin-Sanz P (2003) Selective inhibitors of cyclooxygenase-2 delay the activation of the nuclear factor kappa B and attenuate the expression of inflammatory genes in murine macrophages treated with lipopolysaccharide. *Mol Pharmacol* 63: 671–677
339 Oshima M, Murai N, Kargman S, Arguello M, Luk P, Kwong E, Taketo MM, Evans JF (2001) Chemoprevention of intestinal polyposis in the Apcdelta716 mouse by rofecoxib, a specific cyclooxygenase-2 inhibitor. *Cancer Res* 61: 1733–1740
340 Martinez RV, Reval M, Campos MD, Terron JA, Dominguez R, Lopez-Munoz FJ (2002) Involvement of peripheral cycloxygenase-1 and cyclooxygenase-2 in inflammatory pain. *J Pharm Pharmacol* 54: 405–412
341 Sandrini M, Vitale G, Pini LA (2002) Effect of rofecoxib on nociception and the serotonin system in the rat brain. *Inflamm Res* 51: 154–159
342 El-Shenawy SM, Abdel-Salam OM, Baiuomy AR, El-Batran S, Arbid MS (2002) Studies on the anti-inflammatory and anti-nociceptive effects of melatonin in the rat. *Pharmacol Res* 46: 235–243
343 Hutchins B, Patel H, Spears R (2002) Attenution of pro-inflammatory neuropeptide levels produced by a cyclooxygenase-2 inhibitor in an animal model of chronic temporomandibular joint inflammation. *J Orofac Pain* 16: 312–316
344 Kalgutkar AS, Zhao Z (2001) Discovery and design of selective cyuclooxygenase-2 inhibitors as non-ulcerogenic, anti-inflammatory drugs with potential utility as anti-cancer agents. *Curr Drug Targets* 2: 79–106
345 Dicker AP, Williams TL, Grant DS (2001) Targeting angiogenic processes by combination rofecoxib and ionizing radiation. *Am J Clin Oncol* 24: 438–442
346 Brzozowski T, Konturek PC, Konturek SJ, Pajdo R, Schuppan D, Drozdowicz D, Ptak A, Pawlik M, Nakamura T, Hahn EG (2000) Involvement of cyclooxygenase (COX)-2 products in acceleration of ulcer healing by gastrin and hepatocyte growth factor. *J Physiol Pharmacol* 51: 751–773
347 Laudanno OM, Cesolari JA, Esnarriaga J, Rista L, Piombo G, Maglione C, Aramberry L,

Sambrano J, Godoy A, Rocaspana A (2001) Gastrointestinal damage induced by celecoxib and rofecoxib in rats. *Dig Dis Sci* 46: 779–784
348 Pajdo R, Brzozowski T, Konturek PC, Kwiecien S, Konturek SJ, Sliwowski Z, Pawlik M, Ptak A, Drozdowicz D, Hahn EG (2001) Ischemic preconditioning, the most effective gastroprotective intervention: involvement of prostaglandins, nitric oxide, adenosine and sensory nerves. *Eur J Pharmacol* 427: 263–276
349 Colville-Nash PR, Gilroy DW (2001) Potential adverse effects of cyclooxygenase-2 inhibition: evidence from animal models of inflammation. *BioDrugs* 15: 1–9
350 Simon AM, Manigrasso MB, O'Connor JP (2002) Cyclo-oxygenase 2 function is essential for bone fracture healing. *J Bone Miner Res* 17: 977–978
351 Goodman S, Ma T, Trindade M, Ikenoue T, Matsuura I, Wong N, Fox N, Genovese M, Regula D, Smith RL (2002) COX-2 selective NSAID decreases bone in growth *in vivo*. *J Orthop Res* 20: 1164–1169
352 Cochrane DJ, Jarvis B, Keating GM (2002) Etoricoxib. *Drugs* 62: 2637–2651
353 Riendeau D, Percival MD, Brideau C, Charleson S, Dubé D, Ethier D, Falgueyret J-P, Friesen RW, Gordon R, Greig G et al (2001) Etoricoxib (MK-0663): preclinical profile and comparison with other agents that selectively inhibit cyclooxygenase-2. *J Pharmacol Exp Ther* 296: 558–566
354 Agrawal NG, Porras AG, Matthews CZ, Rose MJ, Woolf EJ, Musser BJ, Dynder AL, Mazina KE, Lasseter KC, Hunt et al (2003) Single- and multiple-dose pharmacokinetics of etoricoxib, a selective inhibitor of cyclooxygenase-2, in man. *J Clin Pharmacol* 43: 268–276
355 Gottesdiener K, Schnitzer T, Fisher C, Bockow B, Markenson J, Ko A, Detora L, Curtis S, Geissler L, Gertz BJ (2002) Results of a randomized, dose-ranging trial of etoricoxib in patients with osteoarthritis. *Rheumatology* 41: 1052–1061
356 Matsumoto AK, Melian A, Mandel DR, McIlwain HH, Borenstein D, Zhao PL, Lines CR, Gertz BJ, Curtis S (2002) A randomized, controlled, clinical trial of etoricoxib in the treatment of rheumatoid arthritis. *J Rheumatol* 29: 1623–1630
357 Collantes E, Curtis SP, Lee KW, Casas N, McCarthy T, Melian A, Zhao PL, Ridgers DB, McCormick CL, Lee M, Lines CR, Gertz BJ (2002) A multinational randomized, controlled, clinical trial of etoricoxib in the treatment of rheumatoid arthritis. *BMC Family Practice* 3: 2–10
358 Sorbera LA, Castaner J, Bayes M, Silvestre JS (2002) Lumiracoxib. *Drugs of the Future* 27: 740–747
359 Reynolds C, Scott G, Looby M, Milosavljev S, Huff JP, Ruff DA, Rordorf C (2003) Pharmacokinetics of lumiracoxib in synovial fluid and plasma of patients with rheumatoid arthritis. European Assoc. Clinical Pharmacology and Therapeutics (EACPT), Istanbul; P-195 (Abstr)
360 Scott G, Branson J, Miloszvljev S, Rordorf C, Haraoui B, Ouellet J-P, Schell E (2003) Lumiracoxib demonstrates dose-proportional and time-independent pharmacokinetics in patients with osteoarthritis of the knee. *Ann Rheum Dis* 62 (Suppl 1): FRI235 (Abstr)
361 Bonner J, Branson J, Milosavljev S, Rordorf C, Scott G (2003) Co-administration of lumiracoxib and warfarin does not alter the pharmacokinetic profile of *R*- or *S*-warfarin. *Ann Rheum Dis* 62 (Suppl 1): FRI0225 (Abstr)
362 Hartman S, Scott G, Rordorf C, Campesrini J, Branson J, Keller U (2003) Lumiracoxib demonstrates high absolute bioavailability in healthy subjects. European Assoc. Clinical Pharmacology and Therapeutics (EACPT), Istanbul; P-199 (Abstr)
363 Wilding IR, Connor AL, Carpenter P, Rordorf C, Branson J, Milosavljev S, Scott G (2003) Lumiracoxib shows similar bioavailability at different sites in the gastrointestinal tract. European Assoc. Clinical Pharmacology and Therapeutics (EACPT), Istanbul; P-196 (Abstr)
364 Grifka J, Zacher J, Brown J, Seriolo B, Lee A, Moore A, Gimona A (2003) Lumiracoxib is effective and well tolerated in patients with osteothritis of the hand: a randmomized placebo-controlled trial. *Ann Rheum Dis* 62 (Suppl 1): FRI0222 (Abstr)
365 Benevolenskaya L, Tüzün S, Hagin E, Moore A, Gimona A (2003) Lumiracoxib is effective in relieving symptoms of osteoarthritis of the hip or knee after 4 weeks of treatment: results from a randomized placebo-controlled trial. *Ann Rheum Dis* 62 (Suppl 1): FRI0246 (Abstr)
366 Fleischmann R, Sheldon E, Maldonado Cocco J, Yu S, Dutta D, Usiskin K (2003) A prospective randomized 13-week study revaluating the efficacy of lumiracoxib in patients with osteoarthritis of the knee. *Ann Rheum Dis* 62 (Suppl 1): FRI0233 (Abstr)

COX-2 Inhibitors
Edited by M. Pairet and J. van Ryn

Efficacy and gastrointestinal safety of selective COX-2 inhibitors

Frank Degner

CDept. Medical Affairs, Boehringer Ingelheim GmbH, Binger Str., D-55216 Ingelheim/Rhein, Germany

Introduction

Nonsteroidal anti-inflammatory drugs (NSAIDs) are among the most commonly prescribed drugs in the world and are used frequently as anti-inflammatory and analgesic agents in the treatment of musculoskeletal conditions. Their use is mainly limited by adverse effects, which are more frequently seen with NSAIDs than with any other prescription drug. These side-effects are of particular importance in the elderly who are more prone to NSAID complications [1].

The background rate of ulcer complications in patients not using NSAIDs varies with risk factors, such as age, but has been estimated to be between 0.2 and one event per 100 patient years. Epidemiological studies have fairly consistently shown a four-fold increase of ulcer complications in patients using NSAIDs compared with control individuals, with an absolute rate in the NSAID group between one and four events per 100 patient years. The fraction of disease attributable to NSAID use has been estimated to be between 25% and 35%, although increased use of low-dose aspirin may have enhanced this. In one study, NSAID use was associated with 1.25 extra hospitalizations per 100 patient years for ulcer complications. Based on figures from the Office of Population Census and Surveys, there are 4000 deaths per annum from peptic ulcer disease in the UK, implying that about 1200 of them are caused by non-aspirin NSAIDs. Recent estimates of 2000 deaths per annum in the UK and 16,500 deaths per annum in the USA are at the upper end of a range of estimates of the mortality attributable each year to these drugs.

The initial hypothesis concerning NSAID action was that they inhibited a single cyclooxygenase (COX) enzyme, and would, therefore, have inseparable efficacy and toxicity [2, 3]. Preliminary data, which suggested a variation in gastrotoxicity, were explained by NSAID use in different disease subsets and age groups, differences in the half-life of NSAIDs and their potency, and fixed-dose prescribing [4].

Subsequent large studies have, however, provided interesting information about relative NSAID toxicity. Singh et al. have developed a "toxicity index" to give an overall measure of the differential adverse effect profile of NSAIDs based on information obtained in patients with rheumatoid arthritis from the Arthritis, Rheumatism and Aging Medical Information System (ARAMIS) database [5]. Langman et al. and Garcia Rodriguez et al. have assessed NSAID use in cases of upper gastrointestinal bleeds and were able to calculate the odds ratios of commonly used NSAIDs for gastrointestinal bleeds [6, 7]. These studies demonstrate that azapropazone, piroxicam and indomethacin have the highest risk of gastrointestinal bleeding, with ibuprofen having a low risk. The risk factors for gastrointestinal adverse events were older age, previous history of gastrointestinal bleed and higher dose of NSAID. Langman et al. noted that peptic ulcer bleeding was strongly associated with the use of NSAIDs in the 3 months prior to admission, and that the risk for most NSAIDs tended to be greatest in those recently starting treatment. This group were also more likely to be on the maximum recommended dose. More recently, MacDonald et al. have demonstrated that gastrointestinal risk from NSAIDs seems to persist with continuous use, and there may even be carry-over toxicity after NSAID cessation [8]. The discovery of the two different forms of COX enzymes (COX-1 and COX-2) expanded on the original concepts of Sir John Vane [9–11], and led to the hypothesis that NSAID toxicity was related to COX-1 inhibition, and that selective COX-2 inhibitors would retain the efficacy of nonselective NSAIDs without their COX-1 mediated side-effects [12]. Many reports on the relative COX-1 and COX-2 inhibition of currently used NSAIDs have subsequently appeared [13], and while their clinical studies demonstrate a wide variation in the incidence of gastrointestinal side-effects with existing NSAIDs, COX-2 selectivity does appear to be associated with a decreased incidence of these side-effects [14–19].

The COX-1-selective drugs indomethacin and piroxicam are associated with high toxicity indices whereas ibuprofen, which is more selective for COX-2, is associated with a low index. Aspirin, however, is the most COX-1-selective drug in this series and has a low toxicity index, suggesting that other factors are also important in determining toxicity. In epidemiological studies, ibuprofen has consistently been associated with lower than average risks of ulcer complications, compared with other nonselective NSAIDs, while other agents, such as azapropazone and piroxicam, have generally been associated with high risks. The safety advantage of ibuprofen is, however, only seen with the lower doses of 1200 mg traditionally used in Europe, rather than the higher doses often used in the USA. Ibuprofen is also a drug with a short serum half-life that persists in joints for much longer, thereby achieving selectively enhanced joint exposure compared with gastric exposure. By contrast, in drugs with a long half-life (such as piroxicam or tenoxicam) no such selectivity is achieved.

Reducing the risks of current NSAIDs use

When prescribing NSAIDs, taking a careful drug history is important to assess the possibility of drug interactions, especially in the elderly. Over-the-counter

NSAIDs are increasingly available and their inadvertent co-use is something that all physicians need to be aware of. Concomitant aspirin, even in low dose, and corticosteroid use increases risk of peptic ulceration with NSAIDs. NSAIDs should be avoided wherever possible in those patients on anticoagulation treatment because of their anti-platelet and protein binding effects. NSAIDs should be used in the lowest possible doses, and it seems prudent initially to use those with lower toxicity indices [20]. There is no significant evidence that controlled release, intramuscular or rectal preparations reduce risks of gastrointestinal adverse effects.

In clinical practice, many patients with risk factors, and those who develop dyspepsia, are prescribed gastroprotective agents (even though the relationship between dyspepsia and peptic ulcer disease is unclear), making the drug combination less cost effective. In chronic NSAID use, standard dose histamine-type 2 (H2)-antagonists reduce the incidence of endoscopic duodenal ulcers, but not gastric ulcers, whereas double-dose H2-antagonists and standard dose proton-pump inhibitors reduce both duodenal and gastric ulcers [21, 22]. Misoprostol 400 mg/day and 800 mg/day reduce gastric and duodenal ulcers, with the higher dose being more effective for reducing gastric ulcers. Notably, only misoprostol at the higher dose has been demonstrated to reduce clinical adverse events, although diarrhoea is a common side-effect at this dose. Sucralfate heals duodenal ulcers as effectively as H2-antagonists, whether or not NSAIDs are continued [23]. Prophylactic gastroprotective agents should be considered in the elderly, those with a history of peptic ulcer disease, those on concomitant corticosteroids, those who develop dyspepsia and those with serious co-morbid conditions.

Efficacy of selective COX-2 inhibitors in rheumatology

Rheumatoid arthritis

Rheumatoid arthritis is a severe inflammatory polyarthritis, with a peak incidence in premenopausal women. It has a prevalence of 0.5–1.0%. Although the causes of rheumatoid arthritis are not fully understood, evidence suggests that pro-inflammatory cytokines play an important role in pathogenesis. It is clear that early on in disease, the major events are cytokine-driven prostaglandin (PG)-mediated inflammatory effects, and later on, cytokine-driven osteoclast activation with subsequent bone damage. In rheumatoid arthritis, COX-2 is markedly up-regulated in inflamed synovial tissue, with resultant increased PG production.

The common feature of disease-modifying anti-rheumatic drugs (DMARDs) is the blockade of cytokines. By doing so, the mediators of bone destruction and of stimulants to the hepatic acute phase response are blocked. With the exception of (nonlicensed) tenidap, the existing NSAIDs inhibit COX alone without any effect on systemic cytokine levels. NSAIDs, however, remain the cornerstone of symptomatic treatment in rheumatoid arthritis. Although clinical experience has shown that NSAIDs do not appear to influence radiological progression, evidence seems to suggest that they improve Health Assessment Questionnaire (HAQ) scores and reduce swollen and tender joint scores in rheumatoid arthritis. Imaging studies, using magnetic resonance imaging and ultrasonography, of early rheumatoid arthritis have also demonstrated the ability of NSAIDs to reduce synovial volume [24].

Selective COX-2 inhibition may well have extra benefits in the control of synovitis. Whether or not selective COX-2 inhibitors given at high doses in early disease could influence disease progression in rheumatoid arthritis remains to be tested in a formal programme. In rheumatoid arthritis, up-regulation of tissue necrosis factor (TNF) is associated with increased COX-2 production. One of the features of anti-TNF therapy is the quick reduction in synovitis. The speed of this response suggests that down-regulation of COX-2 may be involved. Another potential benefit is that COX-2 is expressed in endothelial cells, particularly in situations of shear stress and angiogenesis. If inhibitors have an effect on angiogenesis then clearly there would be a potential role in inhibiting synovial growth.

Osteoarthritis

Osteoarthritis is the most common joint disorder in the world and affects all joints, but most commonly the knees, hips, hands and spine. The prevalence of osteoarthritis increases with age and presents with use-related pain, loss of joint motion and disability. Patients with osteoarthritis have an inflammatory phase with synovitis that is important in the disease progression. Up-regulation of COX-2 has been detected in the synovium from patients with osteoarthritis, although to a lesser extent than in rheumatoid arthritis. There is also evidence of up-regulation of COX-2 and subsequent PG production in osteoarthritic cartilage.

Osteoarthritis presents practical difficulties with current NSAID use in that it is prevalent in that group of patients most at risk of gastrointestinal toxicity, namely the elderly. Patients often find NSAIDs to be the most effective pain relievers in osteoarthritis, creating a major conflict between efficacy and toxicity. It is worth noting that pain is the major reason for joint replacement to be undertaken, and that COX-2 is involved in pain transmission both peripherally and centrally.

Again, there are potential advantages to using selective rather than nonselective NSAIDs in osteoarthritis subgroups with synovitis. Although animal model studies suggest meloxicam, unlike indomethacin, does not inhibit cartilage proteoglycan synthase, the effects of COX-2 inhibition on PG-mediated bone turnover and chondrocyte biology are yet to be determined and further studies will help delineate the role of the COX isoenzymes in bone biology.

Other rheumatological conditions

The potential role of selective COX-2 inhibition in controlling synovitis may also be relevant in systemic lupus erythematosus (SLE), where high-dose inhibition may result in better symptom relief and importantly, a steroid-sparing effect. COX-2 is up-regulated in mesangial cells in lupus nephritis and COX-2 inhibition, in particular without any COX-1 related loss of glomerular filtration rate may well prove beneficial in this group [25].

Up-regulation of COX-2 has also been demonstrated in other inflammatory arthritides, such as ankylosing spondylitis and psoriatic arthritis. In ankylosing spondylitis, adequate analgesia coupled with exercise is the mainstay of treatment to minimise loss of spinal movement. NSAID use has been associated with flares of skin lesions in psoriasis, which is thought to be secondary to an increase in lipoxygenase pathway products as a result of COX inhibition. Selective COX-2 inhibition may change the balance of PGs and lipoxygenase products resulting in fewer exacerbations of skin lesions.

COX-2-selective agents registered for use in rheumatological conditions

Meloxicam is licensed for use in rheumatoid arthritis, osteoarthritis and ankylosing spondylitis at doses of 7.5–15 mg once daily and is superior to placebo and comparable with piroxicam and diclofenac in the treatment of osteoarthritis of the hip and knee [26–28]. In the symptomatic treatment of rheumatoid arthritis, meloxicam is superior to placebo and comparable with naproxen and diclofenac [29–31]. It is also superior to placebo and comparable with piroxicam for symptom relief in ankylosing spondylitis [32].

Celecoxib is licensed for use in osteoarthritis at 100–200 mg once daily and in rheumatoid arthritis at 100–200 mg twice daily, and is also licensed for use in the USA as an adjunct to other prophylactic therapy in familial adenomatous polyposis (FAP). Celecoxib is superior to placebo and as effective as naproxen and diclofenac for symptomatic relief in rheumatoid arthritis [33] and osteoarthritis of the knee [34–37].

Rofecoxib is licensed for use in osteoarthritis at doses of 12.5–25 mg once daily. It is superior to placebo and comparable with ibuprofen and diclofenac for the treatment of osteoarthritis of the hip and knee [38–44], although there are studies showing significantly greater dropouts due to lack of efficacy with rofecoxib than with diclofenac [45]. It is also licensed for use in rheumatoid arthritis and is superior to placebo [46], achieving American College Rheumatology (ACR) 20 Remission in 43.9% and 49.7% of patients taking 25 mg and 50 mg daily, respectively. The drug is not registered for long-term use in daily doses above 25 mg.

Gastrointestinal tolerabiliy and safety of selective COX-2 inhibitors

Although gastrointestinal side-effects of NSAIDs are very important, it should not be forgotten that there are other significant associated problems. Nephrotoxicity can result in acute and chronic renal failure, and hypertensive effects vary widely between patients. PGs in the kidney are important in maintaining glomerular filtration rate and renal blood flow, especially in salt-depleted states [47]. Since glomerular filtration rate is reduced with age, the elderly are more susceptible to these renal adverse effects. This is discussed in more detail in Chapter 8.

Hepatic transaminitis is not uncommon, but is rarely significant after drug cessation. Hypersensitivity reactions may occur in patients with asthma, resulting in bronchospasm and rhinitis. Bleeding complications can result from NSAIDs interfering with platelet function. Confusion and drowsiness have been reported in the elderly and there are rare reports of aseptic meningitis and blood dyscrasias.

Studies in healthy elderly patients suggest that selective COX-2 inhibitors may spare the persistent reduction in glomerular filtration rate seen with conventional NSAIDs, although in patients with salt-depleted states, such as congestive cardiac failure or diuretic use, this benefit seems to be lost. It is postulated that NSAID-associated hypersensitivity and exacerbations of asthma are COX-1 mediated, and this seems to be supported by reports of these patients tolerating selective COX-2 inhibitors.

As with conventional NSAIDs, prescribing physicians need to be aware of drug interactions. Even though selective COX-2 inhibitors have no anti-platelet effect, and their addition leads to little alteration in the international normalized ratio (INR) of patients taking a titrated dose of warfarin, careful monitoring of the INR is advisable especially in the elderly, if an NSAID is indicated. Concomitant aspirin use seems to remove the gastroprotective effect of these drugs, and prophylactic gastroprotective agents may be required as with conventional NSAIDs.

Meloxicam (dose range 7.5–15 mg/day)

Meloxicam was the first of the existing NSAIDs that was found to have COX-2 selectivity. Meloxicam at a dose of 7.5 mg causes little acute gastric mucosal injury. In two large studies involving over 8000 patients, meloxicam 7.5 mg was significantly better tolerated than slow release diclofenac 100 mg or piroxicam 20 mg with regard to symptomatic gastrointestinal adverse events. A combined analysis of studies suggests that the symptom benefits of meloxicam 7.5 mg also applies to meloxicam 15 mg [48].

There have been no prospective randomized trials specifically designed to investigate gastrointestinal complications. In the MELISSA [49] and SELECT [50] studies, however, no ulcer bleeding or perforation was observed with meloxicam, while four such cases were observed each with both diclofenac and piroxicam. Numerically fewer upper gastrointestinal ulcer events were reported with meloxicam compared with diclofenac (5 *versus* 7) or piroxicam (7 *versus* 16). Patients also appeared to spend less time in intensive care units. Analysis of the entire meloxicam database suggests a reported annual incidence of perforations, ulceration and bleeding (PUBs, including uncomplicated upper gastrointestinal ulcers) for meloxicam 7.5 mg and 15 mg of 1.2% *versus* 1.9% for diclofenac 100 mg, 4.8% for piroxicam 20 mg and 1.8% for meloxicam 22.5 mg/day [48]. Postmarketing analyses consistently show a favourable overall and gastrointestinal tolerability for both low risk as well as high risk patients (e.g., those with previous GI history) under actual conditions for meloxicam [51]. Although the effect of meloxicam on outcomes has not been systematically investigated in a prospective way, results of meta-analyses and observational studies are consistent and reassuring. Recently a comparative prescription event monitoring study on 19,087 patients treated with meloxicam and 15,268 patients treated with rofecoxib in England over a period of up to 9 months reported drop-outs for gastrointestinal symptoms in 15.4% with meloxicam and 20.2% patients with rofecoxib [52]. Perforations and gastrointestinal bleeds were observed in 0.4% in both groups. Adverse drug reactions attributed by the physician in the meloxicam group were 1.1% and 2.4% for rofecoxib. A total of 101 (0.66%) thromboembolic events were reported during treatment or within a month of stopping rofecoxib, no comparison data was published for meloxicam.

The differential and dose-dependent effects of meloxicam on platelet aggregation and thromboxane synthesis are remarkable in that studies show inhibition of platelet thromboxane in a dose-dependent fashion over a range of daily doses from 7.5 mg to 30 mg [53]. In these studies, however, it is demonstrated consistently that meloxicam does not inhibit platelet aggregation at doses up to 30 mg. It is under discussion whether small amounts of COX-1 inhibition are harmful or might even have some benefits on the endothelial, renal or cardiovascular system. These are issues that will require future resolution and are further discussed in Chapter 7.

Rofecoxib (dose range 12.5–25 mg/day)

In acute studies, rofecoxib 50 mg caused no depression of PG synthesis and at 250 mg/day caused no increase in acute injury compared with placebo. In endoscopic patient studies, ulcer rates associated with rofecoxib 12.5 mg and 25 mg were no different from those with placebo and were four-fold lower than those of ibuprofen 2.4 g/day [54, 55].

A supratherapeutic dose of rofecoxib 50 mg has been compared with naproxen 1 g/day over a median 9 months in 8076 patients with rheumatoid arthritis in the VIGOR study [56]. This study showed a 54% reduction in clinical ulcers and a 57% reduction in complicated upper gastrointestinal events with rofecoxib compared with naproxen. Rofecoxib 50 mg did not result in an overall reduction in adverse events and the rate of myocardial infarction on rofecoxib was higher than on naproxen. Whether this occurred by chance, because of a harmful effect of COX-2 inhibition or a beneficial effect of COX-1 inhibition with naproxen, or (because it was not seen with other studies involving normal clinical doses of rofecoxib) whether it was dose specific are all currently unclear.

The VIGOR study was large enough to allow identification of subgroups of patients in whom large absolute risk reductions with rofecoxib compared with naproxen were seen. For patients over 75 years of age or those with a past history, risk reductions were sufficiently great that only 10 patients per annum would be treated to avoid one gastrointestinal event on a COX-2 inhibitor compared with a nonselective NSAID.

Celecoxib (dose range 200–400 mg)

In acute studies, celecoxib 800 mg/day was at placebo levels in terms of acute gastric injury [57]. In chronic patient studies, in a dose range of 100–800 mg/day there was likewise no difference from placebo and/or a significant reduction compared with diclofenac or naproxen in five out of six studies [57–62]. In one study, the rate of ulceration on celecoxib did not differ from diclofenac [62].

As with rofecoxib, a supratherapeutic dose of celecoxib (800 mg/day) has been evaluated in the CLASS study [63, 64]. In this study, patients with osteoarthritis and rheumatoid arthritis were enrolled, aspirin less than 325 mg/day was allowed and used by 21% of the patients, and there were differences in trial conduct from rofecoxib. Over the full median 12 months of the study, celecoxib was not significantly different from comparators (ibuprofen or diclofenac) in the rate of ulcer complications, but after 6-month pooled data showed a ~50% reduction in patients receiving celecoxib though this failed to reach significance for each of the individual comparator NSAIDs. In part, this may have been because aspirin was used by 21% of patients, which may have independently increased rates of ulceration in both treatment groups.

Recently Chan et al. compared a strategy of a standard NSAID with a gastroprotective agent *versus* celecoxib alone in high risk arthritic patients who had suffered a recent gastrointestinal hemorrhage on NSAIDs [65]. Ulcer healing was confirmed by endoscopy in 290 subjects who were then randomized to receive celecoxib 400 mg/day alone or diclofenac 150 mg/day with omeprazole 20 mg/day. After 6 months of follow-up, 24 serious GI events occurred. Sixteen cases of recurrent ulcer bleeding occurred (seven with celecoxib and nine with diclofenac + omeprazole). All but one of these was due to recurrent bleeding from the previously identified gastric ulcer site. The 6-month probability of recurrent bleeding was 4.9% and 6.4% for the celecoxib and diclofenac/omeprazole groups respectively (not statistically different). Of the

260 patients who were not taking aspirin, the 6-month probability of recurrent bleeding was 4.5% and 5.6% for the celecoxib and diclofenac/omeprazole groups respectively (not statistically different). The authors conclude that the two strategies for recurrent ulcer prevention are equivalent. It should however, be pointed out that the annulized incidence of NSAID related ulcer bleeding in this study of nearly 10% per year is extremely high. In the MUCOSA study, high-risk patients taking standard NSAIDs without prophylaxis had a 6 month incidence of ulcer complications of 9% [66].

Comparison of randomised clinical trials of meloxicam, rofecoxib and celecoxib

No randomised clinical trials are available to compare adequately the gastrointestinal safety profiles of any of the selective COX-2 inhibitors. Therefore such comparisons can only be done indirectly by using meta-analytical techniques. Table 1 is based on the results of a comprehensive literature database search combined with a vigorous selection process performed by the Canadian Coordinating Office for Health Technology Assessment to summarize randomised clinical trials providing information on clinical ulcer complications of the selective COX-2 inhibitors compared with traditional NSAIDs [67–76]. Table 2 summarizes the results of a meta-analysis performed on these studies regarding clinical ulcer complications and symptomatic ulcers (PUBs). From a gastrointestinal toxicity perspective the data can be summarised that COX-2 selective NSAIDs are safer than naproxen and high-dose ibuprofen and are better tolerated then NSAIDs in general. Nevertheless, data to address the important clinical questions regarding comedication with low-dose aspirin and/or gastroprotective agents in cardiovascular and gastrointestinal risk patients are only beginning to become available and final conclusions are not available at this point in time.

While data from a slightly different set of studies were entered into the analysis performed by the National Institute for Clinical Excellence (NICE), similar results were obtained with regard to the conclusions [77]. NICE recommends restricting the use of selective COX-2 inhibitors to patients at high risk, namely patients with age of 65 years and over; previous clinical history of gastroduodenal ulcer, gastrointestinal bleeding or gastroduodenal perforation; concomitant use of medications that are known to increase the likelihood of upper gastrointestinal adverse events e.g., steroids and anticoagulants; presence of serious co-morbidity, such as cardiovascular disease, renal or hepatic impairment, diabetes and hypertension; requirement for the prolonged use of maximum recommended doses of standard NSAIDs. Existing data suggest that selective COX-2 inhibitors are associated with substantial reductions in risk compared with nonselective NSAIDs in older patients, those with a past history and those using corticosteroids. Although not specifically tested, the use of selective COX-2 inhibitors would seem wise in those with impaired

Table 1. Study characteristics of RCTs with COX-2 selective NSAIDs regarding clinical ulcer complications

Study and study drug	Dose of selective COX-2 inhibitor	NSAID	Number of patients	Mean age (yrs)	Indication	Follow up
Celecoxib						
Silverstein 2000 (CLASS) [63, 64]	400 mg bid	Diclofenac (75 mg bid) Ibuprofen (800 mg tid)	8059	60	OA & RA	4, 13, 26 wks (1 year-FDA)
Goldstein 2000 [69] (Combined analysis study)	25 mg bid to 400 mg bid 7 dose steps	Naproxen (500 mg bid) Diclofenac (75 mg bid) Ibuprofen (800 mg tid)	11008	59	OA & RA	2–24 wks
Success-1 Abstract [70]	100 mg bid 200 mg bid	Naproxen (500 mg bid)	13274	62	OA	6, 12 wks
Rofecoxib						
Bombardier 2000 (VIGOR) [56]	50 mg/day	Naproxen (500 mg bid)	8076	58	RA	6 wks, 4, 8, 12 mo
Langman 1999 [71] (Combined analysis study)	25 mg/day 50 mg/day	Ibuprofen (800 mg tid) Diclofenac (50 mg tid) Nabumetone (1500mg)	5435	63	OA	6wks, 4, 6, 12, 24 mo
Meloxicam						
Dequeker 1998 (SELECT) [50]	7.5 mg/day	Piroxicam	8656	61	OA	4 wks
Hawkey 1998 (MELISSA) [49]	7.5 mg/day	Diclofenac (100 mg/day)	9323	62	OA	4 wks
Yocum 2000 [27]	3.75–15 mg	Diclofenac (100 mg/day)	774	62	OA	12 wks
Goei 1997 [72]	15 mg	Diclofenac (100 mg/day)	258	71	OA	6 wks
Hosie 1997 [73]	15 mg	Piroxicam (20 mg)	455	67	OA	6 mo
Hosie 1996 [74]	7.5 mg	Diclofenac (100 mg/day)	335	64	OA	6 mo
Lindén 1996 [28]	15 mg	Piroxicam 20 mg	260	67	OA	6 wks
Wojtulewski 1996 [31]	7.5 mg	Naproxen 750 mg	379	57	RA	6 mo
Distel 1996 [75, 76] (combined analysis study)	7.5–15 mg	Piroxicam 20 mg/day Naproxen 1000 mg/day Diclofenac 100 mg/day	6129	59	OA & RA	Up to 1 yr

Table 2. Upper gastrointestinal perforations, ulcers and bleeds (PUBs) with COX-2 selective NSAIDs relative to traditional NSAIDs

COX – 2 inhibitor	Number of studies	Number of patients	RR	95% CI	Risk reduction
Celecoxib-6 month	3	30,306	0.41	0.21–0.78	59%
Celecoxib-12 month	3	30,306	0.43	0.21–0.88	57%
Rofecoxib	2	12,997	0.47	0.36–0.63	53%
Meloxicam	8	20,279	0.48	0.26–0.88	52%
Total	13	63,582	0.47	0.38–0.57	53%

coagulation, since they lack the anti-platelet effects of at least some nonselective NSAIDs.

NICE reviewed a total of 53 randomised controlled trials involving 61,731 patients with osteo-and rheumatoid arthritis were reviewed (14 rofecoxib, 15 celecoxib, 3 rofecoxib *versus* celecoxib, 9 meloxicam and 12 etodolac). These trials included both placebo and NSAID comparators (naproxen, diclofenac, ibuprofen, piroxicam, tenoxicam, nabumetone, nimesulide and aspirin). Evidence was found that the COX-2 selective inhibitors are more effective than placebo and of equivalent efficacy to other NSAIDs in their ability to reduce pain (resting and active) and to improve physical and global function in both osteoarthritis and rheumatoid arthritis patients. There was also evidence to suggest that, although the COX-2 selective inhibitors increase the incidence of gastrointestinal adverse events compared to placebo in both sets of patients, the magnitude of this effect is less than that of standard NSAID therapy. Rofecoxib and celecoxib have been shown to be associated with a reduced incidence of gastric erosions on endoscopy compared to standard NSAIDs in patients with arthritis. However, the importance of this finding as a surrogate for the development of peptic ulceration or other major gastrointestinal adverse effects in patients treated with these drugs in routine clinical practice is uncertain. Finally, indirect comparison of the effects of meloxicam, rofecoxib and celecoxib relative to other NSAIDs based on available evidence, does not demonstrate a difference in either efficacy or adverse events between these drugs.

COX-2 inhibition: potential risks

In view of their better safety profile, selective COX-2 inhibitors may well be used more frequently in high-risk patients and subsequent toxicity may not reflect the data from controlled studies. COX-2 is up-regulated in *Helicobacter* gastritis and inflammatory bowel disease, and whether COX-2 inhibition can be harmful in these conditions will depend on the relative contribution of COX-2 to PG synthesis. COX-2 is induced at the rim of ulcers in humans and

COX-2 inhibitors retard ulcer healing in animal studies [78]. It is important, therefore, that these drugs are demonstrated to be safe in this particular subset of patients. Fortunately, there is an increasingly large database of selective COX-2 inhibitor use in unselected patients in whom a favourable gastrointestinal profile is seen.

Selective COX-2 inhibitors will not have the beneficial anti-platelet effects of conventional NSAIDs and this, in combination with the reported role of COX-2 in sustaining vascular protacyclin production, may well have an impact on the incidence of vascular disease [79].

Conclusion

In summary, in the absence of evidence of significant differences in anti-inflammatory efficacy between the COX-2 selective drugs and standard NSAIDs, the avoidance of serious adverse effects, particularly on the gastrointestinal tract, becomes the most relevant factor when considering their use. For the most part the evidence suggests that COX-2 selective agents are less likely to produce symptomatic gastrointestinal tract adverse events. However, there remains some concern regarding the potential cardiovascular risks associated with COX-2 selective agents and therefore caution is needed, as it is for standard NSAID therapy, when prescribing in patients with pre-existing cardiovascular disease.

Overall, the arrival of selective COX-2 inhibitors heralds one of the biggest changes in rheumatological practice and with current knowledge it will be difficult to justify prescribing higher-risk drugs. Prescribers should be aware that gastrointestinal problems are not the only adverse consequences of NSAID use, they should be vigilant for fluid retention and its consequences and should remember that ulcer complications in patients taking a drug may be caused by background problems rather than by the drugs themselves. Our enthusiasm needs to be tempered by the reality that these drugs are still relatively new, and that possible advantages and disadvantages await further evaluation.

References

1 Girgis L, Brooks P (1994) Nonsteroidal anti-inflammatory drugs. Differential use in older patients. *Drugs Aging* 4: 101–112
2 Vane JR (1971) Inhibition of prostaglandin synthesis as a mechanism of action for aspirin-like drugs. *Nature* 231: 232–235
3 Roth GJ, Stanford N, Majerus PW (1975) Acetylation of prostaglandin synthetase by aspirin. *Proc Natl Acad Sci USA* 72: 3073–3076
4 Henry D, Lim LL, Garcia-Rodriguez LA, Perez Gutthann S, Carson JL, Griffin M, Savage R, Logan R, Moride Y, Hawkey C et al (1996) Variability in risk of gastrointestinal complications with individual non-steroidal antiinflammatory drugs: results of a collaborative meta-analysis. *BMJ* 312: 1563–1566
5 Singh G, Ramey DR, Morfeld D, Fries JF (1994) Comparative toxicity of non-steroidal anti-inflammatory agents. *Pharmacol Ther* 62: 175–191

6 Langman MJ (1988) Ulcer complications and nonsteroidal anti-inflammatory drugs. *Am J Med* 84: 15–19
7 Garcia Rodriguez LA, Jick H (1994) Risk of upper gastrointestinal bleeding and perforation associated with individual non-steroidal anti-inflammatory drugs. *Lancet* 343: 769–772
8 MacDonald TM, Morant SV, Robinson GC, Shield MJ, McGilchrist MM, Murray FE, McDevitt DG (1997) Association of upper gastrointestinal toxicity of non-steroidal anti-inflammatory drugs with continued exposure: cohort study. *BMJ* 315: 1333–1337
9 Kujubu DA, Fletcher BS, Varnum BC, Lim RW, Herschman HR (1991) TIS10, a phorbol ester tumor promoter-inducible mRNA from Swiss 3 T3 cells, encodes a novel prostaglandin synthase/cyclooxygenase homologue. *J Biol Chem* 266: 12866–12872
10 O'Banion MK, Sadowski HB, Winn V, Young DA (1991) A serum- and glucocorticoid regulated 4-kilobase mRNA encodes a cyclooxygenase-related protein. *J Biol Chem* 266: 23261–23267
11 Simmons DL, Levy DB, Yannoni Y, Erikson RL (1989) Identification of an ester repressible v-src-inducible gene. *Proc Natl Acad Sci USA* 86: 1178–1182
12 Vane JR, Bakhle YS, Botting RM (1998) Cyclooxygenases 1 and 2. *Annu Rev Pharmacol Toxicol* 38: 97–120
13 Warner TD, Giuliano F, Vojnovic I, Bukasa A, Mitchell JA, Vane JR (1999) Nonsteroidal drug selectivities for cyclo-oxygenase-1 rather than cyclo-oxygenase-2 are associated with human gastrointestinal toxicity: a full *in vitro* analysis. *Proc Natl Aca Sci USA* 96: 7563–7568
14 Brooks P, Emery P, Evans J, Fenner H, Hawkey CJ, Patrono C, Smolen J, Breedveld F, Day R, Dougados M et al (1998) Interpreting the clinical significance of the differential inhibition of cyclooxygenase-1 and cyclooxygenase-2. *Br J Rheumatol* 27 (Suppl. 1): 118
15 Donnelly MT, Hawkey CJ (1997) Review article: COX-2 inhibitors – a new generation of safer NSAIDs? *Aliment Pharmacol Ther* 11: 227–236
16 Emery P (1996) Clinical implications of selective cyclooxygenase-2 inhibition. *Scand J Rheumatol* 102 (Suppl 1): 23–28
17 Hawkey CJ (1999) COX-2 inhibitors. *Lancet* 353: 307–314
18 Hawkey CJ, Lanas AI (2001) Doubt and certainty about nonsteroidal antiinflammatory drugs in the year 2000: a multidisciplinary expert statement. *Am J Med* 110:S79–S100
19 Richardson C, Emery P (1996) The clinical implications of inhibition of the inducible form of cyclo-oxygenase. *Drug Safety* 15: 249–260
20 Rostom A, Wells G, Tugwell P, Welch V, Dube C, McGowan J (2000) The prevention of chronic NSAID induced upper gastrointestinal toxicity: a Cochrane collaboration metaanalysis of randomized controlled trials. *J Rheumatology* 27: 2203–2214
21 Yeomans ND, Tulassay Z, Juhasz L, Racz I, Howard JM, van Rensburg CJ, Swannell AJ, Hawkey CJ (1998) A comparison of omeprazole with ranitidine for ulcers associated with nonsteroidal antiinflammatory drugs. Acid Suppression Trial: Ranitidine *versus* Omeprazole for NSAID-associated Ulcer Treatment (ASTRONAUT) Study Group. *N Engl J Med* 338: 719–726
22 Hawkey CJ, Karrasch JA, Szczepanski L, Walker DG, Barkun A, Swannell AJ, Yeomans ND (1998) Omeprazole compared with misoprostol for ulcers associated with nonsteroidal antiinflammatory drugs. Omeprazole *versus* Misoprostol for NSAID-induced Ulcer Management (OMNIUM) Study Group. *N Engl J Med* 338: 727–734
23 McCarthy DM (1991) Sucralfate. *N Engl J Med* 325: 1017–1025
24 Emery P (1999) COX-1, COX-2: So what? *Scand J Rheumatol* 28: 6–9
25 Tomasoni S, Noris M, Zappella S, Gotti E, Casiraghi F, Bonazzola S, Benigni A, Remuzzi G (1998) Upregulation of renal and systemic cyclooxygenase-2 in patients with active lupus nephritis. *J Am Soc Nephrol* 9: 1202–1212
26 Lund B, Distel M, Bluhmki E (1998) A double-blind, randomized, placebo-controlled study of efficacy and tolerance of meloxicam treatment in patients with osteoarthritis of the knee. *Scand J Rheumatol* 27: 32–37
27 Yocum D, Fleischmann R, Dalgin P, Caldwell J, Hall D, Roszko P (2000) Safety and efficacy of meloxicam in the treatment of osteoarthritis: a 12-week, double-blind, multiple-dose, placebo-controlled trial. *Arch Intern Med* 160: 2947–2954
28 Lindén B, Distel M, Bluhmki E (1996) A double-blind study to compare the efficacy and safety of meloxicam 15 mg with piroxicam 20 mg in patients with osteoarthritis of the hip. *Br J Rheumatol* 35 (Suppl 1): 35–38
29 Furst DE, Kolba KS, Fleischmann R, Silverfield J, Greenwald M, Roth S, Hall DB, Roszko PJ (2002) Dose response and safety study of meloxicam up to 22.5 mg daily in rheumatoid arthritis:

a 12 week multicenter, double blind, dose response study *versus* placebo and diclofenac. *J Rheumatol* 29: 436–446
30 Lemmel EM, Bolten W, Burgos-Vargas R, Platt P, Nissilä M, Sahlberg D, Bjorneboe O, Baumgartner H, Valat JP, Franchimont P et al (1997) Efficacy and safety of meloxicam in patients with rheumatoid arthritis. *J Rheumatol* 24: 282–290
31 Wojtulewski JA, Schattenkirchner M, Barceló P, Le Loët X, Bevis PJR, Bluhmki E, Distel M (1996) A six-month double-blind trial to compare the efficacy and safety of meloxicam 7.5 mg daily and naproxen 750 mg daily in patients with rheumatoid arthritis. *Br J Rheumatol* 35 (Suppl 1): 22–28
32 Dougados M, Gueguen A, Nakache JP, Velicitat P, Veys EM, Zeidler H, Calin A (1999) Ankylosing spondylitis: what is the optimum duration of a clinical study? A one year *versus* a six weeks non-steroidal anti-inflammatory drug trial. *Rheumatology* 38: 235–244
33 Simon LS, Weaver AL, Graham DY, Kivitz AJ, Lipsky PE, Hubbard RC, Isakson PC, Verburg KM, Yu SS, Zhao WW et al (1999) Anti-inflammatory and upper gastrointestinal effects of celecoxib in rheumatoid arthritis: a randomized controlled trial. *JAMA* 282: 1921–1928
34 Bensen WG, Fiechtner JJ, McMillen JI, Zhao WW, Yu SS, Woods EM, Hubbard RC, Isakson PC, Verburg KM, Geis GS (1999) Treatment of osteoarthritis with celecoxib, a cyclooxygenase-2 inhibitor: a randomized controlled trial. *Mayo Clin Proc* 74: 1095–1105
35 McKenna F, Borenstein D, Wendt H, Wallemark C, Lefkowith JB, Geis GS (2001) Celecoxib *versus* diclofenac in the management of osteoarthritis of the knee. *Scand J Rheumatol* 30: 11–18
36 Williams GW, Hubbard RC, Yu SS, Zhao W, Geis GS (2001) Comparison of once-daily and twice-daily administration of celecoxib for the treatment of osteoarthritis of the knee. *Clin Ther* 23: 213–227
37 Zhao S, McMillen JI, Markenson JA, Dedhiya SD, Zhao WW, Osterhaus JT, Yu SS (1999) Evaluation of the functional status aspects of health-related quality of life of patients with osteoarthritis treated with celecoxib *Pharmacotherapy* 19: 1269–1278
38 Acevedo E, Castañeda O, Ugaz M, Beaulieu AD, Pons-Estel B, Caeiro F, Casas N, Garza-Elizondo M, Irazoque F, Hinojosa W et al (2001) Tolerability profiles of rofecoxib (Vioxx) and Arthrotec. A comparison of six weeks treatment in patients with osteoarthritis. *Scand J Rheumatol* 30: 19–24
39 Cannon GW, Caldwell JR, Holt P, McLean B, Seidenberg B, Bolognese J, Ehrich E, Mukhopadhyay S, Daniels B (2000) Rofecoxib, a specific inhibitor of cyclooxygenase 2, with clinical efficacy comparable with that of diclofenac sodium: results of a one-year, randomized, clinical trial in patients with osteoarthritis of the knee and hip. *Arthritis Rheum* 43: 978–987
40 Day R, Morrison B, Luza A, Castaneda O, Strusberg A, Nahir M, Helgetveit KB, Kress B, Daniels B, Bolognese J et al (2000) A randomized trial of the efficacy and tolerability of the COX-2 inhibitor rofecoxib vs ibuprofen in patients with osteoarthritis. *Arch Intern Med* 160: 1781–1787
41 Ehrich EW, Schnitzer TJ, McIlwain H, Levy R, Wolfe F, Weisman M, Zeng Q, Morrison B, Bolognese J, Seidenberg B et al (1999) Effect of specific COX-2 inhibition in osteoarthritis of the knee: a 6 week double blind, placebo controlled pilot study of rofecoxib. *J Rheumatol* 26: 2438–2447
42 Geba GP, Weaver AL, Polis AB, Dixon ME, Schnitzer TJ. Vioxx, Acetaminophen, Celecoxib Trial (VACT) Group (2002) Efficacy of rofecoxib, celecoxib, and acetaminophen in osteoarthritis of the knee: a randomized trial. *JAMA* 287: 64–71
43 Saag K, van der HD, Fisher C, Samara A, DeTora L, Bolognese J, Sperling R, Daniels B (2000) Rofecoxib, a new cyclooxygenase 2 inhibitor, shows sustained efficacy, comparable with other nonsteroidal anti-inflammatory drugs: a 6-week and a 1-year trial in patients with osteoarthritis. *Arch Fam Med* 9: 1124–1134
44 Truitt KE, Sperling RS, Ettinger WH, Jr, Greenwald M, DeTora L, Zeng Q, Bolognese J, Ehrich E (2001) A multicenter, randomized, controlled trial to evaluate the safety profile, tolerability, and efficacy of rofecoxib in advanced elderly patients with osteoarthritis. *Aging Clin Experiment Res* 13: 112–121
45 Li Q (1999) Statistical Review – NDA 21-042 – Rofecoxib. NDA 21-042. *http://www.fda.gov/cder/foi/nda/99/021042_52_Vioxx.htm*. Accessed Dec. 10, 2003
46 Schnitzer TJ, Truitt K, Fleischmann R, Dalgin P, Block J, Zeng Q, Bolognese J, Seidenberg B, Ehrich EW (1999) The safety profile, tolerability, and effective dose range of rofecoxib in the treatment of rheumatoid arthritis. *Clin Ther* 21: 1688–1702
47 Catella-Lawson F, McAdam B, Morrison BW, Kapoor S, Kujubu D, Antes L, Lasseter KC, Quan

H, Gertz BJ, FitzGerald GA (1999) Effects of specific inhibition of cyclooxygenase-2 on sodium balance, hemodynamics, and vasoactive eicosanoids. *J Pharmacol Exp Ther* 289: 735–741
48 Degner F, Richardson B (2001) Review of gastrointestinal tolerability and safety of meloxicam. *Inflammopharmacology* 9: 71–80
49 Hawkey C, Kahan A, Steinbruck K, Alegre C, Baumelou E, Begaud B, Dequeker J, Isomaki H, Littlejohn G, Mau J et al (1998) Gastrointestinal tolerability of meloxicam compared to diclofenac in osteoarthritis patients. Meloxicam Large-Scale International Study Safety Assessment (MELISSA). *Br J Rheumatol* 37: 937–945
50 Dequeker J, Hawkey C, Kahan A, Steinbruck K, Alegre C, Baumelou E, Begaud B, Isomaki H, Littlejohn G, Mau J (1998) Improvement in gastrointestinal tolerability of the selective cyclooxygenase (COX)-2 inhibitor, meloxicam, compared with piroxicam: results of the Safety and Efficacy Large- Scale Evaluation of COX-inhibiting Therapies (SELECT) trial in osteoarthritis. *Br J Rheumatol* 37: 946–951
51 Zeidler H, Kaltwasser JP, Leonard JP, Kohlmann T, Sigmund R, Degner F, Hettich M (2002) Prescription and tolerability of meloxicam in day-to-day practice: postmarketing observational cohort study of 13,307 patients in Germany. *J Clin Rheumatol* 8: 305–315
52 Layton D, Heeley E, Hughes K, Shakir SA (2003) Comparison of the incidence rates of selected gastrointestinal events reported for patients prescribed rofecoxib and meloxicam in general practice in England using prescription-event monitoring data. *Rheumatology* 42: 622–631
53 Rinder HM, Tracey JB, Souhrada M, Wang C, Gagnier RP, Wood CC (2002) Effects of meloxicam on platelet function in healthy adults: a randomized, double-blind, placebo-controlled trial. *J Clin Pharmacol* 42: 881–886
54 Hawkey C, Laine L, Simon T, Beaulieu A, Maldonado-Cocco J, Acevedo E, Shahane A, Quan H, Bolognese J, Mortensen E (2000) Comparison of the effect of rofecoxib (a cyclooxygenase 2 inhibitor), ibuprofen, and placebo on the gastroduodenal mucosa of patients with osteoarthritis: a randomized, double-blind, placebo-controlled trial. The Rofecoxib Osteoarthritis Endoscopy Multinational Study Group. *Arthritis Rheum* 43: 370–377
55 Laine L, Harper S, Simon T, Bath R, Johanson J, Schwartz H, Stern S, Quan H, Bolognese J (1999) A randomized trial comparing the effect of rofecoxib, a cyclooxygenase 2-specific inhibitor, with that of ibuprofen on the gastroduodenal mucosa of patients with osteoarthritis. *Gastroenterology* 117: 776–783
56 Bombardier C, Laine L, Reicin A, Shapiro D, Burgos-Vargas R, Davis B, Day R, Ferraz MB, Hawkey CJ, Hochberg MC et al VIGOR Study Group (2000) Comparison of upper gastrointestinal toxicity of rofecoxib and naproxen in patients with rheumatoid arthritis. *N Engl J Med* 343: 1520–1528
57 Division of Gastrointestinal and Coagulation Drug Products (1998) Medical Officer's Consult Review. *http://www.fda.gov/cder/foi/adcomm/98/adac_12198-consult.pdf.* (Accessed Dec. 10, 2003)
58 Emery P, Zeidler H, Kvien TK, Guslandi M, Naudin R, Stead H, Verburg KM, Isakson PC, Hubbard RC, Geis GS (1999) Celecoxib *versus* diclofenac in long-term management of rheumatoid arthritis: randomised double-blind comparison. *Lancet* 354: 2106–2111
59 Simon LS, Weaver AL, Graham DY, Kivitz AJ, Lipsky PE, Hubbard RC, Isakson PC, Verburg KM, Yu SS, Zhao WW, Geis GS (1999) Anti-inflammatory and upper gastrointestinal effects of celecoxib in rheumatoid arthritis: a randomized controlled trial. *JAMA* 282: 1921–1928
60 Goldstein JL, Correa P, Zhao WW, Burr AM, Hubbard RC, Verburg KM, Geis GS (2001) Reduced incidence of gastroduodenal ulcers with celecoxib, a novel cyclooxygenase-2 inhibitor, compared to naproxen in patients with arthritis. *Am J Gastroenterol* 96: 1019–1027
61 FDA-Study 21 (1998) *http://www.fda.gov/cder/foi/adcomm/98/adac_12198-consult.pdf.* Study 21 (Accessed Dec. 10, 2003)
62 FDA-Study 71 (1998) *http://www.fda.gov/cder/foi/adcomm/98/adac_12198-consult.pdf.* Study 71 (Accessed Dec. 10, 2003)
63 CLASS-12 Month Data (2000) *http://www.fda.gov/ohrms/dockets/ac/01/briefing/367761_05_gi.doc.* (Accessed Dec. 10, 2003)
64 Silverstein FE, Faich G, Goldstein JL, Simon LS, Pincus T, Whelton A, Makuch R, Eisen G, Agrawal NM, Stenson WF et al (2000) Gastrointestinal toxicity with celecoxib vs nonsteroidal anti-inflammatory drugs for osteoarthritis and rheumatoid arthritis: the CLASS study: A randomised controlled trial. Celecoxib Long-term Arthritis Safety Study. *JAMA* 284: 1247–1255
65 Chan FK, Hung LC, Suen BY, Wu JC, Lee KC, Leung VK, Hui AJ, To KF, Leung WK, Wong VW

et al (2002) Celecoxib *versus* diclofenac and omeprazole in reducing the risk of recurrent ulcer bleeding in patients with arthritis. *N Engl J Med* 347: 2104–2110

66 Silverstein FE, Graham DY, Senior J, Davies HW, Struthers BJ, Bittman RM, Geis GS (1995) Misoprostol reduces serious gastrointestinal complications in patients with rheumatoid arthritis receiving nonsteroidal anti-inflammatory drugs. A randomized, doubleblind, placebo-controlled trial. *Ann Intern Med* 123: 241–249

67 Canadian Coordinating Office for Health Technology Assessment (2003) Gastroduodenal ulcers associated with the use of non-steroidal anti-inflammatory drugs: a systematic review of preventive pharmacological interventions. Final Draft January 2003. *http://www.ccohta.ca/publications/pdf/gastroprotective_fr_e.pdf.* (Accessed Dec. 10, 2003)

68 FDA – Study 62 (2000) *http://www.fda.gov/cder/foi/adcomm/98/adac_12198-consult.pdf.* Study 62 (Accessed Dec. 10, 2003)

69 Goldstein JL, Silverstein FE, Agrawal NM, Hubbard RC, Kaiser J, Maurath CJ, Verburg KM, Geis GS (2000) Reduced risk of upper gastrointestinal ulcer complications with celecoxib, a novel COX-2 inhibitor. *Am J Gastroenterol* 95: 1681–1690

70 Singh G, Goldstein J, Bensen W, Agrawal N, Eisen G, Fort J, Bello A, Boots S (2001) SUCCESS-1 in Osteoarthritis (OA) Trial: celecoxib significantly reduces the risk of serious upper GI complications compared to NSAIDs while providing similar efficacy in 13,274 randomized patients. Abstr Perspect 5 (1): *http://www.hopkins-arthritis.som.jhmi.edu/eular2001/oa-treatments-cox2.html* (accessed June 2003)

71 Langman MJ, Jensen DM, Watson DJ, Harper SE, Zhao PL, Quan H, Bolognese JA, Simon TJ (1999) Adverse upper gastrointestinal effects of rofecoxib compared with NSAIDs. *JAMA* 282: 1929–1933

72 Goei Thè HS, Lund B, Distel MR, Bluhmki E (1997) A double-blind, randomized trial to compare meloxicam 15 mg with diclofenac 100 mg in the treatment of osteoarthritis of the knee. *Osteoarthritis Cartilage* 5: 283–288

73 Hosie J, Distel M, Bluhmki E (1997) Efficacy and tolerability of meloxicam *versus* piroxicam in patients with osteoarthritis of the hip or knee. A six-month double-blind study. *Clin Drug Invest* 13: 175–184

74 Hosie J, Distel M, Bluhmki E (1996) Meloxicam in osteoarthritis: a 6-month, double-blind comparison with diclofenac sodium. *Br J Rheumatol* 35 (Suppl 1): 39–43

75 Distel M, Mueller C, Bluhmki E, Fries J (1996) Safety of meloxicam: a global analysis of clinical trials. Br J Rheumatol 35 (Suppl 1): 68–77

76 Distel M, Mueller C, Bluhmki E (1996) Global analysis of gastrointestinal safety of a new NSAID, meloxicam. Inflammopharmacology 4: 71–81

77 National Institute for Clinical Excellence (2001) Guidance on the use of cyclo-oxygenase (COX) II selective inhibitors, celecoxib, rofecoxib, meloxicam and etodolac for osteoarthritis and rheumatoid arthritis. London, UK. *http://www.nice.org.uk/article.asp?a=18030* (accessed June 2003)

78 Wallace JL, McKnight W, Reuter BK, Vergnolle N (2000) NSAID-induced gastric damage in rats: requirement for inhibition of both cyclooxygenase 1 and 2. *Gastroenterology* 119: 706–714

79 McAdam BF, Catella-Lawson F, Mardini IA, Kapoor S, Lawson JA, FitzGerald GA (1999) Systemic biosynthesis of prostacyclin (COX)-2: the human pharmacology of a selective inhibitor of COX-2. *Proc Natl Acad Sci USA* 96: 272–277

COX-2 Inhibitors
Edited by M. Pairet and J. van Ryn

Role of COX-2 in ulcers and ulcer healing

Joanne van Ryn and Michel Pairet

Departments of [1]Drug Discovery Support and [2]Pulmonary Research, Boehringer Ingelheim Pharma GmbH & Co. KG, Birkendorfer Str. 65, D-88397 Biberach, Germany

Introduction

Non steroidal anti-inflammatory drugs (NSAIDs) are among the most frequently prescribed drugs; in the US alone there are over 110 million prescriptions annually [1]. However, their use is limited by their gastrointestinal (GI) side-effects, which is an issue particularly in the elderly where the risk for GI complications is elevated and ~70% take an NSAID (including aspirin) at least once weekly [2–4]. The occurrence of complicated GI events due to NSAIDs in patients with arthritis is 1–1.5% yearly (i.e., only perforation, ulceration and bleeding) [1]. When considering the large number of patients consuming NSAIDs, this translates into a considerable health care problem. Other risk factors associated with NSAID-induced GI complications are history of previous ulcers, concomitant anticoagulation or corticosteroid use and high dose NSAIDs [3]. Use of low-dose aspirin approximately doubles the risk of bleeding in patients taking NSAIDs [3, 4].

It has been demonstrated clinically that COX-2 selective inhibitors result in a reduced number of GI complications as compared to nonselective inhibitors [5–9], presumably by sparing COX-1 inhibition, and thereby maintaining GI mucosal integrity. In addition, the lack of effect of these agents on platelet aggregation should result in less bleeding episodes. Thus, the advantages of COX-2 selective agents are that the anti-inflammatory activity is maintained, while the amount of COX-1 inhibition is reduced. However, with improved pharmacological tools, it is becoming evident that not only COX-1, but also COX-2 is important for gastric mucosal defense.

NSAID-induced gastric injury

Mucosal defense is a term that describes the mechanisms by which the gastric mucosa is protected from injury from noxious agents, either from endogenous secretion of acid, pepsin etc. or from ingested irritants such alcohol or from NSAIDs [10, 11]. A normal epithelial surface without mucosal layer and the epithelial margin of an ulcer with extensive damage are shown in Figure 1.

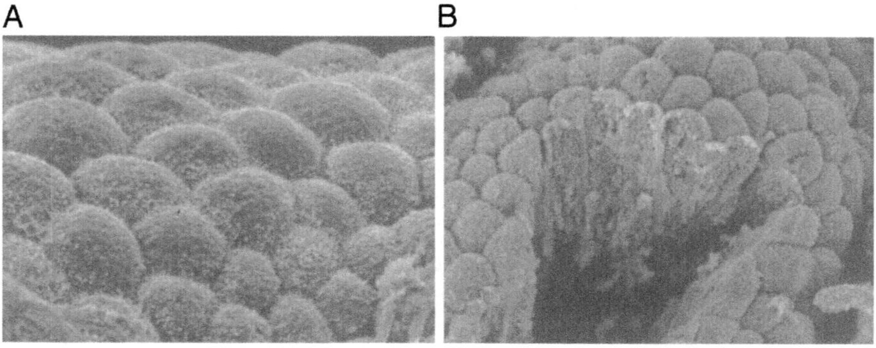

Figure 1. Electron micrograph of healthy human uninjured gastric epithelial cells (A) with the mucus layer removed to allow better visualization of the epithelium. The second panel (B) illustrates the margin of an ulceration. Photographs by Lennart Nielsson, Copyright Boehringer Ingelheim GmbH.

One of the earliest detected forms of damage by particularly aspirin and salicylic acid was the increased gastric permeability and topical injury shortly following administration [11–14]. In addition salicylic acid in concentrations that did not inhibit prostaglandin production could produce gastric injury [12]. The mechanism by which this occurs is still not fully elucidated, whether this involves direct toxic effects on the epithelium, disruption of bicarbonate barrier of the mucosa or disruption of cellular metabolism, such as oxidative phosphorylation remains to be determined [14]. The importance of this effect is also still somewhat controversial, since it is also known that enteric-coated aspirin administration results in similar rates of GI damage as compared to non-coated aspirin [15].

The mucus layer on the luminal side of the gastric epithelial lining provides an important barrier against NSAIDs and other noxious agents and is secreted by protective epithelial cells. Epithelial cells also secrete bicarbonate (Fig. 2), which neutralizes acid in the proximity of the epithelial cells [16] and can undergo rapid responses to mild injury where they proliferate to cover the injured area in an process called restitution [17]. Since the key discovery by Vane that aspirin and NSAIDs inhibit prostaglandins [18], it has become apparent over the years how many gastric protective mechanisms are dependent on prostaglandins and how these mechanisms can be compromised by NSAID intake [10–12, 14]. Prostaglandins are involved in both mucus and bicarbonate production and are both reduced with chronic NSAID administration [19–21].

Another important factor in mucosal integrity is an adequate blood supply. Neuronal input is important in rapid response mechanisms triggering enhanced blood flow that results in removal of toxic products or acid buffering, however, prostaglandins are also important for their vasodilatory properties [10, 11]. NSAID administration has been shown to reduce gastric mucosal blood flow

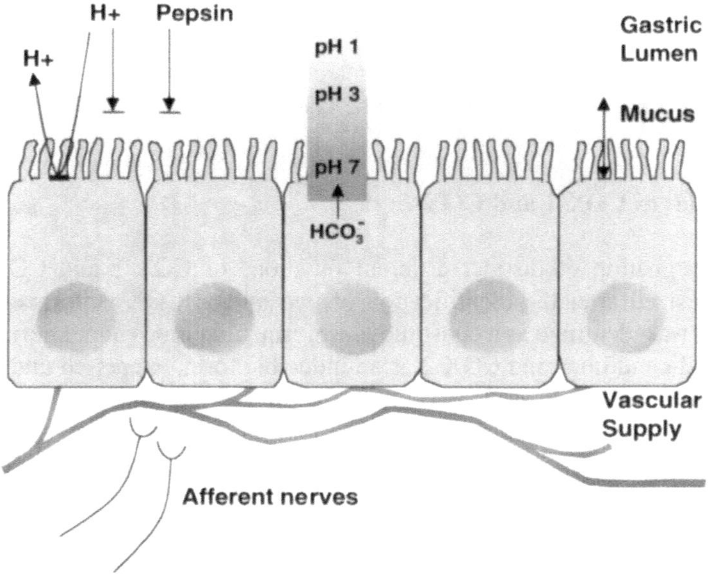

Figure 2. Schematic representation of the mechanisms involved in gastric mucosal defense.

and these effects could be reversed by the administration of prostaglandins [22–24]. This is further exacerbated by neutrophil adhesion to the endothelium in the microcirculation [5, 6, 10]. This results in release of various damaging agents, oxygen-radicals, proteases, cytokines such as tumor necrosis factor α (TNF_{α}) and leukotrienes, which all result in upregulation of adhesion molecules and further neutrophil adhesion [25–28]. This can also result in blockage of blood vessels, leading to a further reduction in blood flow and necrosis. Administration of agents that inhibit neutrophil adherence, prostaglandins, nitric oxide, anti-TNFα antibody or leukotriene inhibitors have all reduced the severity of NSAID-induced damage in animal models [29, 30]. However, unlike animal models, neutrophil and other lymphocyte involvement is not characteristic of GI injury due to NSAID use in humans, thus the relative importance of these findings remains to be determined [6].

It must be stressed that while prostaglandins play a major role in the maintenance of the GI integrity, this is not the only protective mechanism. Nitric oxide duplicates all the protective mechanisms afforded by prostaglandins; it has a role in mucus production [31], gastric mucosal blood flow [32] and inhibition of adhesive molecule expression and thus leukocyte adhesion [33–35].

Though this will not be discussed in detail here, the effect of *Helicobacter pylori* in patients taking NSAIDs is also been extensively studied recently [5, 6]. *H. pylori* infection is the single largest risk factor for ulcer formation in non-NSAID users [6, 36], however, the effect of *H. pylori* in patients taking NSAIDs is controversial and both positive and negative effects on ulcer formation have been reported [37, 38]. It appears that prior infection with H. pylori does not result in an elevated risk of ulcers in NSAID users; in contrast it may even afford some form of protection by elevating gastric adaptation against insult. This issue remains to be clarified.

The roles of COX-1 and COX-2

The designation of distinct different functions of COX-1 and COX-2 was hypothesized after the identification of two unique COX isoforms [39–41]. COX-1 was identified as a constitutive isoform, ubiquitous under normal physiological conditions and COX-2 as an inducible form, expressed under conditions of stimulation, particularly in inflammatory cells [42–44]. This partitioning led to the simple and elegant concept that selective inhibition of COX-2 would result in anti-inflammatory activity, but would not result in the GI and renal side-effects that afflict NSAIDs, because constitutive COX-1 would not be inhibited [45]. While the basic tenets of this concept hold true, selective COX-2 inhibitors do have fewer GI side-effects, they are not devoid of these side-effects.

It has become clear over time that this is not a black and white concept and there are roles for COX-2 in physiological processes just as it appears that COX-1 may also be involved in inflammatory processes. More recent studies identifying COX-1 and -2 in human gastric mucosa [46, 47] have demonstrated the presence of COX-2 in addition to COX-1 in both normal and inflamed gastric tissue, though it appears that basal gastric prostaglandin production is largely dependent on COX-1.

The importance of both COX-1 and COX-2 in maintaining GI mucosa integrity was illustrated by Wallace et al. [48], who used a highly selective COX-1 inhibitor, SC-560 and celecoxib, a selective COX-2 inhibitor, to determine the relative roles of COX-1 and COX-2 inhibition on ulcer formation in healthy rats. SC-560 administration alone resulted in almost complete inhibition of gastric prostaglandin production, consistent with previous reports [49], but did not induce ulceration. Celecoxib did not inhibit gastric prostaglandin production and also did not induce ulceration. These results indicate that COX-2 inhibition alone does not inhibit gastric prostaglandins or induce ulceration; however, surprisingly inhibition of only COX-1 also does not result in ulcer formation, despite inhibition of gastric prostaglandins. When both SC-560 and celecoxib were administered to rats, then ulceration occurred to the same extent as treatment with indomethacin. This clearly demonstrates an additive role for COX-2 inhibition in combination with COX-1 inhibition in ulcer formation. This may also explain why mice lacking the gene for COX-1 do not have spontaneous ulcer formation, despite reduced prostaglandin formation [50].

This phenomenon was explored further by Gretzer et al., who also showed that neither SC-560 nor rofecoxib, a COX-2 inhibitor, induced ulceration when administered to healthy rats. Rats were then exposed to an acid challenge that was shown to elevate COX-2 mRNA levels but not COX-1 mRNA, though this challenge alone was not enough to result in ulceration [51]. When SC-560 was then administered to the acid-challenged rats there was a dose-dependent induction of ulceration. Administration of rofecoxib alone did not induce ulceration, however, the combination of SC-560 and rofecoxib additively induced ulceration as compared to SC-560 alone.

Mechanistically, this may in part be explained by the effect prostaglandins have on gastric blood flow. An important line of defense in the gastric mucosa is the anti-inflammatory reaction, blood flow is increased via afferent nerve stimulation and leukocyte adhesion follows, enhancing gastric defense against toxins and bacteria [52, 53]. Wallace et al. showed that celecoxib or indomethacin increased leukocyte adherence in mesenteric venules in the rat, while SC-560 had no effect [48]. In addition, SC-560 or indomethacin reduced gastric blood flow by ~50%, but celecoxib had no effect. Thus vascularization of the gastric mucosa may be largely dependent on prostaglandins produced via COX-1, while inflammatory processes beginning with leukocyte adherence are regulated by prostaglandins produced by COX-2.

Thus in the normal stomach, both COX-1 and COX-2 inhibition are required to induce ulceration, however, in the presence of a potentially noxious agent, inhibition of COX-1 alone is sufficient to induce these pathological changes, while inhibition of COX-2 under these conditions is not. This illustrates the importance of COX-1 in gastric mucosal protection, but also highlights a role for COX-2. Even mild gastric injury results in the upregulation of COX-2 and the result is a dual resistance against injurious agents.

The importance of COX-1 is illustrated in another study where the ulcerogenic effect of meloxicam, a selective COX-2 inhibitor, was compared to the nonselective NSAID, piroxicam, in equivalent anti-inflammatory doses [54]. Drugs were given subcutaneously to rats 1 h prior to 0.45 N HCl, i.e., doses of acid that induced mild ulcerogenic injury (unpublished data). One hour later, the animals were killed, stomachs removed and analysed for the extent of ulceration. Ulcer area was recorded as a percent of the total glandular area. Meloxicam, at 1.5 and 3 mg/kg s.c., resulted in no additional ulcerogenic damage as compared to placebo animals (14 and 13% compared to 13%, respectively) who received the acid challenge alone. However, pretreatment with the same doses of piroxicam, which inhibits both COX-1 and COX-2 equally, resulted in significantly enhanced ulceration (17 and 22% with 1.5 and 3.0 mg/kg, respectively, Figure 3).

Microvascular rupture associated with ulceration results in blood flow into surrounding tissue with subsequent clot formation and a depletion of nutrients and oxygen in this tissue giving rise to ischemia and necrosis. The subsequent involvement of inflammatory cells is thought to be initiated by activated platelet and thrombin during clotting [55]. Infiltration of monocytes and neutrophils in response to these chemoattractants results in further release of growth factors and cytokines like TNFα, interleukin (IL)-1α and IL-1β [53, 55]. The cytokines

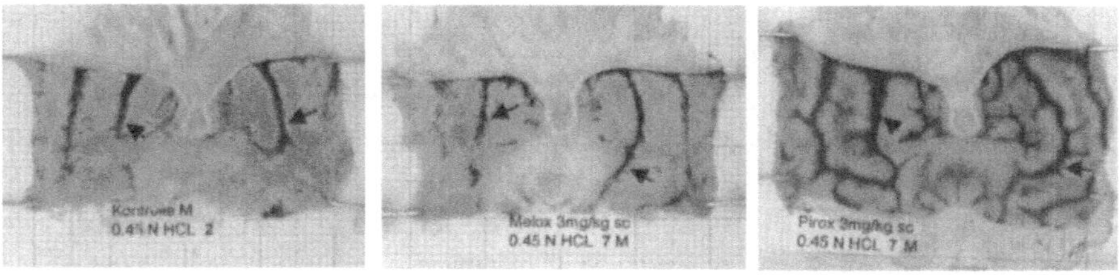

Placebo:	Meloxicam 3 mg/kg:	Piroxicam 3 mg/kg:
13% ulcer area	13% ulcer area	22% ulcer area

Figure 3. Extent of gastric injury in the stomach of rats treated with 0.45 N HCl and either vehicle, meloxicam or piroxicam at the highest dose of 3 mg/kg. The glandular region of the rat stomach is illustrated in the lower portion of each stomach. Areas of ulceration are marked with arrows.

activate endothelial cells, epithelial cells and fibroblasts. This process depends somewhat on the extent of the ulceration, but this inflammatory participation is dependent on COX-2 upregulation in these cells. COX-1 expression remains unchanged after an inflammatory insult, while COX-2 has been localized to monocytes, macrophages, fibroblasts and endothelial cells [46, 47].

There are redundant protective mechanisms in the stomach as described earlier, including nitric oxide and afferent innervation, in addition to COX. It has subsequently been shown that additive inhibition of more protective factors also results in gastric ulceration. Impairment of either COX-2, nitric oxide or afferent denervation in the presence of acid-challenged stomachs did not result in ulceration, however, inhibition of any two mechanisms in stomachs that were also acid-challenged resulted in significant ulceration [31, 56].

The degree of inflammatory burden may also play a role, in rats with adjuvant arthritis the levels of gastric PGE_2 were almost doubled as compared to normal levels [57]. Addition of rofecoxib reduced prostaglandins to levels seen in normal stomachs of non-arthritic rats, implicating COX-2 upregulation in the stomach of arthritic rats [57]. Both rofecoxib and celecoxib induced significant hemorrhagic gastric lesions in these arthritic rats, though the same doses had no injurious effects in normal non-arthritic rats. These data indicate that in the rat in this arthritic model, COX-2 is upregulated and protects against ulceration. When COX-2 is inhibited by selective COX-2 agents, ulceration occurs at doses that would not induce ulceration in the uninjured stomach. Thus COX-2 can play an important role in mucosal defense.

Ulcer healing

An ulcer by definition is a deep defect in the gastric wall resulting from necrosis and involves the entire mucosal thickness, penetrating to the submucosa or deeper [58]. The process of ulcer healing is complex, involving many cell types that are orchestrated by growth factors and cytokines, resulting in cell migration, proliferation, angiogenesis, matrix deposition and eventually scar tissue formation. Ulcers have a distinct ulcer margin composed mostly of epithelial cells. In addition, granulation tissue formed at the base of the ulcer is composed of connective tissue, containing fibroblasts, macrophages and proliferating endothelial cells undergoing angiogenesis [55, 59].

The epithelial cells at the margin of an ulcer dedifferentiate, proliferate and migrate along the surface of either a superficial wound or over the granulation tissue formed at the base a deeper injury, resulting in a continuous layer of epithelium. This is important in protecting the injury site from acid or other irritant exposure. Restitution or re-epithelialization can occur rapidly in response to epidermal growth factor (EGF), transforming growth factor (TGF)-α, basic fibroblast growth factor (bFGF), hepatocyte growth factor (HGF), insulin and platelet-derived growth factor (PDGF) [55, 59]. COX-2 expression is increased at the margins of ulcers and plays an important role in cell proliferation [46, 55, 59], however, this re-epithilization process seems to involve both COX-1 and COX-2. Giap et al. [60] demonstrated *in vitro* that re-epithelialization of a damaged area in a confluent gastric epithelial cell monolayer could be inhibited by indomethacin, but not the COX-2 selective inhibitor NS-398. Stimulation of the epithelial culture by bFGF enhanced repair as compared to basal conditions with no addition of bFGF. This bFGF-enhanced repair could be inhibited by indomethacin and NS-398, this data suggests that both a COX-1- and COX-2-dependent component is required for re-epithelialization.

In addition, epithelial cells from the base of the ulcer margin invade the tissue underlying the ulcer, where they form tubes and branch and form gastric glands. This highly organized event also occurs through growth factor coordination [55, 59]. After several days, adherent inflammatory cells have infiltrated and macrophages become dominant and release a variety of growth factors and cytokines upon activation, including TGFα, EGF, FGF-1, -2 and -4, PDGF, vascular endothelial growth factor (VEGF) and TGFβ [61, 62]. Granulation tissue is present at the base of the ulcer about 2–3 days after ulceration. It consists of proliferating connective tissue cells such as fibroblasts that eventually supply extracellular matrix for the lamina propria, as well as macrophages and endothelial cells [63]. COX-2 expression until day 10 is very high and has been localized to these inflammatory cells, to fibroblasts and to endothelial cells [63, 64]. Inhibition of ulcer healing results in enhanced thickness of the granulation tissue base and resulting scar [63]. COX-1 expression in the area immediately around the ulcer margin was no longer detectable as compared to normal gastric tissue [63]. Inhibition of COX-2 by selective inhibitors or by nonselective inhibitors delayed ulcer healing in experimental models [61–64]. The additional COX-1 inhibitory effect by nonselective inhibitors appears to play no further role. Thus this repair mechanism appears to be dependent on COX-2.

Another important component of wound healing is angiogenesis. Endothelial cell migration, proliferation and microvascular structure formation are triggered by extracellular matrix components (collagens) and growth factors [62, 65]. Two key growth factors are bFGF and VEGF, both of which have been shown to accelerate ulcer healing when administered experimentally [62, 65]. VEGF binds to specific VEGF receptors on endothelial cells, and initiates phosphorylation of numerous cytosolic proteins involved in signal transduction and triggers endothelial cell proliferation [66, 67]. VEGF production is stimulated by a number of growth factors, PDGF, TGF-β1, bFGF, as well as cytokines, nitric oxide, prostaglandins of the E series and importantly, hypoxia. It has also been demonstrated *in vitro* that NSAIDs may interfere with angiogenesis by inhibiting intracellular signaling, suggesting another possible mechanism for delaying ulcer healing [68]. Both NS-398 and indomethacin inhibited *in vitro* angiogenesis in rat aortic cells and human dermal microvascular endothelial cells [68]. This indicates that COX-2 produced prostaglandins are required for angiogenesis and angiogenesis can partially be inhibited by COX-2 inhibition. However, since addition of exogenous prostaglandins did not fully reverse this effect, a further component of NSAID-induced inhibition of angiogenesis may be unrelated to COX inhibition. Higher concentrations of both NS-398 and indomethacin could inhibit endothelial ERK-2, a mitogen-activated protein kinase involved in angiogenesis [68].

Another novel mechanism involving NSAIDs and angiogenesis is the influence of platelet activation on this process. Platelets are involved in wound healing around the ulcer margin and contain a number of growth factors and chemotactic agents which are released with their activation. It has been demonstrated that platelets can modulate angiogenesis by release of angiogenic promoting growth factors such as VEGF [69]. Platelets also contain endostatin, a potent inhibitor of angiogenesis, which is also released when they are activated [69, 70]. This balance of VEGF to endostatin was altered to favor endostatin in rats with delayed ulcer healing after administration of either celecoxib, a selective COX-2 inhibitor or flurbiprofen, a non selective inhibitor. When serum from these rats was added to human umbilical vein cell cultures, cell proliferation was reduced and apoptosis increased. Treatment of rats with a nitric oxide-releasing COX inhibitor did not delay ulcer healing. While increased endostatin levels in the serum of these animals were elevated to a similar extent as with the other drugs, it also increased VEGF levels, thus the ratio of endostatin to VEGF was similar to control levels [70].

Again it must be emphasized that COX is not the only factor involved in promoting ulcer repair and redundant gastric mechanisms such as nitric oxide also play a role in ulcer repair [71, 72]. In addition, the most effective way to negate the damaging effects of NSAIDs on ulcer healing is the addition of a proton pump inhibitor, reducing acid secretion by the gastric mucosa.

Conclusions

It has become apparent in the last few years that COX-2 also plays a role in GI mucosal integrity acting as reinforcement or back up for COX-1 mediated processes and that the original paradigm of "good COX-1" and "bad COX-2" is over simplistic. While it is clear from large, well designed clinical trials that COX-2 inhibitors do result in reduced GI side-effects at least in normal risk arthritic patients, the role of COX-2 inhibitors in patients with existing "silent" ulcers or in patients with an elevated risk of ulceration still needs to be determined. These probably reflect situations of GI inflammation, where COX-2 upregulation is also involved in mucosal integrity, and selective COX-2 inhibition may or may not be of added advantage in these patients.

References

1 Laine L (2003) Gastrointestinal effects of NSAIDs and coxibs. *J Pain Symptom Manage* 25S: S32–S40
2 Talley NJ, Evans JM, Fleming KC, Harmsen WS, Zinsmeister AR, Melton LJ 3rd (1995) Nonsteroidal antiinflammatory drugs and dyspepsia in the elderly. *Dig Dis Sci* 40: 1345–1350
3 Fries JF, Williams CA, Block DA, Michel BA (1991) Non-steroidal anti-inflammatory drug-associated gastropathy: incidence and risk factor models. *Am J Med* 61: 213–222
4 Wolfe MM, Lichtenstein DR, Singh G (1999) Gastrointestinal toxicity of nonsteroidal anti-inflammatory drugs. *N Engl J Med* 340: 1888–1899
5 Hawkey CJ, Langman MJS (2003) Nonsteroidal anti-inflammatory drugs: Overall risks and management. Complementary roles for COX-2 inhibitors and proton pump inhibitors. *Gut* 52: 600–608
6 Chan FKL, Leung WK (2002) Peptic-ulcer disease. *Lancet* 360: 933–941
7 Dequeker J, Hawkey C, Kahan A, Steinbruck K, Alegre C, Baumelou E, Begaud B, Isomaki H, Littlejohn G, Mau J et al (1998) Improvement in gastrointestinal tolerability of the selective cyclooxygenase (COX)-2 inhibitor, meloxicam, compared with piroxicam: results of the Safety and Efficacy Large-scale Evaluation of COX-inhibiting Therapies (SELECT) trial in osteoarthritis. *Brit JRheumatol* 37: 946–951
8 Bombardier C, Laine L, Reicin A, Shapiro D, Burgos-Vargas R, Davis B, Day R, Ferraz MB, Hawkey CJ, Hochberg MC et al VIGOR Study Group (2000) Comparison of upper gastrointestinal toxicity of rofecoxib and naproxen in patients with rheumatoid arthritis. VIGOR Study Group. *N Engl J Med* 343: 1520–1528
9 Silverstein FE, Faich G, Goldstein JL, Simon LS, Pincus T, Whelton A, Makuch R, Eisen G, Agrawal NM, Stenson WF et al (2000) Gastrointestinal toxicity with celecoxib vs nonsteroidal anti-inflammatory drugs for osteoarthritis and rheumatoid arthritis: the CLASS study: A randomized controlled trial. Celecoxib Long-term Arthritis Safety Study. *JAMA* 284: 1247–1255
10 Wallace JL, Bell CJ (1996) Gastromucosal defense. *Curr Opin Gastroenterol* 12: 503–511
11 Halter F, Tarnawski AS, Schmassmann A, Peskar BM (2001) Cyclooxygenase 2- implications on maintenance of gastric mucosal integrity and ulcer healing: controversial issues and perspectives. *Gut* 49: 443–453
12 Hawkey CJ (2000) Nonsteroidal anti-inflammatory drug gastropathy Gastroenterology 119: 521–535
13 Davenport HW (1969) Gastric mucosal hemorrhage in dogs. Effects of acid, aspirin and alcolhol. *Gasteroenterology* 56: 439–449
14 Whittle BJR. COX-1 and COX-2 products in the gut: therapeutic impact of COX-2 inhibitors. *Gut* 47: 320–325
15 Kelly JP, Daufman DW, Jurgelon JM, Sheehan J, Koff RS, Shapiro S (1996) Risk of aspirin-associated major upper-gastrointestinal bleeding with enteric-coated or buffered product. *Lancet* 348: 1413–1416
16 Bickel M, Kauffman GL Jr, (1981) Gastric gel mucus thickness: effect of distention, 16, 16-dimethyl prostaglandin E_2 and carbenoxolone. *Gastroenterology* 80: 770–775
17 Silen W, Ito S (1985) Mechanisms for rapid re-epithelialization of the gastric mucsal surface. *Annu Rev Physiol* 47: 217–229

18 Vane JR (1971) Inhibition of prostaglandin synthesis as a mechanism of action for aspirin-like drugs. *Nature* 231: 232–235
19 Rainsford KD, Willis C (1982) Relationship of gastric mucosal damage induced inpigs by antiinflammatory drugs to their effects on prostaglandin production. *Dig Dis Sci* 27: 624–625
20 Whittle BJ (1981) Temporal relationship between cyclooxygenase inhibition, as measured by prostacyclin biosynthesis, and the gastrointestinal damage induced by indomethacin in the rat. *Gastroenterology* 80: 94–98
21 Uribe A, Johansson C, Rubio C (1987) Cell proliferation of the rat gastrointestinal mucosa after treatment with E2 prostaglandins and indomethacin. *Digestion* 36: 238–245
22 Kitahora T, Guth PH (1987) Effect of aspirin plus hydrochloric acid on the gastric mucosal microcirculation. *Gastroenterology* 93: 810–817
23 Gana TJ, Huhlewych R, Koo J (1987) Focal gastric mucosal blood flow in aspirin-induced ulceration. *Ann Surg* 205: 399–403
24 Rainsford KD (1983) Microvascular injury during gastric damage by anti-inflammatory drugs in pigs and rats. *Agents Actions* 13: 457–460
25 Andrews FJ, Malcontenti-Wilson C, O'Brien PE (1994) Effect of nonsteroidal anti-inflammatory drugs on LFA-1 and ICAM-1 expression in gastric mucosa. *Am J Physiol* 266: 657–664
26 Appleyard CB, McCafferty DM, Tigley AW, Swain MG, Wallace JL (1996) Tumor necrosis factor mediation of NSAID-induced gastric damage: role of leukocyte adherence. *Am J Physiol* 270: 42–48
27 Wallace JL, Reuter B, Cicala C, McKnight W, Grisham MB, Cirino G (1994) Novel nonsteroidal anti-inflammatory drug derivatives with markedly reduced ulcerogenic properties in the rat. *Gastroenterology* 107: 173–179
28 Vaananen PM, Keenan CM, Grisham MB, Wallace JL (1992) Pharmacological investigations of the role of leukotrienes in the pathogenesis of experimental NSAID gastropathy. *Inflammation* 16: 227–240
29 Ham EA, Soderman DD, Zanetti ME, Dougherty HW, McCauley E, Kuehl FA (1983) Inhibition by prostaglandins of leukotriene B_4 release from activated neutrophils. *Proc Natl Acad Sci USA* 80: 4349–4353
30 Gryglewski RJ, Szczeklik A, Wandzilak M (1987) The effect of six prostaglandins, prostacyclin and iloprost on generation of superoxide anions by human oplymorphonuclear leukocytes stimulated by zymosan or formyl-methionyl-leucyl-phenylalanine. *Biochem Pharmacol* 36: 4209–4212
31 Whittle BJR, Lopez-Belmonte J, Moncada S (1990) Regulation of gastric mucosal integrity by endogenous nitric oxide: interactions with prostanoids and sensory neuoropeptides in the rat. *Br J Pharmacol* 99: 607–611
32 Elliot SN, McKnight W, Cirino G, Wallace JL (1995) A nitric oxide-releasing nonsteroidal anti-inflammatory drug accelerates gastric ulcer healing in rats. *Gastroenterology* 109: 524–530
33 Kubes P, Suzuki M, Granger DN (1991) Nitric oxide: an endogenous modulator of leukocyte adhesion. *Proc Natl Acad Sci USA* 88: 4651–4655
34 Kosonen O, Kankaanranta H, Malo-Ranta U, Moilanen E (1999) Nitric oxide-releasing compounds inhibit neutrophil adhesion to endothelial cells. *Eur J Pharmacol* 382: 111–117
35 Kim YM, Talanian RV, Li J, Billiar TR (1998) Nitric oxide prevents IL-1beta and IFN-gamma-inducing factor (IL-18) release from macrophages by inhibiting caspase-1(IL-1beta-converting-enzyme). *J Immunol* 161: 4122–4128
36 Pawlik T, Konturek PC, Konturek JW, Konturek SJ, Brzozowski T, Czesnikiewicz M, Plonka M, Dielanski W, Areny H (2002) Impact of *Helicobacter pylori* and nonsteroidal anti-inflammatory drugs on gastric ulcerogenesis in experimental animals and in humans. *Eur J Pharmacol* 229: 1–15
37 Hawkey CJ, Tulassay Z, Szczepanski L, van Rensburg CJ, Filipowicz-Sosnowska A, Lanas A, Wason CM, Peacock RA, Gillon KRW (2002) Randomised controlled trial of *Helicobacter pylori* eradication in patients on non-steroidal anti-inflammatory drugs: HELP NSAIDs study. *Lancet* 352: 1016–1102
38 Chan FKL, To KF, Wu JCY, Yung MY, Leung WK, Kwok T, Hui Y, Chan HLY, Chan CSY, Hui E et al (2002) Eradication of *Helicobacter pylori* and risk of peptic ulcers in patients starting long-term treatment with non-steroidal anti-inflammatory drugs: a randomised trial. *Lancet* 359: 9–13
39 Xie W, Chipman JG, Robertson DL, Erikson RL, Simmons DL (1991) Expression of a mitogen-responsive gene encoding prostaglandin synthase is regulated by mRNA splicing. *Proc Natl Acad Sci USA* 88: 2692–2696

40 Kujubu DA, Fletcher BS, Varnum BC, Lim RW, Herschman HR (1991) TIS10, a phorbol ester tumor promotor-inducible mRNA from Swiss 3 T3 cells, encodes a novel prostaglandin synthase/cyclooxygenase homologue. J Biol Chem 266: 12866–12872
41 O'Banion MK, Sadowski HB, Winn V, Young DA (1991) A serum- and glucocorticoid-regulated 4-kilobase mRNA encodes a cyclooxygenase-related protein. *J Biol Chem* 266: 23261–23267
42 Fu J-Y, Masferrer JL, Seibert K, Raz A, Needlman P (1990) The induction and suppression of prostaglandin H2 synthase (cyclooxygenase) in human monocytes. *J Biol Chem* 265: 16737–16740
43 O'Neill GP, Ford-Hutchinson AW (1993) Expression of mRNA for cyclooxygenase-1 and cyclooxygenase-2 in human tissues. *FEBS Lett* 330: 156–160
44 Kargman S, Charleson S, Cartwright M, Frank J, Riendeau D, Mancini J, Evans J, O'Neill G (1996) Characterization of prostaglandin G/H synthase 1 and 2 in rat, dog, monkey and human gastrointestinal tracts. *Gastroenterology* 111: 445–454
45 Vane J (1994) Towards a better aspirin. *Nature* 367: 215–216
46 Jackson LM, Wu KC, Mahida YR, Jenkins D, Hawkey CJ (2000) Cyclooxygenase (COX) 1 and 2 in normal, inflamed, and ulcerated human gastric mucosa. *Gut* 47: 762–770
47 To KF, Chan FKL, Cheng ASL, Lee TL, Ng YP, Sung JJY (2001) Up-regulation of cyclooxygenase-1 and -2 in human gastric ulcer. *Aliment Pharmacol Ther* 15: 25–34
48 Wallace JL, McKnight W, Reuter BK, Vergnolle N (2000) NSAID-Induced gastric damage in rats: requirement for inhibition of both cyclooxygenase 1 and 2. *Gastroenterology* 119: 706–714
49 Smith CJ, Zhang Y, Koboldt CM, Muhammad J, Zweifel BS, Shaffer A, Talley JJ, Masferrer JL, Seibert K, Isakson PC (1998) Pharmacological analysis of cyclooxygenase-1 in inflammation. *Proc Natl Acad Sci USA* 95: 13313–13318
50 Langenbach R, Morham SG, Tiano HF, Loftin CD, Ghanayem BI, Chulada PC, Mahler JF, Lee CA, Goulding EH, Kluckman KD et al (1995) Prostaglandin synthase 1 gene disruption in mice reduces arachidonic acid-induced inflammation and indomethacin-induced gastric ulceration. *Cell* 83: 483–492
51 Gretzer B, Maricic N, Respondek M, Schuligoi R, Peskar BM (2001) Effects of specific inhibition of cyclo-oxygenase-1 and cyclo-oxygenase-2 in the rat stomach with normal mucosa and after acid challenge. *Brit J Pharmacol* 132: 1565–1573
52 Asako H, Kubes P, Wallace JL, Wolf RE, Granger DN (1992) Modulation of leukocyte adhesion in rat mesenteric venules by aspirin and salicylate. *Gastroenterology* 103: 146–152
53 Wallace JL, McKnight W, Miyasaka M, Tamatani T, Paulson J, Anderson DC, Granger DN, Kubes P (1993) Role of endothelial adhesion molecules in NSAID-induced gastric mucosal injury. *Am J Physiol* 256: G99–G998
54 Pairet M, van Ryn J, Schierok H, Mauz A, Trummlitz G, Engelhardt G (1998) Differential inhibition of cyclooxygenases-1 and -2 by meloxicam and its 4'-isomer. *Inflamm Res* 47: 270–276
55 Tanawski A (2000) Molecular mechanisms of ulcer healing. *Drug News Perspect* 13: 158–168
56 Maricic N, Ehrlich KH, Schuligoi R, Peskar BM (1999) Cyclooxygenase-2-derived prostaglandins interact with nitric oxide and afferent neurons in the maintenance of rat gastric mucosal integrity. Gastroenterology 116: A280–A281
57 Kato S, Ogawa Y, Kanatsu K, Okayama M, Watanabe T, Arakawa T, Takeuchi K (2002) Ulcerogenic influence of selective cyclooxygenase-2 inhibitors in the rat stomach with adjuvant-induced arthritis. *J Phamacol Exp Ther* 303: 503–509
58 McCarthy DM (2001) Prevention and treatment of gastrointestinal symptoms and complications due to NSAIDs. *Best Practice Res Clin Gastroenterol* 15: 755–773
59 Schmassman A (1998) Mechanisms of ulcer healing and effects of nonsteroidal anti-inflammatory drugs. *Am J Med* 104 (Suppl 3): 43S–51S
60 Giap AQ, Tarnawski A, Hoa NT, Akotia V, Ma TY (2002) NSAID inhibition of RGMI gastric monolayer wound re-epithelialization: comparison of selective COX-2 *versus* non-selective COX inhibitors. *Life Sciences* 70: 3029–3037
61 Shigeta J-I, Takahashi S, Okabe S (1998) Role of cyclooxygenase-2 in healing of gastric ulcers in rats. *J Pharmacol Exp Ther* 286: 1383–1390
62 Szabo S, Folkman J, Vattay P, Morales RE, Pinkus GS, Kato K (1994) Accelerated healing of duodenal ulcers by oral administration of a mutein of basic fibroblast growth factor in rats. *Gastroenterology* 106: 1106–1111
63 Schmassmann A, Peskar BM, Stettler C, Netzer P, Stroff T, Flogerzi B, Halter F (1998) Effects of inhibition of prostaglandin endoperoxide synthase-2 in chronic gastro-intestinal ulcer models in

rats. *Brit J Pharmacol* 123: 795–804
64 Berenguer B, de la Lastra CA, Moreno FJ, Martin MJ (2002) Chronic gastric ulcer healing in rats subjected to selective and non-selective cyclooxygenase-2 inhibitors. *Eur J Pharmacol* 442: 125–135
65 Schmassmann A, Tarnawski A, Peskar BM, Varga L, Flogerzi B, Halter F (1995) Influence of acid and angiogenesis on kinetics of gastric ulcer healing in rats: interaction with indomethacin. *Am J Physiol* 268: G276–G285
66 Leung DW, Cachianes G, Kung WJ, Goeddel DV, Ferrera N (1989) Vascular endothelial growth factor is a secreted angiogenic mitogen. *Science* 246: 1306–1309
67 Szabo S, Folkman J, Vincze A, Sandor ZS, Gombos A (1997) Modulation of vascular factors by VEGF/VPF is sufficient for chronic ulcer healing and acute gastroprotection. *Gastroenterology* 122:A303
68 Jones MK, Wang H, Peskar BM, Levin E, Itani RM, Sarfeh IJ, Tarnawski AS (1999) Inhibition of angiogenesis by nonsteroidal anti-inflammatory drugs: Insight into mechanisms and implications for cancer growth and ulcer healing. *Nature Med* 5: 1418–1423
69 Ma L, Elliott SN, Cirino G, Buret A, Ignarro LJ, Wallace JL (2001) Platelets modulate gastric ulcer healing: Role of endostatin and vascular endothelial growth factor release. *Proc Natl Acad Sci USA* 98: 6470–6475
70 Ma L, del Soldato P, Wallace JL (2002) Divergent effects of new cyclooxygenase inhibitors on gastric ulcer healing: Shifting the angiogenic balance. *Proc Natl Acad Sci USA* 99: 13243–13247
71 Konturek SJ, Brzozowski T, Majka J, Pytoko-Polonczyk J, Stachura J (1993) Inhibition of nitric oxide synthase delays healing of chronic gastric ulcers. *Eur J Pharmacol* 239: 215–217
72 Elliot SN, McKnight W, Cirino G, Wallace JL (1995) A nitric oxide releasing non steroidal anti-inflammatory drug accelerates gastric ulcer healing in rats. *Gastroenterology* 109: 524–530

COX-2 Inhibitors
Edited by M. Pairet and J. van Ryn

COX-2 and the cardiovascular system

Rahul Nayak and Brendan F. McAdam

Division of Cardiovascular Medicine, Vanderbilt University Medical Center, Nashville TN 37232-8802, USA

Introduction

COX-1 derived-prostaglandins mediate physiological responses, such as platelet activation, gastrointestinal (GI) and vascular integrity. Conversely, COX-2 is the isoform responsible for the elaboration of inflammatory prostaglandins and its inhibition is anti-inflammatory [1, 2]. The development of agents selective for COX-2 were expected to maintain the anti-inflammatory efficacy of existing NSAIDs, but have a superior GI profile due to their lack of effect on cytoprotective prostaglandins [3, 4]. This has subsequently been largely verified with clinical studies in target populations and lends credence to the Vane hypothesis that these enzymes serve distinct biological functions [5]. However, studies in animals rendered deficient in both iso-enzymes have shown some surprising deviations from the original Vane paradigm and have suggested that the distinction between mutually exclusive "physiological" and "pathological" COX pathways is somewhat artificial [6–9]. Perhaps even more intriguing was the finding that COX-2 is hydraulically expressed in vascular endothelium *in vivo* and this isoform, and not constitutive COX-1, contributes to the formation of prostacyclin (PGI_2) [10]. It is now known that there are overlaps in expression patterns and functions of both iso-enzymes and this is most apparent in pathological conditions. For example, the expression of both isoforms is increased and co-localized in atherosclerotic lesions, and in myocardial injury and dysfunction from a variety of etiologies [11, 12]. While the role of both isoforms in platelet-vessel wall interactions is expanding, the functional contribution and interplay between these COX pathways in mediating pathobiological responses in atherosclerotic plaque is not well understood.

Prostaglandins in vascular homeostasis

The COX products, thromboxane A_2 (TxA_2) and PGI_2 play an important role in vascular homeostasis by their interaction with cognate cell surface G protein coupled receptors [13–15]. They are formed from the endoperoxide intermediate PGH_2 by the action of thromboxane A_2 and PGI_2 synthases [15] (Fig. 1).

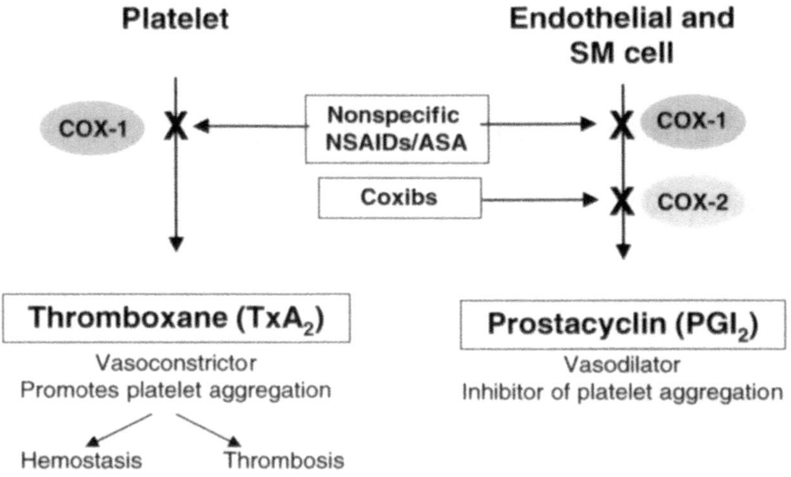

Figure 1. Vascular prostaglandins and their formation by the COX isoforms. The inhibition of endothelial- and smooth muscle (SM) cell-derived PGI_2 and platelet-derived TxA_2 by aspirin, NSAIDs and the COX-2 selective drugs is indicated.

TxA_2 is formed primarily through the activity of platelet COX-1 and is a potent platelet agonist and vasoconstrictor [16]. TxA_2 not only promotes platelet adhesion and aggregation, but acts as an amplifying signal for other agonists, thus playing a central role as a mediator of acute vascular occlusive syndromes [16, 17]. The pro-thrombotic effects of TxA_2 are in part counterbalanced by PGI_2 which has opposite biological effects on platelet function and vascular tone [18]. The interplay between these eicosanoids in the regulation of platelet-vessel wall interactions *in vivo* was confirmed in studies of genetically engineered mice lacking TxA_2 and PGI_2 receptors [19].

PGI_2 can be formed by the activity of both COX-1 and COX-2 in vascular tissue [20]. Although vascular cells under static conditions *in vitro* do not express COX-2, it has been reported that COX-2 can be upregulated in response to hydraulic flow forces of laminar, but not turbulent shear flow conditions [10]. This physiological role for COX-2 *in vivo* has been confirmed by pharmacological studies in healthy subjects, which showed maximal suppression of systemic formation of PGI_2 with selective COX-2 inhibitors, confirming that under normal conditions COX-2 is the main source of PGI_2 *in vivo* [21, 22] (Fig. 2a, 2b).

Increased TxA_2 and PGI_2 generation *in vivo* has been demonstrated in patients with syndromes of platelet activation, not only in severe atherosclerosis, but also in hyperlipidemia and smoking [23, 24]. The increased formation of TxA_2 is probably derived from platelet COX-1, as low-dose aspirin suppresses its formation [25]. Previously, the increased formation of PGI_2 observed in atherosclerosis and in acute coronary syndromes, was thought to reflect a compensatory homeostatic response to increased platelet-vessel wall

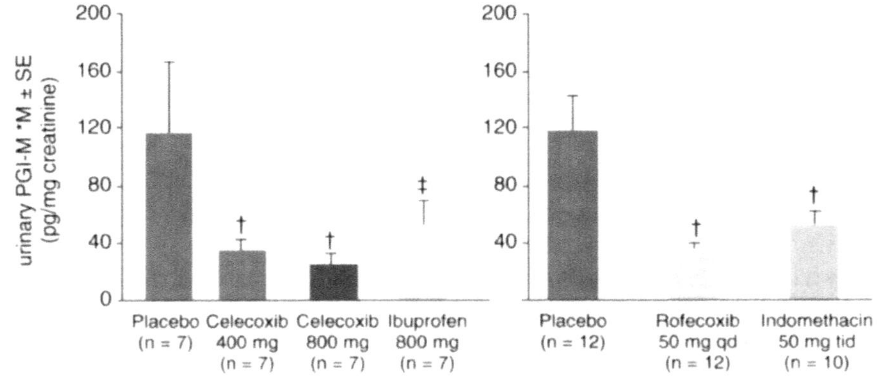

Figure 2 (a) The effect of single dose administration of celecoxib and ibuprofen in healthy volunteers on systemic biosynthesis of PGI_2 as measured by the quantification of its major urinary metabolite PGI-M (2,3 dinor-6-keto $PGF_{1\alpha}$) [21]. (Printed with permission, Copyright National Academy of Sciences, USA). (b) The effect of chronic administration of rofecoxib compared with indomethacin on PGI_2 formation in healthy elderly subjects [22]. (Adapted with permission, Copyright American Society for Pharmacology and Experimental Therapeutics).

interactions [26]. However, recent studies in transgenic mice and in animal models of thrombosis, confirm that this represents a protective response to limit further platelet-dependent thrombosis *in vivo* [19]. Recent reports of thrombosis in patients with lupus with celecoxib and the differential effect of rofecoxib and naproxen on the incidence of myocardial infarction in patients with rheumatoid arthritis has focused attention on the effect of both selective and non-selective agents on vascular eicosanoid balance [27, 28].

Pharmacology of COX inhibition in platelets and in the vasculature

Conventional NSAIDs inhibit COX activity by preventing access of arachidonic acid into the catalytic site of the enzyme and thereby reducing formation of PGH_2, the substrate for TxA_2 and PGI_2 synthases [4, 29]. Traditional NSAIDs do not discriminate between the COX isoforms and by binding reversibly to the active site of COX-1 and COX-2, they concomitantly suppress the formation of both platelet derived TxA_2 and vascular PGI_2 to a similar extent. This occurs in proportion to the plasma concentration and the dosing interval of these drugs [5, 29]. The result of the VIGOR trial showing a difference in the rate of cardiovascular events between naproxen and rofecoxib has prompted renewed interest in whether non-selective inhibitors, particularly naproxen, can exert sustained antithrombotic inhibitory effects that may account for such a difference [28]. There are no prospective controlled studies to specifically address the hypothesis that naproxen is cardioprotective. Unlike NSAIDs, which reversibly inhibit COX, the mode of aspirin is unique as it results in permanent inhibition of the enzyme [30]. Aspirin covalently inhibits COX by irreversibly acetylating a key serine amino acid near the active site of both enzymes (Ser529 in the COX-1 and Ser516 in the COX-2) [31]. In the case of the anucleate mature platelet, which only expresses COX-1 (and cannot synthesize new enzyme), the effect of aspirin on TxA_2 dependent platelet function is permanent for the lifetime of the circulating platelet [32]. The formation of TxA_2 recovers as a function of platelet turnover, requiring the synthesis of new platelets, which are regenerated at a daily rate of 10%. In this way a once daily regimen of aspirin < 100 mg has cumulative effects and completely inhibits TxA_2-dependent platelet function [32].

Thus, the antithrombotic effect of aspirin is primarily derived from its suppression of COX-1 derived TxA_2 dependent platelet aggregation and clinical studies support the marked efficacy of this agent in the secondary prevention of cardiovascular disease [33]. While aspirin is non selective for both isoenzymes, it is more potent in inhibiting COX-1 and low doses can selectively inhibit platelet COX-1 in the portal circulation with a relative sparing of vascular derived PGI_2 [32, 34]. COX-2 is less sensitive to the inhibitory effects of low-dose aspirin and this may account for the substantial residual (COX-2-dependent) PGI_2 biosynthesis *in vivo* at daily doses of aspirin in the range < 100 mg [32, 35]. Moreover, given the short plasma half-life of aspirin, nucleated vascular cells can rapidly re-synthesize the enzyme, thus overcoming the irreversible effect of aspirin on vascular COX [35]. Interestingly, despite development of different formulations of aspirin that result in increased biochemical selectivity for TxA_2 and sparing of PGI_2, it has not been clearly established that more profound suppression of PGI_2 formation by higher doses of aspirin attenuates the beneficial effects of this agent *in vivo* [35]. However, the ACE trial suggested that there might be such an effect, as lower doses of aspirin showed greater efficacy in stroke prevention than higher doses in patients with carotid artery disease [36].

The development of the selective COX-2 inhibitors held the promise of increased anti-inflammatory efficacy and a lower incidence of bleeding (and GI toxicity) related to inhibition of COX-1 [1, 5]. Consistent with their selective effect, clinical studies have confirmed that these agents at therapeutic plasma concentrations have no effect on COX-1 dependent platelet aggregation or on bleeding time [5, 22]. However, they do suppress the formation of systemic PGI_2 to the same extent as traditional NSAIDs in healthy subjects, suggesting that COX-2 is hydraulically induced and is the dominant source of PGI_2 *in vivo* [21, 22]. While this data reflects global vascular synthesis of PGI_2, it is unlikely that this is a uniform pattern of expression. Certain locations in the vasculature are subjected to more turbulent flow conditions where local COX-2 may not be upregulated. This may be relevant to sites such as vascular bifurcations that are predisposed to focal atherogenesis. However, other factors such as inflammatory stimuli are known to increase the expression of COX-2 at these local sites and net local formation of PGI_2 may be dependent on the balance between these and other factors. Although an inflammatory effect for PGI_2 has been reported, the alterations in PGI_2 likely reflect a protective response that limits the consequences of platelet activation [19, 37].

Since selective COX-2 inhibitors have no effect on platelet function, the biological consequences of an isolated, yet incomplete reduction in PGI_2 on platelet-vessel wall interactions *in vivo* has been the subject of much discussion. Mice rendered deficient in both copies of the PGI_2 receptor (IP) have a normal phenotype and do not exhibit spontaneous thrombosis, but do so only in response to a thrombotic stimulus [37]. Pharmacological reduction of systemic biosynthesis of PGI_2 by selective inhibitors is not comparable to mice that are homozygous for IP receptor deletion. It is not known if the heterozygous mouse (which may be equivalent to the 60% reduction in systemic biosynthesis that is observed with selective inhibitors) is protected from risk of thrombosis. However, these mice are protected from the development of vascular lesions when crossed with a mouse model of atherosclerosis [38]. Selective inhibition of PGI_2 in healthy subjects appears to have no clinical sequelae and indeed short-term treatment with celecoxib did not alter endothelial function in volunteers [22, 39]. Moreover, the clinical experience with these agents (at recommended doses with prolonged exposure) in patients particularly with osteoarthritis and otherwise healthy older populations with Alzheimer's disease, have confirmed their cardiovascular safety [5, 40].

Is selective inhibition of COX-2 prothrombotic in "at risk" subjects?

The more important and intensely controversial issue is whether these selective agents, by reducing PGI_2, have the potential to increase the likelihood of arterial thrombotic events in populations who may be at increased risk, or indeed do conventional NSAIDs have cardioprotective effects (Fig. 3). Many of the patients who are candidates for longer-term treatment with these agents are also at increased risk for or have established atherosclerosis (either documented or clinically silent). The biosynthesis of both TxA_2 and PGI_2 is increased not only in syndromes of altered platelet activation such as acute coronary syndromes, but also in patients with extensive atherosclerosis, in diabetics and in smokers [23, 24, 26, 41]. Suppression of PGI_2 without coincident inhibition of platelet COX-1-dependent TxA_2 formation may be undesirable in these clinical conditions. Most of the clinical studies on the longer-term effect of these agents have been performed in relatively healthy subjects and the available data on the cardiovascular effects of both selective and traditional NSAIDs is derived from recent clinical studies with the two new coxibs [5, 28].

Studies in experimental models suggest that there may be thrombotic risk. Transgenic mice that lack PGI_2 receptors have an increased risk of thrombosis when subjected to appropriate stimuli [37]. Importantly, PGI_2 modulated the prothrombotic actions of TxA_2 in these mutant mice, which confirmed a critical vaso-protective role for this autacoid under thrombotic conditions [19]. Hennan et al. showed in a canine model of coronary thrombosis that celecox-

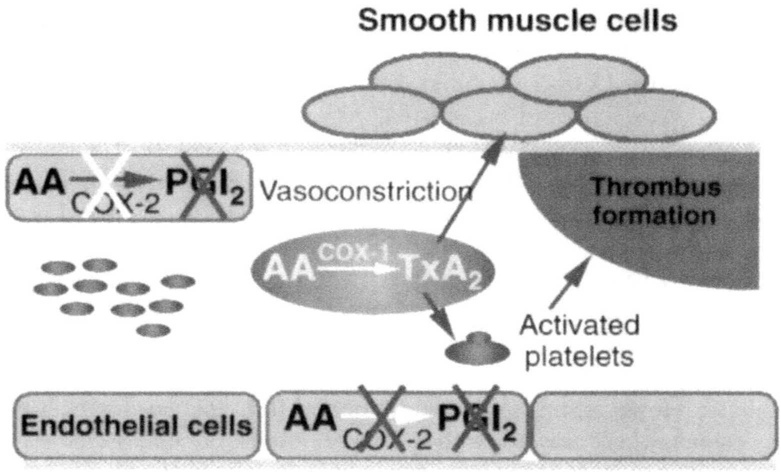

Figure 3. Vascular expression of COX and potential pro-thrombotic effects of selective COX-2 inhibitors.

ib attenuated the beneficial effects of aspirin on time to vessel occlusion and prolongation of bleeding time when it was administered as a high dose followed by a recovery period (permitting regeneration of vascular COX) [42]. Interestingly, administration of celecoxib alone had a similar effect to high-dose aspirin and did not alter time to coronary vessel occlusion despite inhibiting vasodilator responses suggesting that inhibition of COX-2 derived PGI_2 alone did not accelerate the process of thrombosis in this model of vessel injury. These data also demonstrate the complex temporal interplay between platelet-vessel wall interactions in the setting of sequential and selective inhibition of COX activity.

Possible thrombotic hazard with coxibs was reported in four patients with lupus anticoagulant when vascular thrombotic events occurred after the initiation of therapy with celecoxib [27]. The potential importance of this finding was further suggested by the results of the VIGOR trial, which examined the long-term effects of rofecoxib and naproxen in patients with rheumatoid arthritis [28]. This study was designed to explore efficacy and demonstrate superior GI safety of the coxib over naproxen in patients with rheumatoid arthritis. This multi-center study enrolled over 8,000 mostly women (mean age 58 years) with median follow up of 9 months. Aspirin use was not permitted to allow specific evaluation of the effects of this agent on GI toxicity. There was no difference in overall mortality, but the incidence of vascular thrombotic events diverged significantly between the treatment groups, 1.7% in the rofecoxib treatment group compared with 0.7% in the naproxen arm. This represented a five-fold difference in rates of myocardial infarction and this effect appeared within 6–8 weeks after the start of the study (Fig. 4) [28, 43]. Patients with

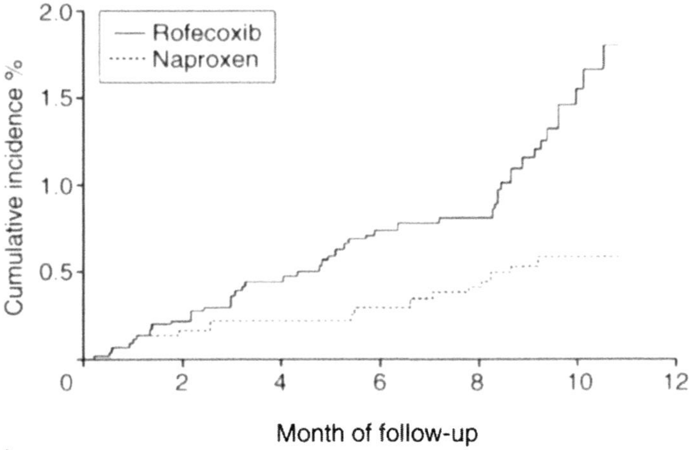

No. at risk							
Rofecoxib	4047	3643	3405	3177	2806	1067	531
Naproxen	4029	3647	3395	3172	2798	1073	514

Figure 4. Kaplan-Meier analysis of the time to cardiovascular events with rofecoxib and naproxen in the VIGOR trial [51].

rheumatoid arthritis have a well-documented increased risk of cardiovascular events, but it is unknown if this is related to increased platelet activation and/or to other inflammatory mechanisms that may impact on vascular dysfunction [44, 45]. Despite exclusion of patients with prior "cardiovascular disease" in this study, a review of the patient demographics showed that 40% had a "history of some form of vascular pathology" and 4% (?) met FDA criteria for the use of aspirin for secondary prevention [28, 43]. *Post hoc* analysis showed that this subset of subjects accounted for the excess of myocardial infarctions (>30%) and after re-analysis excluding this group, there was no significant difference in the event rates between naproxen and rofecoxib.

These adverse vascular effects were not observed in the other large arthritis study (CLASS) which was designed to examine the effect of celecoxib in 8,059 patients (mean age 60 years) with osteoarthritis and rheumatoid arthritis compared with ibuprofen and diclofenac, with follow-up over 6 months. Again most of the patients were females (average age 60 years). Aspirin use was permitted for those with an indication for such treatment (22%) [46]. There were no differences in cardiovascular outcomes at 6 months between the treatment groups and the overall event rates were low (<1% after 6 months of follow-up) [46, 47].

Although these studies were not designed or powered to demonstrate adverse thrombotic events (a pre-specified end point) between these agents and comparator drugs, and while the absolute numbers were small, the differences in the rates of myocardial infarction were significant in VIGOR. However, there was no placebo arm to adjudicate background rates, and the doses used were higher than those currently recommended [28, 46].

Several mechanisms have been proposed to explain the differential rate of myocardial infarction in the VIGOR study between naproxen and rofecoxib and why such an effect was not observed in the CLASS trial with a similarly selective agent.

Naproxen effect or rofecoxib effect on vascular eicosanoids?

Assuming that the difference between naproxen and rofecoxib was not due to chance, two mechanisms have been proposed based on known effects of these inhibitors on vascular eicosanoids: either an anti-platelet effect of naproxen or a possible pro-thrombotic effect of rofecoxib or both.

Unlike aspirin, traditional NSAIDs inhibit both COX-1 derived TxA_2 and COX-2 derived PGI_2 to varying degrees for the duration of the dosing interval. Their effects on platelet-vessel wall function *in vivo* at anti-inflammatory dosing schedules are not well characterized. Knijff-Dutmer et al. reported that treatment with naproxen 500 mg bid for 2 weeks in patients with rheumatoid arthritis resulted in significant inhibition of platelet aggregation *ex vivo* when compared with nabumetone, but there was no effect on bleeding time [48]. The exact timing of the sampling for platelet aggregation studies in relation to the administration of naproxen was not reported. Van Hecken et al. showed that treatment with ibuprofen 800 mg tid and naproxen sodium 500 mg bid for 5 days resulted in >90% inhibition in the capacity of platelet COX-1 to generate TxA_2 and inhibited platelet aggregation for up to 8 h [49]. However, in both studies the comparative effects of these agents on urinary prostaglandins was not reported. Prolonged functional inhibition of TxA_2 dependent platelet aggregation requires >95% sustained suppression of platelet capacity to generate TxA_2. It is not clear whether this reversible pharmacodynamic effect that has been reported with naproxen is similar to the irreversible effect of aspirin. It remains to be determined whether this effect is sufficient to translate into cardiovascular benefit and is currently the subject of significant debate [17].

There is some indirect evidence that other NSAIDs can have protective cardiovascular effects [50]. In the absence of prospective controlled data on possible cardio-protective effects of naproxen, several recent case control and observational studies have yielded either conflicting results or suggested modest benefits with the acknowledged limitations of such analyses [51–54]. However, a much larger cohort study in subjects with an increased incidence of coronary artery disease by Ray et al. demonstrated that there was no protective effect of naproxen and other traditional NSAIDs on the risk of cardiac events in 532,634 person years of follow-up [55]. However, the significant difference in myocardial infarction between groups (despite the small numbers of events) in VIGOR exceeds the modest and variable benefit reported in the case control studies and indeed the maximum benefit of aspirin in placebo-controlled trials, suggesting that other mechanisms may be operant [56].

The difference between the groups may also reflect the expected lack of effect of rofecoxib on platelet COX-1 and/or be linked with the second hypothesis that PGI_2 is important in restraining platelet activation and that its suppression would lead to the unopposed effects of TxA_2. To address this hypothesis, we examined the short-term effect of COX-2 inhibition in smoking, a well-documented paradigm of platelet activation [41]. The increased biosynthesis of TxA_2 in smoking, as measured by its urinary metabolite, 11-dehydro-TxA_2, was reduced to the level of non-smokers with rofecoxib without alteration in serum TxB_2. (McAdam BF, unpublished work) These data suggest that COX-2 and not COX-1 is responsible for the increased TxA_2 biosynthesis in smoking and this may reflect induction of COX-2 in inflammatory cells [57]. Interestingly, Patrono et al. has previously suggested that the aspirin insensitive increase in TxA_2 formation in unstable coronary disease may be derived from COX-2 in activated circulating monoyctes [58]. Another possible explanation for this finding relates to the expression of COX-2 in immature platelets as reported by Rocca et al., although increased platelet turnover has not been reported in smoking [59].

COX inhibition in the "inflamed" atheromatous plaque?

Despite the well-documented effect of selective inhibitors on the vascular formation of eicosanoids, there is very limited data on the effects of therapeutic doses of COX inhibitors on the prostaglandin formation in the vessel wall itself. It is possible that these agents, either selective or non selective, may have effects within the vascular wall depending on degree of tissue penetration. Consistent with the new perspective of inflammatory signaling in atherosclerosis, both COX enzymes are expressed in several cells within the atheromatous plaque and elevated plasma C reactive protein (CRP) levels have been reported to reflect arterial wall inflammation [11, 60, 61]. COX-2 derived PGE_2 in inflammatory cells modulates the expression of matrix metalloproteinases, which regulate structural content of the fibrous cap of atheroma and may play a role in the biology of plaque rupture [57, 62]. In this microenvironment, prostaglandins can signal not only through cell surface receptors, but are also able to regulate the transcription of inflammatory genes, such cytokines (and interestingly COX-2 itself) by binding to nuclear receptors, PPARs. Thus, by modulating upstream elements of several intracellular signaling pathways, prostaglandins may act as an alternative stimulus for cytokine release and subsequent increased formation of CRP [63, 64]. While NSAIDS have been shown to act as ligands for these nuclear receptors, it is unknown if the selective inhibitors can transduce biological effects by this pathway [65]. Moreover, it is unknown if naproxen could exert a cardioprotective effect as a result of its anti-inflammatory activity in the vessel wall. However, the effect of COX inhibition within the plaque is likely to be complex and may vary as a function of time and the degree of inflammation. It remains to be determined how COX derived prostanoids alter plaque-platelet and leukocyte interactions and how this might impact on the progression of this disease or on its acute manifestations.

Can an effect on hypertension lead to vascular events in VIGOR?

The increased incidence of hypertension and fluid retention observed with the 50 mg dose of rofecoxib has been raised as a possible explanation for the increased in myocardial infarction and stroke observed with rofecoxib in VIGOR [28]. More patients developed hypertension in the rofecoxib arm than in the naproxen arm (4.6 mmHg compared with 1.0 mmHg in mean systolic BP). While this may have long-term consequences on risk of vascular events, it is unlikely to explain the differences between these agents after 6–8 weeks of treatment [43]. There were no differences between celecoxib and the comparator NSAIDs in the CLASS study on this parameter [47].

Disparity in cardiovascular event rates between CLASS and VIGOR?

The apparent divergent results on rate of vascular thrombotic events with two selective inhibitors that produce similar suppressive effects on vascular PGI_2 without altering platelet function, is initially surprising. However, there are important differences between the studies that may in part account for the disparate findings. Perhaps, of most relevance were the study populations; in VIGOR, all the subjects had rheumatoid arthritis but only 28% of patients in CLASS had this diagnosis. Although overall thrombotic event rates were small and similar in both studies, patients with rheumatoid arthritis have an increased risk of adverse vascular events compared with those with osteoarthritis and the general population [44, 45]. Aspirin use was not permitted in the VIGOR study, whereas 22% of patients were on this agent in CLASS, however, subsequent analysis in those subjects not taking aspirin showed no increased event rates [28, 46]. Finally, different NSAIDs did not permit useful comparisons between these studies as differences in their vascular pharmacology have now been reported.

A recent widely reported meta-analysis comparing the cardiovascular event rates from four trials (including CLASS and VIGOR) with those from the placebo group of a meta-analysis of data from aspirin prevention trials, heightened the controversy and suggested that both celecoxib and rofecoxib were pro-thrombotic. However, this analysis had significant limitations in methodology and used statistical comparisons that were not valid between the studies with different populations [66]. Konstam et al. performed a meta-analysis of all trials with rofecoxib and did not observe increased cardiovascular events when compared with placebo, however, most of the subjects in these combined studies were at lower risk for events [40].

Other selective agents, such as meloxicam and nimesulide have been in use for several years without apparent adverse CV events based on post market surveillance. Though it must be noted there are significant limitations in accurate assessment of causal relationships of these clinically important adverse events, particularly since this potential effect was not appreciated until recently. These agents were also never formally studied in large prospective trials with long-term follow-up.

Not a selective COX-2 class effect but a single compound effect?

Are there differences between COX-2 selective agents to account for the differential effect on vascular outcomes between the CLASS and VIGOR trials? A trend for adverse vascular events was also observed in two recent smaller studies in the same population with rheumatoid arthritis with a more potent selective inhibitor, etoricoxib, when compared with placebo or naproxen over a short treatment period. The absolute event rate numbers were small (3 in >600 patients), but this occurred in a population that was already carefully screened for cardiovascular disease and aspirin use was permitted [67, 68]. This effect was not seen in trials in the same patient population when celecoxib (200–800 mg daily) was compared to naproxen over a 12-week period [69]. It is possible that there are differences in the pharmacokinetic or pharmacodynamic properties of both coxibs that account for a drug-specific effect rather than a class effect.

While all the coxibs are lipophilic and have a similar volume of distribution, rofecoxib and etoricoxib exhibit higher specificity for COX-2 and have a slower dissociation from the active site (i.e., almost irreversible binding) and have longer half-lives compared to celecoxib and valdecoxib [5, 70]. Whether these characteristics contribute to possible differences in clinical outcome remains to be determined as surprisingly little is known about the effects of these agents on tissue prostaglandin formation in inflamed synovial tissue (or in vascular tissue). Thus, for example, a coxib with a shorter half-life and/or with quicker dissociation from the COX active site might permit recovery of endothelial PGI_2 formation and thereby not predispose to continuous unopposed platelet COX-1 dependent TxA_2 formation. As there are no head to head studies at comparable doses in the same population to assess the impact of these effects on anti-inflammatory end points, renal and hypertensive effects and on potential vascular adverse effects, definitive conclusions cannot be drawn.

The initiation of atherothrombosis involves a process of platelet adhesion, activation and finally aggregation. Some of the selective agents have a modest inhibitory effect on the capacity of platelet COX-1 to generate TxA_2 without altering aggregation responses (due to the non-linear relationship between inhibition of aggregation *ex vivo* and of the capacity to generate TxA_2) [17, 22, 49]. It is not known if a modest effect on TxA_2 reduces the release of soluble mediators that influences platelet-vessel interactions, such as expression of adhesion molecules and activation of the tissue factor pathway with CD40 ligand [71]. These possibilities are somewhat speculative but merit further investigation.

Apart from recent intense interest in the vascular effects, COX inhibitors appear to have effects in other aspects of cardiovascular biology, such as atherogenesis and cardiac function.

Do COX pathways modulate atherosclerosis?

Inflammatory signaling is now considered central to the pathobiology of atherosclerosis and increased expression of both COX enzymes has been reported in atherosclerotic plaque [11, 60]. In addition to their vasoactive properties and effects on hemostasis, both PGI_2 and TxA_2 have been shown *in vitro* to have opposing actions on mitogenic responses in cells critical to the atherosclerotic process. Pharmacological TxA_2 receptor antagonism reduced atherosclerosis in the apolipoprotein E knockout mouse and deletion of both copies of the IP receptor accelerated lesion formation in the low-density lipoprotein receptor knockout mouse [38, 72]. What is the functional contribution of COX to this process and what are the implications for pharmacologically induced sustained 60% reduction in PGI_2 on the progression of atherosclerosis?

Two groups reported divergent results on the effects of pharmacologically defined inhibition of both COX isoenzymes using selective and non-selective inhibitors in the same LDL receptor deficient mouse model [73, 74]. An impressive reduction in the burden of plaque was observed with 6 weeks of treatment with rofecoxib that was not reproduced with the less selective COX-2 inhibitor, nimesulide, when administered over a longer period. In both studies, the non-selective indomethacin also exhibited a marked inhibitory effect [73, 74]. Thus, it remains to be defined if this protective effect is a COX-1 or COX-2 mediated response, although genetic manipulation of macrophage COX-2 in these animals resulted in a significant reduction in the extent of progression in the early stages of atherosclerosis [74]. It is not known whether pharmacological differences between the COX inhibitors in the mouse or whether activation of PPARs are responsible for the disparate data in this model [65].

While useful insights into COX biology can be obtained in these animal models, there are limitations in extrapolating these findings to human disease, as the process of plaque development in the mouse is significantly different. Periods of intermittent plaque rupture and fissuring that are mechanistically important in human disease are not features of this process in these mouse models [75].

Does COX modulate cardiac function?

Although COX-1 is the only isoform expressed in normal cardiomyocytes, recently the expression of COX-2 has been shown in cardiac tissue from patients with advanced cardiac dysfunction and also in myocardial infarction [12, 76]. Inhibition of this isoform was found to improve myocardial function in a rat model of infarction [76]. These beneficial effects in infarction contrast with the role of COX-2 in ischemic preconditioning in rabbits and anthracycline induced cardiac toxicity in rodents, when inhibition of COX-2 derived prostaglandins was deleterious [77, 78].

Recently a small pilot, open-label, randomized, prospective study examined the effect of the selective COX-2 inhibitor, meloxicam, in patients with acute coronary syndromes (who also were given aspirin) and reported that there was a trend towards benefit in these subjects, perhaps reflecting an anti-inflammatory effect on the vasculature [79]. These data suggest that the expression of COX-2 plays an important but variable biological role in cardiac disease and further studies are necessary to elucidate the role of COX across the spectrum of cardiac disease.

Unresolved issues in the use of selective inhibitors in higher risk patients

Many questions remain about therapeutic options in elderly patients who have co-existing rheumatic complaints and possible risk factors for atherosclerosis. Thus, when considering treatment options in patients with rheumatic disorders, the decision to treat with traditional NSAIDs or selective agents will require at least some assessment of both the risk of GI adverse events and risk of cardiovascular events.

If there is pro-thrombotic potential with these agents, the intuitive response would be to administer aspirin, however, the advantage of superior GI tolerability may be lost when aspirin is co-administered with the coxib relative to those patients who received standard NSAIDs as noted in the CLASS study [46]. Whether this combination maintains an advantage over nonselective NSAIDs remains to be determined and the cost effectiveness of this strategy will require further investigation. However, the same issue pertains to those patients at risk who may be administered an NSAID. Co-administration of aspirin with NSAIDs such as ibuprofen may attenuate the antithrombotic effect of aspirin by preventing its access to the catalytic pocket of COX-1 [80]. This effect has not been observed with the rofecoxib or meloxicam [81]. Low doses of aspirin also partially inhibit PGI_2 formation and it is unclear if this effect may accentuate the effect of the selective inhibitor on this eicosanoid [35, 36].

Interestingly administration of celecoxib (and valdecoxib) after aspirin attenuates its prolongation of bleeding time [82]. This is similar to the effect seen in a dog model of coronary thrombosis when the beneficial effects of aspirin on vessel occlusion were lost after celecoxib administration [42]. While this raises some theoretical concerns about co-administration of aspirin with these agents, no adverse effects were noted in the subset of subjects from CLASS who were given aspirin with celecoxib [46, 47].

Conclusions

COX-2 inhibitors were introduced on the premise that they would have an improved safety profile owing to fewer toxic GI effects that had complicated the use of traditional NSAIDs. Recent studies have largely confirmed their superiority in a broad range of patients with rheumatic disorders and their cardiovascular safety profile in normal subjects and those at low risk for events has been reassuring. However, the unexpected findings of a potential for increased cardiovascular thrombosis in certain subsets of higher risk patients raises important concerns over the use of these agents. There is limited clinical experience with the longer-term effect of this class of drugs in higher risk subjects such as those with hyperlipidemia, diabetes and those with unrecognized atherosclerosis, which are characterized by increased TxA_2 generation. Certainly more information is required on the effects of these agents in these higher risk subjects as well as those with hypertension and renal disease. Perhaps the results of the VIGOR trial highlight the potential for adverse interactions in higher risk subsets as COX expression is modified by pathophysiological circumstance as noted in arteriosclerosis. This is perhaps consistent with the findings in IP knockout mice, whose cardiovascular phenotype under physiological conditions is benign but in the setting of adverse vascular stimuli, exhibits deleterious responses.

Presently, it is not clear if there are one or more mechanisms that account for such a possible thrombotic hazard. Is it mediated by an alteration in vascular eicosanoids, through an effect on COX pathways in atherosclerotic plaque or is it related to specific kinetic or pharmacodynamic features of the individual agents or does all of the above play a role. Several investigators have called for studies to specifically examine possible hazards with existing agents, particularly in patients at increased risk for vascular events. However, at risk patients, like those in VIGOR, would not likely be enrolled without concomitant aspirin, which would limit the ability of clinical trials to resolve this cardiovascular controversy. Based on the studies with aspirin and the differences between the rates of events in both CLASS and VIGOR, it is estimated that several thousand subjects with extended follow up would be required to perform a trial that would have the power to ascertain differences. Such a study is underway with lumiracoxib and as aspirin use is permitted, the cardiovascular dilemma will probably not be adequately addressed despite its large size. It will, however, yield information on the effect of aspirin co-therapy on GI outcomes when administered with this coxib.

The original Vane paradigm continues to undergo modification as new clinical and experimental studies reveal novel insights into the biology of COX expression and its functional contribution not only in the vasculature, but also across the spectrum of cardiovascular disease. While the risk of vascular thrombosis with the use of these selective inhibitors in "at risk" elderly populations with sub-clinical atherosclerosis needs to be clarified, by comparison, the role of COX and its inhibition may yet be a novel new target for anti-inflammatory therapy in the earlier stages of this disease.

References

1 Vane J (1994) Towards a better aspirin. *Nature* 367: 215–216
2 Masferrer JL, Zweifel BS, Manning PT, Hauser SD, Leahy KM, Smith WG, Isakson PC, Seibert K (1994) Selective inhibition of inducible cyclooxygenase 2 *in vivo* is anti-inflammatory and nonulcerogenic. *Proc Natl Acad Sci USA* 91: 3228–3232
3 Kurumbail RG, Stevens AM, Gierse JK, McDonald JJ, Stegeman RA, Pak JY, Gildehaus D, Miyashiro JM, Penning TD, Seibert K et al (1996) Structural basis for selective inhibition of cyclooxygenase-2 by anti-inflammatory agents. *Nature* 384: 644–648
4 Vane JR, Botting R (1996) Overview-Mechanisms of action of anti-inflammatory drugs. In: JR Vane, J Botting, R Botting (eds): *Improved Non steriodal anti-inflammatory drugs: COX-2 enzyme Inhibitors*, Kluwer Academic Publishers, Dordrecht, 1–27
5 FitzGerald GA, Patrono C (2001) The coxibs, selective inhibitors of cyclooxygenase-2. *N Engl J Med* 345: 433–442
6 Morham SG, Langenbach R, Loftin CD, Tiano HF, Vouloumanos N, Jennette JC, Mahler JF, Kluckman KD, Ledford A, Lee CA (1995) Prostaglandin synthase 2 gene disruption causes severe renal pathology in the mouse. *Cell* 83: 473–482
7 Langenbach R, Morham SG, Tiano HF, Loftin CD, Ghanayem BI, Chulada PC, Mahler JF, Lee CA, Goulding EH, Kluckman KD (1995) Prostaglandin synthase 1 gene disruption in mice reduces arachidonic acid-induced inflammation and indomethacin-induced gastric ulceration. *Cell* 83: 483–492
8 Smith, WL, Langenbach, R (2001) Why there are two cyclooxygenase isozymes. *J Clin Invest* 107: 1491–1495
9 McAdam BF, Mardini I, Habib A, Burke A, Lawson JA, Kapoor S, FitzGerald, GA (2000) Effect of regulated expression of human cyclooxygenase isoforms on eicosanoid and isoeicosanoid production in inflammation *J Clin Invest* 105: 1473–1482
10 Topper JN, Cai J, Falb D, Gimbrone MA Jr, (1996) Identification of vascular endothelial genes differentially responsive to fluid mechanical stimuli: cyclooxygenase-2, manganese superoxide dismutase, and endothelial cell nitric oxide synthase are selectively up-regulated by steady laminar shear stress. *Proc Natl Acad Sci USA* 93: 10417–10422
11 Schonbeck U, Sukhova GK, Graber P, Coulter S, Libby P (1999) Augmented expression of cyclooxygenase-2 in human atherosclerotic lesions. *Am J Pathol* 155: 1281–1291
12 Wong Y, Fukuchi M, Melnyk P, Rodger IW, Giaid A (1998) Induction of COX-2 and activation of NFκB in myocardium of patients with CHF. *Circulation* 98: 100–103
13 Smith WL (1992) Prostanoid biosynthesis and mechanism of action. *Am J Physiol* 263: F181–F191
14 Dubois RN, Abramson SB, Crofford L, Gupta RA, Simon LS, Van De Putte LB, Lipsky PE (1998) Cyclooxygenase in biology and disease. *FASEB J* 2: 1063–1073
15 FitzGerald GA, Patrono C (1998) Antiplatelet drugs. In: M Verstraete, V Fuster, E Topol (eds): *Cardiovascular thrombosis- thrombocardiology, thromboneurology*. Lippincott-Raven, Philadelphia, 121–139
16 Hamberg M, Svensson J, Samuelsson B (1975) Thromboxanes: a new group of biologically active compounds derived from prostaglandin endoperoxides. *Proc Natl Acad Sci USA* 72: 2294–2298
17 FitzGerald GA (1991) Mechanisms of platelet activation: TxA_2 as amplifying signal for other ago-

nists. *Am J Cardiology* 68: 11B–15B
18 Moncada S, Gryglewski R, Bunting S, Vane JR (1976) An enzyme isolated form arteries transforms prostaglandin endoperoxides to an unstable substance that inhibits platelet aggregation. *Nature* 263: 663–665
19 Cheng Y, Austin SC, Rocca B, Koller BH, Coffman TM, Grosser T, Lawson JA, FitzGerald GA (2002) Role of prostacyclin in the cardiovascular responses to thromboxane A_2. *Science* 296: 539–541
20 Habib A, Creminon C, Frobert Y, Grassi J, Pradelles P, Maclouf J (1994) Demonstration of an inducible COX in human endothelial cells using antibodies raised against the carboxyl-terminal region of COX-2. *J Biol Chem* 268: 23448–23454
21 McAdam BF, Catella-Lawson F, Mardini IA, Kapoor S, Lawson JA, FitzGerald GA (1999) Systemic biosynthesis of prostacyclin by cyclooxygenase (COX)-2: the human pharmacology of a selective inhibitor of COX-2. *Proc Natl Acad Sci USA* 96: 272–277
22 Catella-Lawson F, McAdam B, Morrison BW, Kapoor S, Kujubu D, Antes L, Lasseter KC, Quan H, Gertz BJ, FitzGerald GA (1999) Effects of specific inhibition of COX-2 on sodium balance, hemodynamics and vasoactive eicosanoids. *J Pharm and Exper Therap* 289: 735–741
23 Fitzgerald DJ, Roy L, Catella F, FitzGerald GA (1986) Platelet activation in unstable coronary disease *N Engl J Med* 315: 983–989
24 Davi G, Gresele P, Violi F, Basili S, Catalano M, Giammarresi C, Volpato R, Nenci GC, Ciabattoni G, Patrono C (1997) Diabetes mellitus, hypercholesterolemia and hypertension but not vascular disease *per se* are associated with persistent platelet activation *in vivo*. *Circulation* 96: 69–75
25 Patrignani P, Filabozzi P, Patrono C (1982) Selective cumulative inhibition of platelet thromboxane production by low-dose aspirin in healthy subjects. *J Clin Invest* 69: 1366–1372
26 FitzGerald, GA, Smith B, Pedersen AK, Brash AR (1984) Increased prostacyclin biosynthesis in patients with severe atherosclerosis and platelet activation. *N Engl J Med* 310: 1065–1068
27 Crofford LJ, Oates JC, McCune WJ, Gupta S, Kaplan MJ, Catella-Lawson F, Morrow JD, McDonagh KT, Schmaier AH (2000) Thrombosis in patients with connective tissue diseases treated with specific COX-2 inhibitors. A report of four cases. *Arthritis Rheum* 48: 1891–1896
28 Bombardier, C, Laine L, Reicin A, Shapiro D, Burgos-Vargas R, Davis B, Day R, Ferraz M, Hawkey CJ, Hochberg MC et al (2000) Comparison of upper gastrointestinal toxicity of rofecoxib and naproxen in patients with rheumatoid arthritis: VIGOR study group. *N Engl J Med* 343: 1520–1528
29 Vane JR (1971) Inhibition of prostaglandin synthesis as a mechanism of action for aspirin-like drugs. *Nature New Biol* 231: 232–235
30 Roth GJ, Stanford N, Majerus PW (1975) Acetylation of prostaglandin synthase by aspirin. *Proc Natl Acad Sci USA* 72: 3073–3077
31 Loll PJ, Picot D, Garavito RM (1995) The structural basis of aspirin activity inferred from the crystal structure of inactivated prostaglandin H synthase. *Nat Struct Biol* 2: 637–643
32 Patrono C (1994) Aspirin as an anti-platelet drug. *N Engl J Med* 330: 1287–1294
33 Antithrombotic Trialists' Collaboration (2002) Collaborative meta-analysis of randomised trials of antiplatelet therapy for prevention of death, myocardial infarction, and stroke in high-risk patients. *Brit Med J* 324: 71–86
34 Pedersen AK, FitzGerald GA (1984) Dose related kinetics of aspirin. Pre-systemic acetylation of platelet cyclooxygenase. *N Engl J Med* 311: 1206–1211
35 Clarke RJ, Mayo G, Price P, FitzGerald GA (1991) Suppression of thromboxane A_2 but not systemic prostacyclin by controlled-release aspirin. *N Engl J Med* 325: 1137–1141
36 Ferguson G, Eliasziw M, Barr HW, Clagett GP, Barnes RW, Wallace MC, Taylor DW, Haynes RB, Finan JW, Hachinski VC et al (1999) Low-dose and high-dose acetylsalicylic acid for patients undergoing carotid endarterectomy: a randomized controlled trial. *Lancet* 353: 2179–2184
37 Murata T, Ushikubi F, Matsuoka T, Hirata M, Yamasaki A, Sugimoto Y, Ichikawa A, Aze Y, Tanaka T, Yoshida N et al (1997) Altered pain perception and inflammatory response in mice lacking prostacyclin receptor. *Nature* 388: 678–682
38 Egan KM, Austin SC, Smyth EM, FitzGerald GA (2000) *Circulation* 102, 234 (abstr.)
39 Verma S, Raj SR, Shewchuk L, Mather KJ, Anderson TJ (2001) Cyclooxygenase-2 blockade does not impair endothelial vasodilator function in healthy volunteers: Randomized evaluation of rofecoxib *versus* naproxen on endothelium-dependent vasodilatation. *Circulation* 104: 2879–2882
40 Konstam MA, Weir MR, Reicin A, Shapiro D, Sperling RS, Barr E, Gertz BJ (2002) Cardiovascular thrombotic events in controlled, clinical trials of rofecoxib. *Circulation* 104:

2280–2288
41 Nowak J, Murray JJ, Oates JA, FitzGerald GA (1987) Biochemical evidence of a chronic abnormality in platelet and vascular function in healthy individuals who smoke cigarettes. *Circulation* 76: 6–14
42 Hennan JK, Huang J, Barrett TD, Driscoll EM, Willens DE, Park AM, Crofford LJ, Lucchesi BR (2001) Effects of selective cyclooxygenase-2 inhibition on vascular responses and thrombosis in canine coronary arteries. *Circulation* 104: 820–825
43 Food and Drug Administration (2001) Arthritis Advisory Committee Open Panel Meeting 2/8/01: review of VIGOR, Cardiovascular Safety Review of Rofecoxib. Rockville MD, FDA. URL: *www.fda.gov/ohrms/dockets/ac/01/briefing/3677b2_06/cardio.pdf*
44 Wallberg-Jonsson S, Ohman ML, Dahlqvist SR (1997) Cardiovascular morbidity and mortality in patients with seropositive rheumatoid arthritis in northern Sweden. *J Rheumatol* 24: 445–451
45 De Maria, M (2002) Relative risk of CV events in patients with rheumatoid arthritis *Am J Card* 89S: 33D–38D
46 Silverstein FE, Faich G, Goldstein JL, Simon LS, Pincus T, Whelton A, Makuch R, Eisen G, Agrawal NM, Stenson WF et al (2000) Gastrointestinal toxicity with celecoxib vs nonsteroidal anti-inflammatory drugs for osteoarthritis and rheumatoid arthritis: the Celecoxib Long-Term Arthritis Safety Study (CLASS): a randomized controlled trial. *J Am Med Assoc* 284: 1247–1255
47 White WB, Faich G, Whelton A, Maurath C, Ridge NJ, Verburg KM, Geis GS, Lefkowith JB (2002) Comparison of thromboembolic events in patients treated with celecoxib, a cyclooxygenase-2 specific inhibitor, *versus* ibuprofen or diclofenac. *Am J Cardiol* 89: 425–430
48 Knijff-Dutmer EA, Martens A, vd Laar MA (1999) Effects of nabumetone compared with naproxen on platelet aggregation in patients with rheumatoid arthritis. *Ann Rheum Dis* 58: 257–259
49 Van Hecken A, Schwartz JI, Depre M, De Lepeleire I, Dallob A, Tanaka W, Wynants K, Buntinx A, Arnout J, Wong PH et al (2000) Comparative inhibitory activity of rofecoxib, meloxicam, diclofenac, ibuprofen, and naproxen on COX-2 *versus* COX-1 in healthy volunteers. *J Clin Pharmacol* 40: 1109–1120
50 Brochier ML (1993) Evaluation of flurbiprofen for prevention of re-infarction and reocclusion after successful thrombolysis or angioplasty in acute myocardial infarction: the Flurbiprofen French Trial. *Eur Heart J* 14: 951–957
51 Rahme E, Pilote L, LeLorier J (2002) Association between naproxen use and protection against acute myocardial infarction. *Arch Int Med* 162: 1111–1115
52 Solomon DH, Glynn RJ, Levin R, Avron J (2002) Nonsteroidal anti-inflammatory drug use and acute myocardial infarction. *Arch Intern Med* 162: 1099–1104
53 Watson, DJ, Rhodes T, Cai B, Guess HA (2002) Lower risk of thromboembolic cardiovascular events with naproxen among patients with rheumatoid arthritis. *Arch Intern Med* 162: 1105–1110
54 Garcia Rodriguez LA, Varas C, Patrono C (2000) Differential effects of aspirin and non-aspirin nonsteroidal anti inflammatory drugs in the primary prevention of myocardial infarction in postmenopausal women. *Epidemiology* 11: 382–387
55 Ray WA, Stein CM, Hall K, Daugherty JR, Griffin MR (2002) NSAIDs and risk of serious coronary heart disease: an observational cohort study. *Lancet* 359: 118–123
56 Sanmuganathan PS, Ghahramani P, Jackson PR, Wallis EJ, Ramsay LE (2001) Aspirin for primary prevention of coronary heart disease: safety and absolute benefit related to coronary risk derived from meta-analysis of randomized trials. *Heart* 85: 265–271
57 Mertz PM, DeWitt DL, Steller-Stevenson WG, Wahl LM (1994) IL-10 suppression of monocyte COX-2. Mechanism of inhibition of prostaglandin dependant matrix metalloproteinases production. *J Biol Chem* 269: 21322–21329
58 Cipollone F, Patrignani P, Greco A, Panara MR, Padovano R, Cuccurullo F, Patrono C, Rebuzzi AG, Liuzzo G, Quaranta G et al (1997) Differential suppression of thromboxane biosynthesis by indobufen and aspirin in patients with unstable angina. *Circulation* 96: 1109–1116
59 Rocca B, Secchiero P, Ciabattoni G, Ranelletti FO, Catani L, Guidotti L, Melloni E, Maggiano N, Zauli G, Patrono C (2002) Cyclooxygenase-2 expression is induced during human megakaryopoiesis and characterizes newly formed platelets. *Proc Natl Acad Sci USA* 99: 7634–7639
60 Ross R (1999) Atherosclerosis: an inflammatory disease. *N Engl J Med* 340: 115–126
61 Ridker PM, Hennekens CH, Buring JE, Rifai N (1997) Inflammation, aspirin, and the risk of cardiovascular disease in apparently healthy men. *N Engl J Med* 336: 973–979
62 Cipollone F, Prontera C, Pini B, Marini M, Fazia M, De Cesare D, Iezzi A, Ucchino S, Boccoli G, Saba V et al (2001) Over expression of functionally coupled COX-2 and PGE synthase in symp-

tomatic atherosclerotic plaques as a basis of PGE_2 dependent plaque instability. *Circulation* 104: 921–927
63 Ricote M, Huang J, Fajas L, Li A, Welch J, Najib J, Witztum JL, Auwerx J, Palinski W, Glass CK (1998) Expression of the peroxisome proliferator-activated receptor γ (PPARγ) in human atherosclerosis and regulation in macrophages by colony stimulating factors and oxidized low density lipoprotein. *Proc Natl Acad Sci USA* 95: 7614–7619
64 Inoue H, Tanabe T, Umesono K (2000) Feedback control of cyclooxygenase-2 expression through PPARγ. *J Biol Chem* 275: 28028–28032
65 Lehmann JM, Lenhard JM, Oliver BB, Lenhard JM, Oliver BB, Ringold GM, Kliewer SA (1997) Peroxisome proliferator-activated receptors α and γ are activated by indomethacin and other non-steroidal anti-inflammatory drugs. *J Biol Chem* 272: 3406–3410
66 Mukherjee D, Nissen SE, Topol EJ (2001) Risk of cardiovascular events associated with selective COX-2 inhibitors. *J Am Med Assoc* 286: 954–959
67 Matsumoto AK, Melian A, Mandel DR, McILwain H, Borenstein D, Zhao PL, Lines CR, Gertz BJ, Curtis S (2002) A randomized, controlled trial of etoricoxib in the treatment of rheumatoid arthritis. *J Rheumatol* 29: 1623–1630
68 Collantes E, Curtis SP, Lee KW, Casas N, McCarthy T, Melian A, Zhao PL, Rodgers DB, McCormick CL, Lee M et al (2002) A multinational randomized controlled clinical trial of etoricoxib in the treatment of rheumatoid arthritis. *BMC Fam Pract* 3: 1–10
69 Simon LS, Weaver Al, Graham DY, Kivitz AJ, Lipsky PE, Hubbard RC, Isakson PC, Verburg KM, Yu SS, Zhao WW et al (1999) Anti-inflammatory and upper gastrointestinal effects of celecoxib in rheumatoid arthritis: a randomized controlled trial. *J Am Med Assoc* 282: 1921–1928
70 McMurray RW, Hardy KJ (2002) COX-2 inhibitors: today and tomorrow. *Am J Med Sci* 323: 181–189
71 Libby P (2001) Current concepts of the pathogenesis of the acute coronary syndromes. *Circulation* 104: 365–372
72 Cayette AJ, Du Y, Oliver-Krasinki J, Lavielle G, Verbeuren TJ, Cohen RA (2000) The thromboxane receptor antagonist S1 8886 but not aspirin inhibits atherogenesis in apo E deficient mice: evidence that eicosanoids other than thromboxane contribute to atherosclerosis. *Arterioscler Thromb Vasc Biol* 20: 1724–1728
73 Pratico D, Tillmann C, Zhang ZB, Li H, FitzGerald GA (2001) Acceleration of atherogenesis by COX-1-dependent prostanoid formation in low density lipoprotein receptor knockout mice. *Proc Natl Acad Sci USA* 98: 3358–3363
74 Burleigh ME, Babaev VR, Oates JA, Harris RC, Gautam S, Riendeau D, Marnett LJ, Morrow JD, Fazio S, Linton MF (2002) Cyclooxygenase-2 promotes early atherosclerotic lesion formation in LDL receptor-deficient mice. *Circulation* 105: 1816–1823
75 Fuster V, Badimon L, Badimon JJ, Chesebro JH (1992) The pathogenesis of coronary artery disease and acute coronary syndromes. *N Engl J Med* 326: 310–318
76 Saito T, Rodger IW, Hu F, Shennib H, Giaid A (2000) Inhibition of COX-2 improves cardiac function in myocardial infarction. *Biochem Biophys Res Commun* 273: 772–775
77 Shinmura K, Tang XL, Wang Y, Xuan YT, Liu SQ, Takano H, Bhatnagar A, Bolli R (2000) Cyclooxygenase-2 mediates the cardioprotective effects of the late phase of ischemic preconditioning in conscious rabbits. *Proc Natl Acad Sci USA* 97: 10197–10202
78 Dowd NP, Scully M, Adderley SR, Cunningham AJ, Fitzgerald DJ (2001) Inhibition of cyclooxygenase-2 aggravates doxorubicin-mediated cardiac injury *in vivo*. *J Clin Invest* 108: 585–590
79 Altman R, Luciardi HL, Muntaner J, Del Rio F, Berman SG, Lopez R, Gonzalez C (2002) Efficacy assessment of meloxicam, a preferential COX-2 inhibitor, in acute coronary syndromes without ST segment elevation: the NSAIDs in Unstable Angina Treatment 2 (NUT-2) pilot study. *Circulation* 106: 191–195
80 Catella-Lawson F, Reilly MP, Kapoor SC, Cucchiara AJ, DeMarco S, Tournier B, Vyas SN, FitzGerald GA (2001) Cyclooxygenase inhibitors and the antiplatelet effects of aspirin. *New Engl J Med* 345: 1809–1817
81 van Ryn J, Kink-Eiband M, Kuritsch I, Feifel U, Hanft G, Wallenstein G, Trummlitz G, Pairet M (2003) Meloxicam does not interfere with the antiplatelet effect of aspirin in healthy volunteers. EULAR 2003, Lisbon, Portugal, Abstract THU114
82 Gotta AW (2002) Valdecoxib (Pharmacia). *Curr Opin Investig Drugs* 3: 240–245

COX-2 Inhibitors
Edited by M. Pairet and J. van Ryn

Effects on the kidney: role of COX-2 inhibitors

Dirk O. Stichtenoth

Institute of Clinical Pharmacology, Medizinische Hochschule Hannover, D-30623 Hannover, Germany

Introduction

Renal side-effects of non-steroidal anti-inflammatory drugs (NSAIDs) range from mild to life threatening and a life-long necessity for dialysis. They can be divided into three categories [1–3]: (1) Renal side-effects of NSAIDs due to inhibition of prostanoid synthesis. Most of the unwanted renal effects of NSAIDs are related to this mechanism, including reduction in renal blood flow and glomerular filtration rate, sodium and water retention, and hyperkalemia. (2) Analgesic nephropathy. Habitual use of NSAIDs and non-NSAID analgesics can cause analgesic nephropathy with the characteristic papillary necrosis. Most likely chronic hypoperfusion of the renal medulla caused by NSAIDs leads to ischaemic damage [1]. For the non-NSAID analgesics the mechanism of damage is unknown [4]. (3) Acute interstitial nephritis, which is a cell-mediated immune response to NSAIDs and many other drugs [3].

The incidence of all NSAID renal side-effects can be as high as 18%; severe renal side-effects with clinically apparent symptoms occur in 1% of individuals taking NSAIDs [5]. Though this percentage is small, the absolute numbers are high, due to the large exposure to NSAIDs [3]. More than 10% of the population in developed countries ingest NSAIDs intermittently or routinely per year. As a result, 3% of all cases of acute renal failure are caused by NSAIDs and these drugs are implicated as causal agents in up to 30% of patients with endstage renal failure [6, 7]. Thus, any effort to avoid renal side-effects of NSAIDs is of great epidemiological importance.

The value of selective COX-2 inhibitors in terms of renal safety will be reviewed here, beginning with a discussion of physiological mechanisms in the kidney affected by NSAIDs and the roles of the COX isoenzymes in the kidney.

Prostanoid functions in the kidney and NSAID effects

Prostaglandin E_2 (PGE_2), prostacyclin (PGI_2) and to a minor degree $PGF_{2\alpha}$ are the predominant prostanoids in the kidney [8, 9]. Renomedullary interstitial cells synthesize mainly PGE_2 and renal vasculature and glomeruli produce predominantly PGI_2 [10]. The effects of these prostanoids are mediated via G-protein-coupled receptors. Many receptor subtypes and a specific pattern of distribution are responsible for the multiple and at times apparently opposing functional effects of a single prostaglandin [11].

Regulation of renal hemodynamics

Infusion of PGE_1 into renal arteries of dogs produced an increase in renal blood flow [12]. The same effect was also shown for PGE_2, PGD_2 and PGI_2 [13]. In conscious dogs COX inhibition by indomethacin had no effect on renal blood flow [14]. In contrast, in anesthetized, laparotomized dogs indomethacin led to striking reductions of renal blood flow, pointing to the important role of other mediators which, when unopposed by renal prostanoids, cause pronounced vasoconstriction [15].

Since these initial studies, much has been learned about the crucial role of prostanoids in pathophysiological states for maintenance of renal blood flow and glomerular filtration rate and we are now aware of the risk factors for acute renal failure induced by NSAIDs (Tab. 1) [2]. Accordingly, indomethacin

Table 1. Risk factors for renal side-effects of NSAIDs [2]. Edema, hypertension and hyperkalemia are listed when they may occur as isolated symptoms, however, a reduction of GFR/RBF can also cause edema, hypertension and hyperkalemia.

Risk factor	NSAID renal side-effect
Preexisting renal impairment	reduction of GFR/RBF, acute renal failure, edema, hypertension, hyperkalemia
Congestive heart failure	reduction of GFR/RBF, acute renal failure, edema, hypertension
Sodium depletion by diet or diuretics	reduction of GFR/RBF, acute renal failure
Dehydratation	reduction of GFR/RBF, acute renal failure
Liver cirrhosis with ascites	reduction of GFR/RBF, acute renal failure, edema, hypertension
Lupus nephritis	reduction of GFR/RBF, acute renal failure, edema, hypertension, hyperkalemia
Status post surgery	reduction of GFR/RBF, acute renal failure
Co-medication with nephrotoxic drugs (e.g., aminoglycosides, ciclosporine)	reduction of GFR/RBF, acute renal failure, edema, hypertension
Co-medication with ACE inhibitors, angiotensin receptor antagonists	reduction of GFR/RBF, acute renal failure, hyperkalemia
Co-medication with potassium sparing diuretics	hyperkalemia

GFR = Glomerular filtration rate. RBF = Renal blood flow

caused a striking decrease in glomerular filtration rate in patients with disorders characterized by a low effective plasma volume like liver cirrhosis with ascites, heart failure and nephrotic syndrome [16–18]. In all these diseases vasoconstrictor effects of renal nerve activity, catecholamines, angiotensin II and vasopressin play an important role. Other risk factors for acute renal failure by inhibition of prostanoid synthesis are preexisting chronic renal insufficiency, dietary sodium depletion, therapy with diuretics, and systemic lupus erythematosus with renal involvement [2].

The renal medulla has a very low oxygen tension, because blood flow is provided exclusively by the juxtamedullary nephrons. Here, prostanoids play a critical role in protecting the renal medulla from ischaemic damage [19]. Thus, one important factor in the development of analgesic nephropathy is the redistribution of blood flow away from the renal medulla caused by the habitual use of NSAIDs.

Electrolyte excretion

PGE_2 reduces sodium reabsorption in a segment beyond the proximal tubule, primarily in the juxtamedullary nephrons [20, 21]. Sodium retention is difficult to demonstrate in normal volunteers [22], however, in patients with post malignant hypertension on a 100 mEq sodium diet, inhibition of renal prostanoid production by indomethacin 25 mg three times daily (TID) caused sodium retention and weight gain [22]. In contrast, sodium retention is easily observed after prostanoid inhibition in patients with congestive heart failure, liver cirrhosis or nephrotic syndrome and may lead to significant clinical deterioration (Tab. 1) [16–18]. It must be stressed that the simultaneous decreases of renal blood flow, glomerular filtration rate and sodium excretion make it difficult to localize the point of action of the drugs. However, sodium retention is more severe than can be explained solely by a reduction in the filtered load and an increased tubular reabsorption is most likely also involved.

NSAIDs also have a direct effect on potassium excretion. In the macula densa PGI_2 and PGE_2 mediate renin release [23, 24]. PGI_2 synthesized in the afferent arterioles directly stimulate the juxtaglomerular cells to release renin, whereas PGE_2 from tubular cells stimulate renin release via the macula densa [24]. We showed that indomethacin completely abrogates the increase in plasma renin activity seen within minutes after giving furosemide intravenously to healthy volunteers [22]. In sodium depleted volunteers inhibition of renin release by NSAIDs can be observed only by inhibiting sympathetic stimulation of renin release, i.e., by beta receptor blockade [25]. In patients with normal renal function hyperkalemia is not a common problem. However, hyperkalemia may become a severe threat to patients with renal disease, which may in itself can cause potassium retention or when drugs that cause potassium retention are taken simultaneously (Tab. 1).

Antidiuretic hormone

Antidiuretic hormone (ADH) stimulates intrarenal synthesis of PGE_2 and $PGF_{2\alpha}$, which attenuates the hydroosmotic effect of ADH as part of a negative feedback loop [26]. NSAIDs interrupt this negative feedback mechanism. Indomethacin led to an increase of urine osmolality in response to ADH in hypophysectomized dogs [27]. In humans, infusion of hypertonic saline increases urine osmolality and free water reabsorption; both effects are enhanced by indomethacin, showing that endogenous ADH, released in large amounts by hypertonic saline, will become more effective when renal prostanoid production is suppressed [28].

COX-1 and COX-2 in the kidney

In the late 1970s Smith and coworkers located COX-1 in arterial vascular endothelial cells, collecting tubules and medullary interstitial cells of five mammalian species, including rat, rabbit, guinea pig, sheep and cow [29]. This was performed using polyclonal COX antibodies, which were retrospectively classified as COX-1 antibodies. COX antigenicity was not detectable in the macula densa or in the adjoining juxtaglomerular cells, proximal or distal convoluted tubules, or Henle's loop in any species [29]. After discovery of COX-2, Harris et al. confirmed these results in the rat kidney [30]. In addition, a new surprising finding was that COX-2 mRNA and protein was detected under control conditions at locations where no COX-1 was detectable: in the interstitial cells of the papilla, macula densa and the adjacent cortical thick ascending limb [30]. Furthermore, cortical COX-2 expression under a normal diet was restricted to a few macula densa, whereas chronic salt depletion increased COX-2 expression strikingly, both in terms of frequency and number of cells within the macula densa [30].

Yang et al. corroborated these findings and were able to demonstrate in the rat kidney a divergent regulation of COX-2 in the macula densa and renal medulla by changing concentrations of dietary salt [31]. Medullary COX-2 mRNA and protein was increased by high dietary sodium intake, whereas COX-2 expression in macula densa was decreased by this treatment [31]. In contrast, sodium depletion induced a three-fold increase in COX-2 in the macula densa and tended to reduce COX-2 mRNA in the inner medulla [31]. COX-2 expression in the macula densa could also be stimulated by inhibition of salt transport in the loop of Henle [32]. In this study, treatment of rats with furosemide for 1 week led to a three-fold increase in COX-2 protein in the macula densa, whereas hydrochlorothiazide had no effect on COX-2 mRNA or immunoreactivity [32].

COX-2 also seems to play an important role in mediating prostanoid-ADH antagonism. Addition of a hypertonic medium to collecting duct cell cultures induced COX-2 and increased PGE_2 synthesis. This could be completely blocked by the COX-2 inhibitor NS-398 or by indomethacin [33]. Accordingly, in the medulla of the rat kidney, dehydration led to enhanced COX-2 expression in collecting ducts [33].

In the human kidney both COX-isoforms are expressed constitutively. Low levels of both COX-1 and COX-2 mRNA were observed in the kidneys of victims of accidents [34]. Subsequently, Kömhoff et al. confirmed the constitutive expression of COX-2 in similar amounts as COX-1 [35]. COX-1 mRNA and protein were localized in endothelial and smooth muscle cells of arteries and veins, collecting duct cells, and in interstitial cells of the renal cortex and medulla [35]. For COX-2 the same pattern of vascular expression was shown as with COX-1. COX-2 was abundantly expressed intraglomerularly in podocytes, suggesting that COX-2 may contribute to the regulation of renal perfusion and filtration properties [35]. The same study also investigated the localization of COX-1 and COX-2 in the human fetal kidney, because animal data point to participation of COX-2 in renal development [36–39]. COX-2 was detected only at the endstage of renal development in podocytes, whereas COX-1 was expressed abundantly with a similar localization as in adults and additionally in podocytes [35].

Further studies yielded divergent results, which may be explained by different methodological approaches and functional state of the kidneys investigated. While Khan et al. found the same restriction of COX-2 to blood vessels and podocytes [40], Nantel et al. revealed that COX-2 is also expressed in the human macula densa, thick ascending limb of Henle's loop and medullary interstitial cells [41]. Moreover, in human fetal kidneys marked COX-2 levels in macula densa and thick ascending limb of Henle's loop were demonstrated [42]. This COX-2 expression declines during nephron maturation and COX-2 was no longer detected in the adult kidney [42].

Thus, COX-2 is expressed in the human kidney with a pattern similar to the known intrarenal localization in other mammals: Vascular endothelium and smooth muscle cells, podocytes, macula densa, thick ascending limb of Henle's loop and medullary interstitial cells. In the fetal kidney COX-2 was detected in macula densa, thick ascending limb of Henle's loop and at the endstage of renal development in podocytes. Thus, one might expect that COX-2 is involved in all prostanoid-controlled mechanisms in the human kidney and further in renal development. However, before drawing any definite conclusions, the studies on the renal effects of COX-2 inhibitors in animals and humans *in vivo* must be reviewed.

Animal experimental work with selective COX-2 inhibition *in vivo*

As far as renal hemodynamics are concerned, experiments in sodium and volume depleted conscious dogs, volume depleted rats and rats with liver cirrhosis and ascites indicate that COX-1 is the critical enzyme [43–46]. All of these pathophysiological conditions are risk factors for the deterioration of renal blood flow and glomerular filtration rate by NSAIDs. In these studies, COX-2 inhibitors had no effects on renal hemodynamics, whereas nonselective NSAIDs or the selective COX-1 inhibitor SC-560 in comparable doses reduced renal perfusion markedly [43–46].

In contrast, for sodium and water excretion and the control of renin release it has repeatedly been shown in rats and mice that COX-2 is the responsible isoenzyme. Treatment of rats with the COX-2 inhibitor celecoxib for 3 weeks caused significant weight gain (20%) and an increase of systolic blood pressure of 30 mm Hg relative to vehicle-treated controls [47]. In hypertensive rats, the same treatment schedule with celecoxib led to a mean increase in systolic blood pressure of 33 mm Hg [47].

The COX-2 inhibitor NS-398 blocked the increase in renal renin content in sodium depleted mice [48]. Also in rats on a low sodium diet the furosemide stimulated increase of plasma renin activity was blocked by the COX-2 inhibitor SC-58125 to the same degree as indomethacin [49]. Interestingly, urinary PGE_2 excretion was markedly reduced by indomethacin but unaffected by SC-58125, suggesting, that PGE_2 in the normal kidney is synthesized largely by COX-1, whereas COX-2 appears to contribute qualitatively but not quantitatively [49]. COX-2 in the macula densa and renin release are coordinately upregulated in the rat model of renovascular hypertension [50, 51]. Administration of celecoxib almost normalized plasma renin activity and lowered mean arterial blood pressure, thereby proving the functional linkage between COX-2 and renin release [51]. Mice genetically devoid of COX-1 have normal plasma renin activity, the increase in plasma renin activity induced by captopril could be blocked, again, by celecoxib [52].

A detailed insight into the roles of the COX-isoenzymes in renal development is provided by studies with COX-1 and COX-2 knockout mice. Renal malformations primarily due to a block in the development of glomeruli occurred in COX-2, but not in COX-1 knockout mice. Most of the animals died within 16 weeks of birth [36–38]. In the rat, COX-2 mRNA and immunoreactivity were detected in a subset of the epithelial cells of the developing thick ascending limb of Henle [39]. Both the time course and magnitude of COX-2 expression support the conclusion that COX-2 plays an important role in rat nephron development. More relevant to the therapeutic situation, Kömhoff et al. demonstrated in mice and rats that treatment with celecoxib from postnatal day 0 to day 21, when COX-2 expression peaked in renal cortex, severely reduced glomerular diameter. In contrast, treatment limited to pregnancy duration did not affect glomerular size [53].

In the inflamed kidney COX-2 expression is induced and COX-2 is the responsible enzyme for synthesis of pro-inflammatory prostanoids. Seibert et al. [54] investigated the contribution of COX-1 and COX-2 to prostanoid production in the inflamed kidney using a model of unilateral ureteral obstruction, which leads to hydronephrosis and inflammatory infiltration. The 10-fold higher PGE_2 excretion in the venous effluent of the hydronephrotic kidney was inhibited by the COX-2 inhibitor DuP-697 and by indomethacin, dexamethasone and cycloheximide, an inhibitor of protein biosynthesis [54]. In the normal kidney, only indomethacin was able to reduce PGE_2 production, indicating that the majority of PGE_2 in the normal kidney is produced by COX-1 [54]. Several other studies revealed enhanced COX-2 expression in experimental glomerulonephritis [55–57], nephrotoxic nephritis [58], passive Heymann nephritis [59], subtotal renal ablation [60, 61] and lupus nephritis in mice and man [62]. Treatment with COX-2 inhibitors reduced inflammation, proteinuria and retarded progression of renal injury in these models [59, 61].

Clinical experience with selective COX-2 inhibition

The first experiences with COX-1 sparing drugs were made before the discovery of the COX-isoenzymes. Glucocorticoids potently suppress COX-2 expression but have no inhibitory activity on preformed COX [63, 64]. Dexamethasone at dosages that induce a Cushingoid appearance had no effect on renal and total body prostanoid production in healthy volunteers [65] rabbits [66] and rats [67]. Clinically, glucocorticoids have impressive anti-inflammatory actions but they are devoid of the NSAID typical side-effects. The absence of unwanted renal effects of glucocorticoids shows that prostanoids synthesized by an inducible COX-2 are of no importance for renal physiology, otherwise glucocorticoids should have detrimental effects on renal function. It must be stressed that this holds true only for the inducible COX-2, and that there is also a constitutive pool of COX-2 as described above, however, its expression seems to be insensitive to glucocorticoids [68].

Salicylic acid is the first NSAID. It is the active substance present in herbal remedies that have been used by humans for more than 3500 years. Surprisingly, salicylic acid and other non-acetylated salicylates are selective COX-2 inhibitors and suppress the induction of COX-2 [69, 70]. In healthy volunteers, sodium salicylate in the anti-inflammatory dose of 53 mg/kg body weight did not inhibit renal PGE_2 synthesis, whereas the equivalent dose of acetylsalicylic acid led to a marked decrease [71]. Salsalate, the prodrug of salicylic acid, is a potent NSAID, but does not cause reduction in renal blood flow and glomerular filtration rate, sodium and water retention, or a decrease in lithium clearance [72–74]. Unfortunately, the therapeutic value of non-acetylated salicylates is undeservedly forgotten in many countries.

Preferential COX-2 inhibitors

Etodolac, meloxicam, nimesulide and non-acetylated salicylates are classified as preferential COX-2 inhibitors, i.e., NSAIDs with a moderate COX-2 selectivity [69, 75]. Amongst them, meloxicam is the best documented. In healthy volunteers we demonstrated that meloxicam at the therapeutic dose of 7.5 mg OD had no effects on platelet aggregation and thromboxane formation in contrast to a marked inhibition by indomethacin 25 mg TID [76]. Thus, meloxicam at this dose level is COX-1 sparing in humans *in vivo*. In the same study, renal PGE_2 synthesis remained unaffected by meloxicam but was markedly inhibited by indomethacin [76]. Thus, under physiological conditions the majority of renal PGE_2 is synthesized by COX-1. This is in agreement with the animal experimental findings [49, 54].

This selectivity observed in normal volunteers translates into improved clinical safety as far as renal hemodynamics is concerned. Meloxicam in the maximum recommended dose of 15 mg OD for 28 days did not cause further deterioration of renal function in 25 patients with mild renal impairment, a risk factor for NSAID-induced renal failure [77]. Convincingly, in patients with chronic heart failure, another risk factor for deterioriation of renal function by NSAIDs, treatment with meloxicam 15 mg OD for 7 days caused no renal side-effects nor attenuated the diuretic effect of furosemide [78].

Two prospective, large scale, double-blind randomized trials support the findings of the above mentioned studies. In the MELISSA (n = 9323) and SELECT (n = 8656) trial meloxicam 7.5 mg was compared to diclofenac 100 mg slow release and piroxicam 20 mg, respectively, for 28 days in patients with osteoarthritis [79, 80]. While the clinical efficacy of meloxicam was equivalent to the comparator drugs in both trials, renal toxicity of meloxicam, measured as increases in creatinine and urea in serum, was significantly less as compared to diclofenac and piroxicam [79, 80]. The results of MELISSA and SELECT show a reduced risk of renal side-effects with meloxicam, it is obvious, however, that they were not eliminated.

To address the question, whether renin release is COX-2 dependent in humans, we undertook a randomized cross-over study in healthy volunteers comparing the effects of meloxicam in the COX-1 sparing dosage of 7.5 mg/d to indomethacin 25 mg TID on plasma renin activity [81]. Meloxicam abolished the furosemide stimulated plasma renin as effectively as indomethacin, suggesting that the COX-isoenzyme responsible for renin release in man is COX-2 (Fig. 1) [81]. Thus, inhibition of renin release with subsequent hyperkalemia must be considered as a side-effect of COX-2 inhibition. We also determined renal PGE_2 synthesis, which was, in accordance with our previous finding, not inhibited by meloxicam whereas indomethacin led to a significant reduction [76, 81].

First generation COX-2 inhibitors

Preferential COX-2 inhibitors like meloxicam can exert at the highest therapeutic dosage clinically relevant COX-1 inhibition. Compounds with a higher selectivity for COX-2 are classified as specific or highly selective COX-2 inhibitors [69, 75]. The first specific COX-2 inhibitors were celecoxib and rofecoxib.

As far as renal perfusion is concerned, even two-fold the therapeutic dosage of celecoxib or rofecoxib (Fig. 2) caused no decrease of glomerular filtration

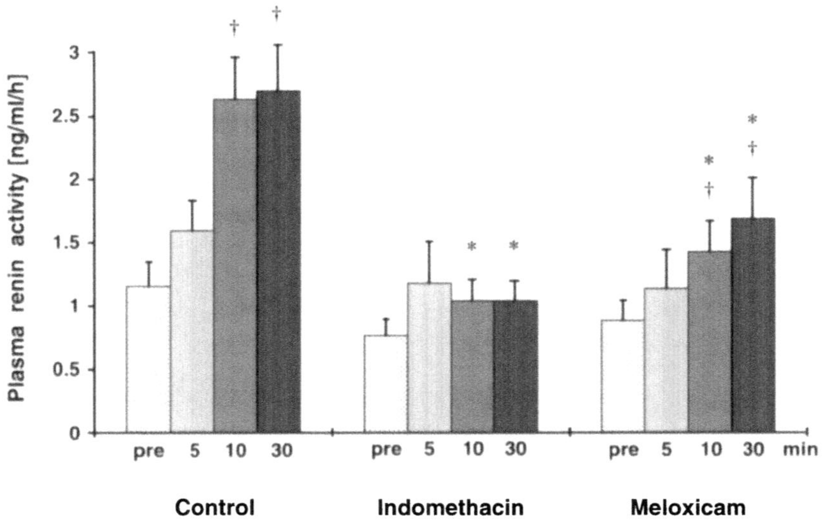

Figure 1. Effects of meloxicam and indomethacin on plasma renin activity before (pre) and 5, 10 and 30 min after furosemide 20 mg i.v. Mean ± SE of plasma renin activity in 15 healthy volunteers on control day or on the last day of treatment with meloxicam 7.5 mg once daily (OD) for 6 days or indomethacin 25 mg TID for 3 days in a randomized, cross-over study design. * = $p < 0.05$ *versus* control. † = $p < 0.05$ *versus* pre furosemide. (Figure reproduced from [81] with the kind permission of the *Journal of Investigative Medicine*.)

rate in healthy elderly volunteers and normotensive salt-depleted young healthy volunteers [82–85]. Both groups are at moderate risk for impairment of renal function by NSAIDs. However, when a second risk factor is present, such as older age plus sodium and volume depletion, rofecoxib at therapeutic dosage led to the same reduction of glomerular filtration rate as indomethacin 50 mg TID (Fig. 3) [86]. In addition, a number of case reports of acute renal failure caused by celecoxib or rofecoxib alert us, that the gain of renal safety in terms of renal perfusion by use of specific COX-2 inhibitors is small [87–89]. Thus, the same precautions as for conventional NSAIDs must be used.

While the clinical data support the experimental results in animals that COX-2 is only partly involved in the regulation of renal hemodynamics, they clearly show that COX-2 is the critical COX-isoenzyme for sodium and water excretion. Several double blind, placebo-controlled and randomized studies provide detailed data on this issue [82–86]. In healthy, elderly volunteers on a controlled normal sodium diet, intake of rofecoxib 50 mg once daily as well as indomethacin 50 mg TID for 14 days, each reduced sodium excretion by 20% during the first 3 days of administration, thereafter returning to normal [82]. No edema occurred, body weight and diastolic blood pressure remained unchanged. Similar effects were found for celecoxib in a study in normotensive salt-depleted young healthy volunteers [83]. Celecoxib was given for 7 days in a dosage of 200 mg BID and 400 mg BID, the latter is twice the maximum recommended dosage in rheumatoid arthritis, the comparators were

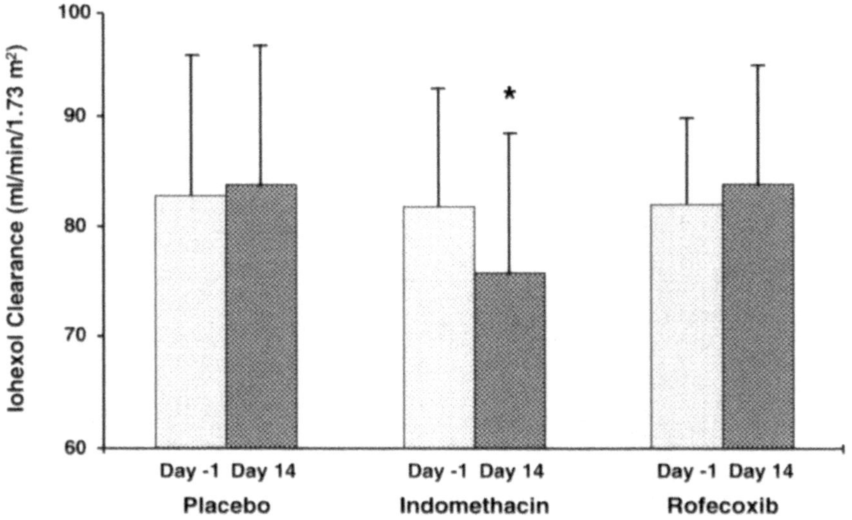

Figure 2. Effects of rofecoxib and indomethacin on glomerular filtration rate. Mean ± SD of iohexol clearance in 36 healthy elderly volunteers treated with either rofecoxib 50 mg OD, indomethacin 50 mg TID or placebo for 14 days in a double-blind, randomized, parallel-group study design. *$p < 0.05$ before (day -1) *versus* treatment day 14. Data from [82].

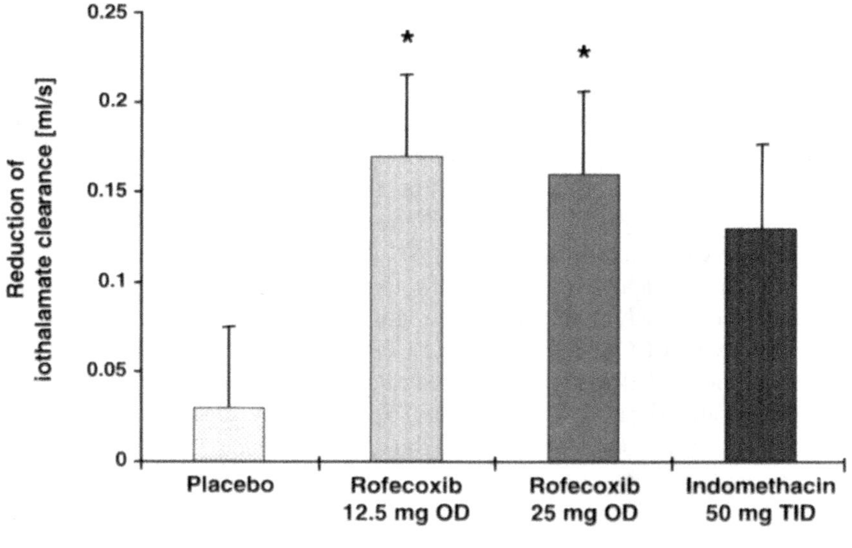

Figure 3. Effects of rofecoxib and indomethacin on glomerular filtration rate. Mean ± SE of iothalamate clearance in 60 healthy elderly volunteers receiving a low-sodium diet on the 6th day of treatment with either rofecoxib 12.5 mg OD, rofecoxib 25 mg OD, indomethacin 50 mg TID or placebo in a double-blind, randomized, parallel-group study design. * $p < 0.05$ *versus* placebo. Data from [86].

naproxen 500 mg BID for 7 days and placebo [83]. Celecoxib caused sodium retention to the same degree as the nonselective naproxen, even under the low sodium diet used in this study. The reduction of urinary sodium excretion was most pronounced 2–3 h after the first dose, but still present on the last day of treatment [83]. Endogenous lithium clearance, reflecting tubular sodium reabsorption, was decreased accordingly (Fig. 4). Remarkably, urine flow was markedly reduced for several hours after intake of celecoxib and naproxen; the effect was attenuated on the last day of treatment but still significant for the two doses of celecoxib [83]. This observation strongly suggests medullary COX-2 as the responsible isoenzyme for ADH antagonism, since reduction in urine output is a well known renal effect of NSAIDs and caused by inhibition of prostanoid-ADH antagonism.

The analysis of the safety data from the large clinical trials with celecoxib and rofecoxib show that as anticipated from the studies in healthy volunteers, the incidence of peripheral edema and hypertension occurring with celecoxib or rofecoxib was significantly higher than with placebo and comparable to traditional NSAIDs [90–92]. Furthermore, as demonstrated for rofecoxib, there is a clear dose-dependency with highest incidence of edema in patients treated with 50 mg rofecoxib once daily, a dosage twice the maximum recommended dose for rheumatoid arthritis but still COX-1 sparing [93]. In a 6 week head-to-head trial rofecoxib 25 mg OD was compared to celecoxib 200 mg OD in 810 older hypertensive osteoarthritis patients. The incidence of peripheral edema and

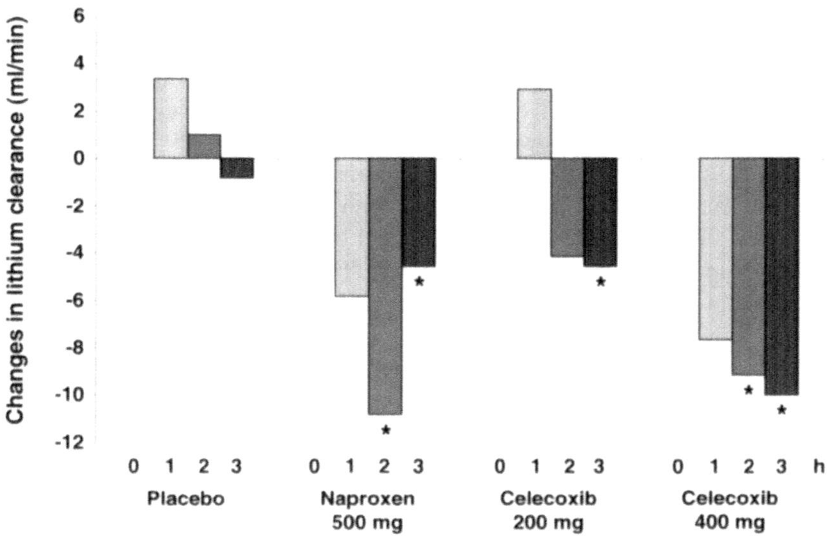

Figure 4. Effects of celecoxib and naproxen on lithium clearance. Mean of 40 healthy volunteers receiving a low-sodium diet after the first administration of either celecoxib 200 mg, celecoxib 400 mg, naproxen 500 mg or placebo in a double-blind, randomized, parallel-group study design. *$p < 0.05$ *versus* placebo. Data from [83].

increase in systolic blood pressure was higher with rofecoxib (9.5% and 17%, respectively) than with celecoxib (4.9% and 11%, respectively) [94]. However, the dose of celecoxib 200 mg OD is less antiphlogistically potent than rofecoxib 25 mg OD; a further bias was the non comparable pharmacokinetics of once daily celecoxib with a half-life of 10 h *versus* rofecoxib with 17 h. Another recently published head-to-head comparison of rofecoxib 12.5 and 25 mg OD to celecoxib 200 mg OD and acetaminophen 1 g QID in 382 osteoarthritis patients without concurrent diseases demonstrated superior efficacy of rofecoxib 25 mg OD over the other treatments and no difference in the incidence of edema and hypertension, which were low (1–4%) in this study population [95].

With regard to renin release, we recently confirmed our previous results obtained with meloxicam [81] in a double-blind, randomized three-way crossover study in healthy volunteers comparing celecoxib 200 mg BID with indomethacin 50 mg TID and placebo. The unpublished results clearly demonstrate a decrease of furosemide stimulated and unstimulated plasma renin activity by celecoxib. Thus, the COX-isoenzyme responsible for renin release in humans is most likely COX-2.

Second generation COX-2 inhibitors

Valdecoxib, its water-soluble prodrug parecoxib, etoricoxib and lumiracoxib represent the second generation of COX-2 inhibitors. The COX-2 selectivity of valdecoxib and parecoxib in the whole blood assay is comparable to rofecoxib, whereas etoricoxib and lumiracoxib possess a several-fold higher COX-2 selectivity than rofecoxib [96]. To date there is only limited experience on the renal effects of these compounds.

For valdecoxib incidence of water retention and edema is evident, 12% with the 80 mg OD dose, 9% with 40 mg OD and 10% with 10 mg OD was observed [97]. Thus, with the therapeutic doses of valdecoxib, 10 and 20 mg OD, the rate of renal side-effects is similar to naproxen, ibuprofen or diclofenac. The available data from clinical trials for lumiracoxib show a frequency of moderate to severe edema of 1%. The corresponding incidence rates were 3% for diclofenac and 0% for placebo [98].

Although experience with the renal safety of the second generation of COX-2 inhibitors is very limited, the speculation is justified that their higher COX-2 selectivity does not translate into renal sparing effects.

References

1 Schlöndorff D (1993) Renal complications of nonsteroidal anti-inflammatory drugs. *Kidney Int* 44: 643–653

2 Stichtenoth DO, Frölich JC (2000) COX-2 and the kidneys. *Curr Pharm Des* 6: 1737–1753

3 Whelton A (1999) Nephrotoxicity of nonsteroidal anti-inflammatory drugs: physiologic foundations and clinical implications. *Am J Med* 106: 13S–24S

4 De Broe ME, Elseviers MM (1998) Analgesic nephropathy. *N Engl J Med* 338: 446–452
5 Murray MD, Brater DC, Tierney WM, Jui SL, McDonald CJ (1990) Ibuprofen-associated renal impairment in a large general internal medicine practice. *Am J Med Sci* 299: 222–229
6 Kleinknecht D (1993) Diseases of the kidney caused by non-steroidal antiinflammatory drugs. In: JH Stewart (ed.): *Analgesic and NSAID-induced kidney disease*. Oxford University Press, Oxford, 160–179
7 Schwarz A, Offermann G, Keller F (1992) Analgesic nephropathy and renal transplantation. *Nephrol Dial Transplant* 7: 427–432
8 Larsson C, Anggard E (1973) Regional differences in the formation and metabolism of prostaglandis in the rabbit kidney. *Eur J Pharmacol* 21: 30–36
9 Whorton AR, Smigel M, Oates JA, Frölich JC (1978) Regional differences in prostaglandin formation by the kidney: Prostacyclin is a major prostaglandin of renal cortex. *Biochim Biophys Acta* 529: 176–180
10 Garcia-Perez A, Smith WL (1984) Apical-basolateral membrane assymetry in canine cortical collecting tubule cells: bradykinin, arginine vasopressin, prostaglandin E_2 interrelationships. *J Clin Invest* 74: 63–74
11 Breyer MD, Jacobson HR, Breyer RM (1996) Functional and molecular aspects of renal prostaglandin receptors. *J Am Soc Nephrol* 7: 8–17
12 Johnston HH, Herzog JP, Lauler DP (1967) Effect of prostaglandin E_1 on renal hemodynamics, sodium and water excretion. *Am J Physiol* 213: 939–946
13 Lifschitz MD (1981) Prostaglandins and renal blood flow. *Kidney Int* 19: 781–785
14 Kirschenbaum MA, Stein JH (1976) Effect of inhibition of prostaglandin synthesis on urinary sodium excretion in conscious dog. *J Clin Invest* 57: 517–521
15 Aiken JW, Vane JR (1973) Intrarenal prostaglandin release attenuates the renal vasoconstrictor activity of angiotensin. *J Pharmacol Ther* 184: 678–687
16 Arisz L, Donker AJM, Brentjens JRH, van der Hem GK (1976) The effect of indomethacin on proteinuria and kidney function in the nephrotic syndrome. *Acta Med Scand* 199: 121–125
17 Walshe JJ, Venuto RC (1979) Acute oliguric renal failure induced by indomethacin: possible mechanism. *Ann Intern Med* 91: 47–49
18 Antillon M, Cominelli F, Lo S Moran M, Somberg K, Reynolds T, Zipser RD (1990) Effects of oral prostaglandins on indomethacin induced renal failure in patients with cirrhosis and ascites. *J Rheumatol Suppl* 20: 46–49
19 Brezis M, Rosen S (1995) Hypoxia of the renal medulla – its implications for disease. *N Engl J Med* 332: 647–655
20 Kauker ML (1977) Prostaglandin E_2 effect from the luminal side on renal tubular ^{22}Na efflux: tracer microinjection studies. *Proc Soc Exp Biol Med* 154: 274–277
21 Stokes JB, Kokko JP (1977) Inhibition of sodium transport by prostaglandin E_2 across the isolated, perfused rabbit collecting tubule. *J Clin Invest* 59: 1099–1104
22 Frölich JC, Hollifield JW, Dormois JC Frölich BL, Seyberth H, Michaelis AM, Oates JA (1976) Suppression of plasma renin activity by indomethacin in man. *Circ Res* 39: 447–452
23 Whorton AR, Misono K, Hollifield J, Frölich JC, Inagami T, Oates JA (1977) Prostaglandins and renin release: I. Stimulation of renin release from rabbit renal cortical slices by PGI_2. *Prostaglandins* 14: 1095–1104
24 Ito S, Carretero OA, Abe K, Beierwaltes WH, Yoshinaga K (1989) Effect of prostanoids on renin release from rabbit isolated afferent arterioles with and without attached macula densa. *Kidney Int* 35: 1138–1144
25 Frölich JC, Hollifield JW, Michelakis AM Vesper BS, Wilson JP, Shand BG, Seyberth HJ, Frölich WH, Oates JA (1979) Reduction of plasma renin activity by inhibition of the fatty acid cyclooxygenase in human subjects: independence of sodium retention. *Circ Res* 44: 781–787
26 Walker LA, Whorton AR, Smigel M, France R, Frölich JC (1978) Antidiuretic hormone increases renal prostaglandin synthesis *in vivo*. *Am J Physiol* 235: F180–F185
27 Anderson RJ, Berl T, McDonald KM, Schrier RW (1975) Evidence for an *in vivo* antagonism between vasopressin and prostaglandin in the mammalian kidney. *J Clin Invest* 56: 420–426
28 Kramer HJ, Backer A, Hinzen S, Dusing R (1978) Effects of inhibition of prostaglandin-synthesis on renal electrolyte excretion and concentrating ability in healthy man. *Prostaglandins Med* 1: 341–349
29 Smith WL, Bell TG (1978) Immunohistochemical localization of the prostaglandin forming cyclooxygenase in renal cortex. *Am J Physiol* 235: F451–F457
30 Harris RC, McKanna JA, Akai Yg, Jacobson HR, Dubois RN, Breyer MD (1994)

Cyclooxygenase-2 is associated with the macula dense of rat kidney and increases with salt restriction. *J Clin Invest* 94: 2504–2510
31 Yang T, Singh I, Pham H, Sun D, Smart A, Schnerman JB, Briggs JP (1998) Regulation of cyclooxygenase expression in the kidney by dietary salt intake. *Am J Physiol* 274: F481–F489
32 Mann B, Hartner A, Jensen BL, Kammerl M, Krämer BK, Kurtz A (2001) Furosemide stimulates macula densa cyclooxygenase-2 expression in rats. *Kidney Int* 59: 62–68
33 Yang T, Schnermann JB, Briggs JP (1999) Regulation of cyclooxygenase-2 expression in renal medulla by tonicity *in vivo* and *in vitro*. *Am J Physiol* 277: F1–F9
34 O'Neill GP, Ford-Hutchinson AW (1993) Expression of mRNA for cyclooxygenase-1 and 2 in human tissues. *FEBS Lett* 330: 156–160
35 Kömhoff M, Gröne H-J, Klein T, Seyberth HW, Nüsing RM (1997) Localization of cyclooxygenase-1 and -2 in adult and fetal human kidney: implication for renal function. *Am J Physiol* 272: F460–F468
36 Dinchuk JE, Car BD, Focht RJ, Johnston JJ, Jaffee BD, Covington MB, Contel NR, Eng VM, Collins RJ, Czerniak PM et al (1995) Renal abnormalities and an altered inflammatory response in mice lacking cyclooxygenase II. *Nature* 378: 406–409
37 Morham SG, Langenbach R, Loftin CD, Tiano HF, Vouloumanos N, Jennette JC, Mahler JF, Kluckman KD, Ledford A, Lee CA (1995) Prostaglandin synthase 2 gene disruption causes severe renal pathology in the mouse. *Cell* 83: 473–482
38 Langenbach R, Morham SG, Tiano HF, Loftin CD, Ghanayem BI, Chulada PC, Mahler JF, Lee CA, Goulding EH, Kluckman KD (1995) Prostaglandin synthase 1 gene disruption in mice reduces arachidonic acid-induced inflammation and indomethacin-induced gastric ulceration. *Cell* 83: 483–492
39 Zhang M-Z, Wang J-L, Cheng H-F, Harris RC, McKanna JA (1997) Cyclooxygenase-2 in rat nephron development. *Am J Physiol* 273: F994–F1002
40 Khan KN, Venturini CM, Bunch RT, Brassard JA, Koki AT, Morris DL, Trump BF, Maziasz TJ, Alden CL (1998) Interspecies differences in renal localization of cyclooxygenase isoforms: implications in nonsteroidal antiinflammatory drug-related nephrotoxicity. *Toxicol Pathol* 26: 612–620
41 Nantel F, Meadows E, Denis D, Connolly B, Metters KM, Giaid A (1999) Immunolocalization of cyclooxygenase-2 in the macula densa of human elderly. *FEBS Letters* 457: 475–477
42 Koki A, Seibert K, Khan NK, Masferrer JL (1999) Cyclooxygenase-2 expression in normal human tissues and pathologic disease. Abstract book of Poster Presentations Third International Workshop on COX-2, Hawaii, Abstr 46
43 Gans KR, Galbraith W, Roman RJ, Haber SB, Kerr JS, Schmidt WK, Smith C, Hewes WE, Ackerman NR (1990) Anti-inflammatory and safety profile of DuP 697, a novel orally effective prostaglandin synthesis inhibitor. *J Pharmacol Exp Ther* 254: 180–187
44 Bosch-Marce M, Claria J, Titos E, Masferrer JL, Altuna R, Poo JL, Jimenez W, Arroyo V, Rivera F, Rodes J (1999) Selective inhibition of cyclooxygenase-2 spares renal function and prostaglandin synthesis in cirrhotic rats with ascites. *Gastroenterology* 116: 1167–1175
45 Black SC, Brideau C, Cirino M, Belley M, Bosquet J, Chan CC, Rodger IW (1998) Differential effect of a selective cyclooxygenase-2 inhibitor *versus* indomethacin on renal blood flow in conscious, volume-depleted dogs. *J Cardiovasc Pharmacol* 32: 686–694
46 Lopez-Parra M, Claria J, Planaguma A, Titos E, Masferrer JL, Woerner BM, Koki AT, Jimenez W, Altuna R, Arroyo V et al (2002) Cyclooxygenase-1 derived prostaglandins are involved in the maintenance of renal function in rats with cirrhosis and ascites. *Br J Pharmacol* 135: 891–900
47 Muscara MN, Vergnolle N, Lovren F, Triggle CR, Elliott SN, Asfaha S, Wallace JL (2000) Selective cyclo-oxygenase-2 inhibition with celecoxib elevates blood pressure and promotes leukocyte adherence. *Br J Pharmacol* 129: 1423–30
48 Harding P, Sigmon DH, Alfie ME, Huang PL, Fishman MC, Beierwaltes WH, Carretero OA (1997) Cyclooxygenase-2 mediates increased renal renin content induced by low sodium diet. *Hypertension* 29: 297–302
49 Pairet M, Churchill L, Engelhardt G (1996) Differential inhibition of cyclooxygenase 1 and 2 by NSAIDs. In: N Bazan, J Botting, J Vane (eds): *New targets in inflammation: Inhibitors of COX-2 or adhesion molecules.* Kluwer Academic Publishers, London, 23–37
50 Hartner A, Goppelt-Struebe M, Hilgers KF (1998) Coordinate expression of cyclooxygenase-2 and renin in the rat kidney in renovascular hypertension. *Hypertension* 31: 201–205
51 Wang J-L, Cheng H-F, Harris RC (1999) Cyclooxygenase-2 inhibition decreases renin content and lowers blood pressure in a model of renovascular hypertension. *Hypertension* 34: 96–101

52 Cheng HF, Wang SW, Zhang MZ, McKanna JA, Breyer R, Harris RC (2002) Prostaglandins that increase renin production in response to ACE inhibition are not derived from cyclo-oxygenase-1. *Am J Physiol Regul Integr Comp Physiol* 283: R638–R646

53 Kömhoff M, Wang JL, Cheng HF, Langenbach R, McKanna JA, Harris RC, Breyer MD (2000) Cyclooxygenase-2-selective inhibitors impair glomerulogenesis and renal cortical development. *Kidney Int* 57: 414–422

54 Seibert K, Masferrer JL, Needleman P, Salvemini D (1996) Pharmacological manipulation of cyclo-oxygenase-2 in the inflamed hydronephrotic kidney. *Br J Pharmacol* 117: 1016–1020

55 Chanmugam P, Feng L, Liou S, Jang BC, Boudreau M, Yu G, Lee JH, Kwon HJ, Beppu T, Yoshida M et al (1995) Radicicol, a protein tyrosine kinase inhibitor, suppresses the expression of mitogen-inducible cyclooxygenase in macrophages stimulated with lipopolysaccharide and in experimental glomerulonephritis. *J Biol Chem* 270: 5418–5426

56 Hirose S, Yamamoto T, Feng L, Yaoita E, Kawasaki K, Goto S, Fujinaka H, Wilson CB, Arakawa M, Kihara I (1998) Expression and localization of cyclooxygenase isoforms and cytosolic phospholipase A2 in anti-Thy-1 glomerulonephritis. *J Am Soc Nephrol* 9: 408–416

57 Schneider A, Harendza S, Zahner G, Jocks T, Wenzel U, Wolf G, Thaiss F, Helmchen U, Stahl RA (1999) Cyclooxygenase metabolites mediate glomerular monocyte chemoattractant protein-1 formation and monocyte recruitment in experimental glomerulonephritis. *Kidney Int* 55: 430–441

58 Khan KN, Alden CL, Gleissner SE, Gessford MK, Maziasz TJ (1998) Effect of papillotoxic agents on expression of cyclooxygenase isoforms in the rat kidney. *Toxicol Pathol* 26: 137–142

59 Heise G, Grabensee B, Schror K, Heering P (1998) Different actions of the cyclooxygenase 2 selective inhibitor flosulide in rats with passive Heymann nephritis. *Nephron* 80: 220–226

60 Wang J-L, Cheng H-F, Zhang M-Z, McKanna JA, Harris RC (1998) Selective increase of cyclooxygenase-2 expression in a model of renal ablation. *Am J Physiol* 275: F613–F622

61 Sanchez PL, Salgado LM, Ferreri NR, Escalante B (1999) Effect of cyclooxygenase-2 inhibition on renal function after renal ablation. *Hypertension* 34: 848–853

62 Tomasoni S, Noris M, Zapella S (1998) Upregulation of renal and systemic cyclooxygenase-2 in patients with active lupus nephritis. *J Am Soc Nephrol* 9: 1202–1212

63 Fu JY, Masferrer JL, Seibert K, Raz A, Needleman P (1990) The induction and suppression of prostaglandin H2 synthase (cyclooxygenase) in human monocytes. *J Biol Chem* 265: 16737–16740

64 Masferrer JL, Seibert K, Zweifel B, Needleman P (1992) Endogenous glucocorticoids regulate an inducible cyclooxygenase enzyme. *Proc Natl Acad Sci USA* 89: 3917–3921

65 Rosenkranz B, Náray-Fejes-Tóth A, Fejes-Tóth G, Fischer C, Sawada M, Frölich JC (1985) Dexamethasone effect on prostanoid formation in healthy man. *Clin Sci* 68: 681–685

66 Naray-Fejes-Toth A, Fejes-Toth G, Fischer C, Frölich JC (1984) Effect of dexamethasone on *in vivo* prostanoid production in the rabbit. *J Clin Invest* 74: 120–123

67 Nasjletti A, Erman A, Cagen LM, Baer PC (1984) Plasma concentrations, renal excretion, and tissue release of prostaglandins in the rat with dexamethasone-induced hypertension. *Endocrinology* 114: 1033–1040

68 Vio CP, Cespedes C, Gallardo P, Masferrer JL (1997) Renal identification of cyclooxygenase-2 in a subset of thick ascending limb cells. *Hypertension* 30: 687–692

69 Frölich JC (1997) A classification of NSAIDs according to the relative inhibition of cyclooxygenase isozymes. *Trends Pharmacol Sci* 18: 30–34

70 Wu KK, Sanduja R, Tsai AL, Ferhanoglu B, Loose-Mitchell DS (1991) Aspirin inhibits interleukin-1-induced prostaglandin H synthase expression in cultured endothelial cells. *Proc Natl Acad Sci USA* 88: 2384–2387

71 Rosenkranz B, Fischer C, Frölich JC (1986) Effects of salicylic and acetylsalicylic acid alone and in combination on platelet aggregation and prostanoid synthesis in man. *Br J Clin Pharmacol* 21: 309–317

72 Bombardier C, Peloso PMJ, Goldsmith CH, and the Salsalate-Diclofenac Study Group (1995) Salsalate, a nonacetylated salicylate, is as efficacious as diclofenac in patients with rheumatoid arthritis. *J Rheumatol* 22: 617–624

73 Clive DM, Stoff JS (1984) Renal syndromes associated with nonsteroidal antiinflammatory drugs. *N Engl J Med* 310: 563–572

74 Carmichael J, Shankel S (1985) Effects of nonsteroidal anti-inflammatory drugs on prostaglandins and renal function. *Am J Med* 78: 992–1000

75 The International COX-2 Study Group (1998) The classification of cyclooxygenase inhibitors. *J Rheumatol* 25: 2298–2303

76 Stichtenoth DO, Wagner B, Frölich JC (1997) Effects of meloxicam and indomethacin on cyclooxygenase pathways in healthy volunteers. *J Investig Med* 45: 44–49
77 Bevis PJR, Bird HA, Lapham G (1996) An open study to assess the safety and tolerability of meloxicam 15 mg in subjects with rheumatic disease and mild renal impairment. *Br J Rheumatol* 35 (Suppl.1): 56–60
78 Müller FO, Middle MV, Schall R, Terblanche J, Hundt HK, Groenewoud G (1997) An evaluation of the interaction of meloxicam with frusemide in patients with compensated chronic cardiac failure. *Br J Clin Pharmacol* 44: 393–398
79 Hawkey C, Kahan A, Steinbrück K, Alegre C, Baumelou E, Begaud B, Dequeker J, Isomaki H, Littlejohn G, Mau J, Papazoglou S (1998) Gastrointestinal tolerability of meloxicam compared to diclofenac in osteoarthritis patients. *Br J Rheumatol* 37: 937–945
80 Dequeker J, Hawkey C, Kahan A, Steinbruck K, Alegre C, Baumelou E, Begaud B, Isomaki H, Littlejohn G, Mau J, Papazoglou S (1998) Improvement in gastrointestinal tolerability of the selective cyclooxygenase (COX)-2 inhibitor, meloxicam, compared with piroxicam: results of the safety and efficacy large-scale evaluation of COX-inhibiting therapies (SELECT) trial in osteoarthritis. *Br J Rheumatol* 37: 946–951
81 Stichtenoth DO, Wagner B, Frölich JC (1998) Effect of selective inhibition of the inducible cyclooxygenase on renin release in healthy volunteers. *J Investig Med* 46: 290–296
82 Catella-Lawson F, McAdam B, Morrison BW, Kapoor S, Kujubu D, Antes L, Lasseter KC, Quan H, Gertz BJ, FitzGerald GA (1999) Effects of specific inhibition of cyclooxygenase-2 on sodium balance, hemodynamics, and vasoactive eicosanoids. *J Pharmacol Exp Ther* 289: 735–741
83 Rossat J, Maillard M, Nussberger J, Brunner HR, Burnier M (1999) Renal effects of selective cyclooxygenase-2 inhibition in normotensive salt depleted subjects. *Clin Pharmacol Ther* 66: 76–84
84 Whelton A, Schulman G, Wallemark C, Drower EJ, Isakson PC, Verburg KM, Geis GS (2000) Effects of celecoxib and naproxen on renal function in the elderly. *Arch Int Med* 160: 1465–1470
85 Schwartz JI, Vandormael K, Malice MP, Kalyani RN, Lasseter KC, Holmes GB, Gertz BJ, Gottesdiener KM, Laurenzi M, Redfern KJ et al (2002) Comparison of rofecoxib, celecoxib, and naproxen on renal function in elderly subjects receiving a normal salt diet. *Clin Pharmacol Ther* 72: 50–61
86 Swan SK, Rudy DW, Lasseter KC, Ryan CF, Buechel KL, Lambrecht LJ, Pinto MB, Dilzer SC, Obrda O, Sundblad KJ et al (2000) Effect of cyclooxygenase-2 inhibition on renal function in elderly persons receiving a low salt diet. *Ann Intern Med* 133: 1–9
87 Perazella MA, Eras J (2000) Are selective COX-2 inhibitors nephrotoxic? *Am J Kidney Dis* 35: 937–940
88 Stafford C, Bestoso JT (2000) Celecoxib-induced acute renal failure. *J Am Soc Nephrol* 11: 134A
89 Gadalean F, Barreto G, Epstein M (2001) Acute renal failure induced by COX-2 specific inhibitors. *J Am Soc Nephrol* 12: 779A
90 Briefing document for Celebrex™ (celecoxib) (2001) FDA Arthritis Advisory Committee, 7.2.2001, NDA 20–998
91 Briefing document for Vioxx™ (rofecoxib) (2001) FDA Arthritis Advisory Committee, 8.2.2001, NDA 21-042
92 Whelton A, Maurath CL, Verburg KM, Geis GS (2000) Renal safety and tolerability of celecoxib, a novel cyclooxygenase-2 inhibitor. *Am J Ther* 7: 159–175
93 Scott LJ, Lamb HM (1999) Rofecoxib. *Drugs* 58: 499–505
94 Whelton A, Fort JG, Puma JA, Normandin D, Bello AE, Verburg KM for the SUCCESS VI Study Group (2001) Cyclooxygenase-2-specific inhibitors and cardiorenal function: A randomized, controlled trial of celecoxib and rofecoxib in older hypertensive osteoarthritis patients. *Am J Ther* 8: 85–95
95 Geba GP, Weaver AL, Polis AB, Dixon ME, Schnitzer TJ for the VACT Group (2002) Efficacy of rofecoxib, celecoxib, and acetaminophen in osteoarthritis of the knee – a randomized trial. *J Am Med Assoc* 287: 64–71
96 Stichtenoth DO, Frölich JC (2003) The second generation of COX-2 inhibitors: What advantages do the newest offer? *Drugs* 63: 33–45
97 Product Information Bextra™, valdecoxib (2001) Pharmacia Corporation, Chicago, IL
98 Schnitzer T, Geusens P, Hasler P, Patel SK, Poor G, Senftleber J, Jayawardene S, Sloan VS, Macerata RS (2000) Efficacy and safety of COX-189 in osteoarthritis: a multinational study. *Arthritis Rheum* 43 (Suppl.): S336

Edited by M. Pairet and J. van Ryn

Effects on the lungs: role of COX-2 inhibitors

R. Stokes Peebles, Jr.[1] and Koichi Hashimoto[2]

[1] *Division of Allergy, Pulmonary and Critical Care Medicine, Vanderbilt University School of Medicine, Nashville, TN 37232-2650, USA*
[2] *Department of Microbiology, School of Medicine, Fukushima Medical University, 1st Hikariga-oka Fukushima, Fukushima, Japan 960-1295*

Introduction

Intervention trials using cyclooxygenase (COX)-2 inhibitors in both humans and animals reveal that this enzyme has important regulatory roles in myriad facets of lung biology. These trials also indicate that COX-2 inhibitors have tremendous potential to treat several diseases for which there are currently suboptimal therapeutic options. In this chapter, we will first outline data regarding COX-2 expression in the lung. We will then focus on various lung disease models in which COX-2 expression is increased and is thought to have a pathogenic role, and then review investigations in which COX-2 inhibition has altered these models (Tab. 1).

COX-2 expression in the lung begins in the fetus [1]. Immunostaining of the lungs of 16 to 32 week gestational age human fetuses that were aborted for

Table 1. Pharmacologic agents detailed in this review

Pharmacologic agent	References
COX-2 inhibitors	
JTE-522	[16, 52, 63]
NS-398	[17, 18, 52, 62, 64, 65, 67, 68, 83]
Nimesulide	[20]
Celecoxib	[21, 24, 25, 45, 59, 81]
Rofecoxib	[53, 60]
SC-58236	[51, 63, 69]
DUP-697	[63]
FK3311	[74, 75]
Nonspecific COX inhibitors]	
Aspirin	[17, 18, 23, 24]
Indomethacin	[27–29, 46–48, 61, 64]
Thromboxane receptor antagonist	
SQ29.548	[67, 68]

fetal anomalies, but that had normal lungs at autopsy, revealed that COX-2 expression was present in the epithelial cells of the alveoli and bronchi with no expression in the interstitium or endothelium [1]. Bronchial staining was localized to the ciliated epithelial cells and alveolar staining occurred in cells resembling type II pneumocytes and cuboidal cells [1]. In pre-term infants (gestational age 23 to 30 weeks) who died of immaturity and respiratory distress syndrome with hyaline membranes, alveolar staining was greatest for COX-2 in cells resembling type II pneumocytes as well as in the areas of the hyaline membranes [1]. Compared to pre-term infants, there was a similar distribution of COX-2 staining in term infants with gestational age of 38 to 42 weeks who died of cardiac abnormalities; however, the staining was less intense [1)]. Thus, there seems to be a negative correlation between the intensity of COX-2 staining and gestational age. These results imply biologic function for COX-2 in the developing human lung that may be regulated by alveolar type II pneumocytes. In adults, COX-2 expression is negligible in lung epithelium from normal controls [2, 3].

Lung cancer

Prostaglandins affect cell proliferation and alter the immune system response to malignant cells, therefore the COX enzymes may directly affect tumorigenesis. For instance, the COX product prostaglandin E_2 (PGE_2) is suspected to promote the development of tumors by augmenting angiogenesis and decreasing immune surveillance [2]. There are several lines of evidence that PGE_2 may have a role in the regulation of lung cancers. First, increased levels of PGE_2 are present in bronchoalveolar lavage (BAL) fluid in patients with either primary lung cancer or with extrathoracic cancer with lung metastases compared to BAL fluid from patients with extrathoracic cancer without lung metastases or patients with benign lung disease [4]. In addition, microsomal PGE_2 synthase is upregulated in non small cell lung cancers [5]. *In vitro* studies examining the regulation of PGE_2 production in tumor cell lines reveal that COX-2, but not COX-1, inhibition decreases tumor related PGE_2 production [6]. Additionally, the unstable COX product intermediary, PGH_2, is able to isomerize to form malondialdehyde, a potent mutagen, by both enzymatic and nonenzymatic reactions [7]. Finally, peroxidative activity of COX-2 can augment free radical damage by oxidating aromatic and heterocyclic amines and dihydrodiol derivatives [8].

While COX-1 expression in neoplastic tissue is reported to be unchanged from its level of constitutive expression in normal lung [2, 9], there is variable expression of COX-2 dependent upon tumor type and histologic grade. For instance, COX-2 expression seems to be greatest in adenocarcinoma of the lung, with much lower COX-2 expression in squamous cell lung carcinomas and almost absent expression in small cell lung cancers [9–11]. COX-2 expression is reported to be increased in epithelial lung tissue immediately adjacent to neoplastic tissue, while not present in lung epithelium distant to the neoplasm, suggesting possible paracrine regulation [2]. There has been conflicting data on the correlation of COX-2 expression in adenocarcinoma of the lung and histologic grade of the neoplasm. While some report that COX-2 mRNA expression by Northern Blot analysis and COX-2 immunostaining of histologic sections was greatest in well-differentiated adenocarcinomas with much lower expression in poorly differentiated adenocarcinoma [11], others have found consistently high level of COX-2 expression within lung adenocarcinomas irrespective of histologic grade [2, 11]. High levels of COX-2 expression have been noted in invasive adenocarcinomas, with even greater expression in lymph node metastases than in the corresponding primary tumor [2, 10].

The prognostic significance of elevated COX-2 expression in lung cancers has also been investigated by several groups. Achiwa et al. found that there was no relationship between increased COX-2 expression in tumors and clinical outcome in a cohort of 130 patients with resected lung adenocarcinoma. However, after analyzing the prognostic significance of increased COX-2 expression after stratification according to disease stage, a significant relationship between elevated COX-2 expression and shortened survival was observed only in a cohort of patients with stage I disease [12]. Similar results were published by Khuri et al. who found that in patients with stage I non-small cell lung cancer, the strength of COX-2 expression correlated with a significantly worsened survival rate. In this series, the median survival times for patients whose tumors exhibited either strong, intermediate or weak, and no COX-2 expression was 1.04, 5.50, and 8.54 years, respectively [13]. Brabender et al. also found that in curatively resected non-small cell lung cancer, high COX-2 expression in the tumor tissue was significantly associated with poor survival with multivariate analysis revealing that high COX-2 expression was an independent predictor of decreased survival in this patient population [14]. However, others have found no relationship between COX-2 expression and survival [15].

There is a wide range of evidence suggesting that COX-2 inhibitors might be useful in the treatment of lung cancers (Tab. 2). The COX-2 selective inhibitor JTE-522 inhibited growth of the human lung adenocarcinoma cell line ACC-LC-319 *in vitro* as a single agent and also improved the antitumor effect of the anti-cancer drugs docetaxel, vinorelbine, and amrubicin [16]. This same COX-2 inhibitor significantly inhibited the growth of ACC-LC-319 cells xenografted into athymic nude mice [16]. In this *in vivo* model, JTE-522 enhanced the tumor inhibitory effects of docetaxel and vinorelbine, but not amrubicin [16]. JTE-522 also had antiangiogenic effects by reducing the microvessel count within tumors *in vivo* [16].

The COX-2 inhibitor NS-398, as well as aspirin, significantly inhibited tumor multiplicity in a murine model of lung tumorigenesis caused by administration of the tobacco-specific nitrosamine 4-(methylnitrosamino)-1-(3-pyridyl)-1-butanone given in the drinking water for 7 weeks [17]. In this

Table 2. Studies examining the effects of COX-2 inhibitors in cancer models

COX-2 inhibitor	Effect	Reference
JTE-522	inhibited growth of lung adenocarcinoma cell line ACC-LC-319 *in vitro*	[16]
JTE-522	inhibited the growth of ACC-LC-319 cells xenografted into athymic nude mice	[16]
JTE-522	reducing the microvessel count within tumors *in vivo*	[16]
NS-398	inhibited tumor multiplicity in a murine model of lung tumorigenesis	[17]
nimesulide	decreased proliferation of non-small cell lung cancer cell lines *in vitro*	[20]
celecoxib	reduced the growth of Lewis lung carcinoma cells *in vivo*	[21]
celecoxib	reduced growth of Lewis lung carcinoma cells *in vivo*	[22]
celecoxib	not effective in preventing the development of lung tumors	[23]
celecoxib	did not inhibit tumor multiplicity caused by urethane	[23]

model, plasma PGE_2 levels were increased four-fold, but were reduced by both NS-398 and aspirin [17]. In addition, a concentration of either aspirin or NS-398 greater than 100 μM inhibited the proliferation of both 82-132 and LM2 murine lung tumor cells by 50% [17]. In this same model, aspirin and NS-398 significantly increased the apoptotic index, upregulated Fas-L and BAD expression, and decreased expression of cyclin B2 [18].

NS-398 also regulates secretion of cytokines by tumor cells. A549 non-small cell lung cancer cells upregulate COX-2 expression and secrete PGE_2 in response to IL-1β, effects that are blocked by NS-398 [19]. NS-398 inhibited the ability of IL-1β-stimulated A549 cells to induce IL-10 in lymphocytes and macrophages and reversed the tumor-derived PGE_2-dependent inhibition of macrophage IL-12 production when whole blood was cultured in tumor supernatant [19]. Another COX-2 inhibitor, nimesulide, decreased proliferation of non-small cell lung cancer cell lines in a dose-dependent fashion *in vitro* by inducing apoptosis [20]. In this study, there was a substantial relationship between the expression of COX-2 within the tumor cell lines and the effectiveness of nimesulide in inhibiting the growth of the individual tumor lines [20].

Celecoxib, another selective COX-2 inhibitor also has significant anti-tumor effects *in vivo* in some murine models, but not all [21]. Celecoxib administration from the time of implantation of Lewis lung carcinoma cells into the paws of C57/Bl6 mice significantly reduced the growth of these tumors in a dose-dependent fashion (Fig. 1) [21]. In this same experiment, at higher doses celecoxib also reduced the number and size of metastatic nodules in the lung [21]. In a different murine model in which Lewis lung carcinoma cells are implanted subcutaneously, celecoxib reduced tumor growth *in vivo* and also inhibited *in vitro* fibroblast production of vascular endothelial growth

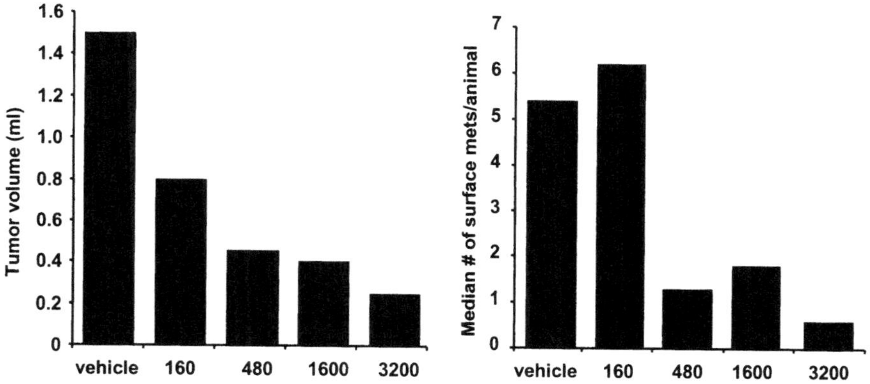

Figure 1. Celecoxib potently inhibited tumor growth and metastasis in the syngeneic Lewis lung model. Lewis lung carcinoma (10^6 cells) was implanted into the paws of C57/Bl6 mice to develop tumors. Celecoxib was administered to mice (n = 20/treatment arm) at 0, 160, 480, 1600, and 3200 ppm in their food from the day of implant. Celecoxib dose-dependently inhibited primary tumor growth (left panel), and number of surface metastatic lesions (right panel).

factor (VEGF) by 92% [22]. However, neither celecoxib nor aspirin were effective in preventing the development of lung tumors induced in a two stage protocol in which dosing of 3-methylcholanthrene precedes chronic butylated hydroxytoluene administration [23]. In another model, celecoxib did not inhibit tumor multiplicity caused by urethane [23]. In fact, the urethane-induced lung tumors were larger in the mice fed chow in which celecoxib was added, even though the tumors in these mice produced 60% less PGE_2 compared to mice fed regular chow [23]. Therefore, COX-2 inhibitors are effective in preventing tumor growth in some, but not all, murine models of lung neoplasia.

Although there have been no studies examining the protective effect of COX-2 inhibitor use in human lung cancer, a prospective study revealed that COX inhibition might be effective in preventing tumor development. For instance, regular users of aspirin had a significantly lower incidence of lung cancer 0.68 (95% CI, 0.49–0.94) [24].

Radiation injury

Given the protective effect of COX-2 inhibitors in some models of lung cancer, it is important to understand the interaction of these agents with forms of cancer therapies that might be used to treat this disease. Some prostaglandins have protective effects against ionizing radiation and inhibit radiation injury [25, 26]. In fact, prostaglandin production by neoplasms is linked to resistance to radiotherapy [25, 27] and inhibition of prostaglandin synthesis by indomethacin increases radiosensitivity [27–29]. Selective COX-2 inhibition also increases the response to radiation. In an *in vivo* model, NS-398 augmented the effect of radiation in delaying growth of human lung cancer cells that expressed COX, NCI-H460, implanted into the left thigh of athymic nude mice [30]. NS-398 did not have this same effect on HCT-116 human colon cancer cells which did not express COX-2, suggesting that the NS-398 enhanced radiosensitivity was dependent upon the cell line's characteristic of expressing COX-2 [30]. Radiation induced apoptosis was also augmented by NS-398 treatment in both H460 cells and rat intestinal epithelial cells stably transfected with COX-2 DNA, but not in either rat intestinal epithelial cells stably transfected with antisense COX-2 DNA nor in HCT-116 cells [30]. These limited results infer that COX-2 inhibitors augment the effect of radiation on neoplasms that express COX-2, but have no effect on tumors that do not express COX-2. Thus, COX-2 inhibitors could have a high potential for increasing the therapeutic ratio of radiotherapy.

Asthma

There have been conflicting reports regarding the expression of COX-2 in the lungs from asthmatic subjects. While a four-fold increase in bronchial epithelial COX-2 immunostaining in asthmatic subjects compared to non-asthmatic controls has been reported [31], another group found no significant differences in the level of immunostaining among asthmatics, chronic bronchitics, or controls with no lung disease [32]. Redington et al. found that COX-2 mRNA expression and immunoreactive protein was increased in the airway epithelium of asthmatics that had not been treated with corticosteroids compared with non-asthmatic controls, while asthmatics receiving corticosteroid therapy had decreased COX-2 expression [33].

COX products have diverse effects on the inflammatory cascade and are presumed to have a pathogenic role in asthma [34, 35]. PGD_2 is increased in the BAL fluid of asthmatic subjects while both PGD_2 and $PGF_{2\alpha}$ contract human airways smooth muscle *in vitro* and are potent bronchoconstrictors *in vivo* [36, 37]. Thromboxane also causes human airway smooth muscle contraction *in vitro* [38]. On the other hand, PGI_2 causes relaxation of isolated precontracted human bronchi, but has minimal effects on airway function *in vivo* [37, 39]. There are several lines of evidence that suggest that PGE_2 has important immunomodulatory effects in human allergic airway inflammation. For instance, PGE_2 inhalation inhibits the pulmonary immediate and late phase responses to inhaled allergen [40, 41]. Compared to vehicle inhalation, inhaled PGE_2 also decreases the change in methacholine airway reactivity and reduces the number of eosinophils after inhaled allergen challenge [40]. There is also evidence that PGE_2 generated during viral respiratory tract infections modulates allergic inflammation in the lung [42, 43]. The results from these studies suggest that PGE_2 has a greater immunomodulatory effect than direct effect on airway caliber. This notion is supported by the fact that PGE_2 inhalation prior to segmental allergen challenge significantly reduced the BAL levels of PGD_2, an important product of mast cell degranulation, and the cysteinyl leukotrienes [44].

To our knowledge, at this time there have been no published studies investigating the effect of selective COX-1 or COX-2 inhibitors on allergic inflammation or asthma in humans; however, in a small study of eight subjects, celecoxib had no effect on pulmonary function, bronchial responsiveness, or cough reflex sensitivity in asthmatics [45]. The role of the COX enzymes in allergic bronchospasm has been examined by treating subjects with indomethacin, a non-selective COX inhibitor that blocks both COX-1 and COX-2, prior to allergen challenge. Indomethacin had no effect on baseline pulmonary function in either allergic asthmatics or nonasthmatic subjects with allergic rhinitis [46]. However, after indomethacin treatment, nonasthmatic subjects with allergic rhinitis had greater sensitivity to allergen challenge as measured by change in the forced expiratory volume in 1 s (FEV_1) and specific airway conductance [46]. Indomethacin treatment prior to allergen challenge caused a small but significant decrease in specific airway conductance in the allergic asthmatic subjects, yet indomethacin had no effect on allergen-induced changes in FEV_1 [46]. Other investigators found that indomethacin treatment had no significant effect on baseline airway responsiveness to histamine or any effect on either the immediate or late phase pulmonary response to allergens in allergic asthmatics [47, 48].

Animal studies suggest that COX-2 is an important modulator of allergic inflammation in the lung. Mice deficient in the COX-1 and COX-2 synthase genes had heightened pulmonary eosinophilia and serum IgE levels compared to wild type mice, while airway hyperresponsiveness was present only in the COX-1 deficient mice [49]. As COX-1 critically affects the transition from $CD4^-CD8^-$ cells to $CD4^+CD8^+$ cells and COX-2 has a very important role in early thymocyte proliferation and differentiation and subsequent maturation of the $CD4^+$ T-lymphocyte lineage [50], experiments were performed using pharmacologic inhibitors of COX-1 and COX-2 to avoid any confounding factors of developmental T-lymphocyte abnormalities in COX-1 and COX-2 deficient mice. Mice treated with a COX-1 inhibitor (SC58560), a COX-2 inhibitor (SC58236), or indomethacin during the development of allergic airway disease all had augmented lung levels of IL-13, increased lung eosinophilia, and airway responsiveness compared to vehicle-treated mice [51]. Since COX inhibition during the development of allergic airway disease resulted in increased allergic inflammation and airway responsiveness, these results suggest that a COX product may restrain allergic inflammation and might be a therapeutic target for the treatment of asthma. It is very important to recognize that in these two murine studies, COX inhibition was present throughout the entire development of allergic disease, from the initial stage of antigen presentation and throughout all airway allergen challenges. In addition, the antigen sensitization procedure consisted of an injection of intraperitoneal ovalbumin formulated with aluminum hydroxide. Different results were obtained in a study performed in guinea pigs that utilized an allergen sensitization protocol of two ovalbumin aerosols administered a week apart [52]. In this study, selective COX-2 inhibitors (NS-398 and JTE-522) administered only during the period of airway challenge each inhibited allergic inflammation in the lung, yet had no effect on airway responsiveness to histamine [52].

Approximately 10% of asthmatics will experience an asthma exacerbation after using aspirin or other NSAIDS [53]. Patients with aspirin intolerant asthma (AIA) have marked overexpression of leukotriene (LT) C_4 synthase, an enzyme that converts LTA_4 to the cysteinyl leukotrienes LTC_4, LTD_4, and LTE_4 [54, 55]. Some investigators theorize that the increase in the cysteinyl leukotrienes, mediators that are known to have a pathogenic role in asthma, may be critical in producing AIA [56]. Inhaled PGE_2 attenuated the decrease in pulmonary function that occurs when subjects with AIA are challenged with inhaled lysine acetylsalicylate and also decreased urinary LTE_4 excretion, a marker of LTC_4 synthase activity [57]. *In vitro* assays confirm this protective effect of PGE_2 in AIA. Although there was no difference in cysteinyl leukotriene production in peripheral blood leukocytes from subjects with AIA and either asthmatics who were tolerant of asthma or non-asthmatic controls, the peripheral leukocytes from the AIA subjects had a significant increase in cysteinyl leukotrienes after aspirin stimulation that was not present in the leukocytes from the other two groups [58]. When PGE_2 was added to the aspirin stimulated cells from the AIA subjects, there was a significant decrease in cysteinyl leukotriene levels [58]. This suggests that there is a significant inhibitory effect of PGE_2 on the production of the cysteinyl leukotrienes in AIA subjects. Recent studies showed that AIA subjects do not develop bronchospasm when they take the selective COX-2 inhibitors celecoxib [59] and rofecoxib [53, 60]. These studies suggest that COX-1 inhibition may be critical for aspirin intolerance [53]. Thus, a COX-1 product, possibly PGE_2, may be critical in restraining the activation of cells responsible for the overproduction of bronchoconstrictor leukotrienes.

Sepsis

Sepsis induces pulmonary inflammation with augmented production of COX products. To our knowledge, there are no clinical trials using specific COX-2 inhibitors in sepsis. However, a non-selective COX inhibitor, ibuprofen, did not reduce the incidence or duration of shock or the acute respiratory distress syndrome and did not significantly improve the rate of survival at 30 days in septic patients compared to placebo treatment [61]. Several animal models have been developed to determine the contribution of COX-2 to sepsis-induced lung injury through the use of selective COX-2 inhibitors (Tab. 3). Most of these models utilize lipopolysaccharide (LPS) administration as the experimental correlate of sepsis-induced injury.

LPS has also been shown to increase COX-2 expression in isolated perfused rat lungs when subtoxic doses of LPS were administered through the pul-

Table 3. Studies examining the effects of COX-2 inhibitors in models of sepsis

COX-2 Inhibitor	Effect	Reference
NS-398	prevented bronchoconstriction induced by LPS and GM-CSF	[62]
NS-398	prevented LPS-induced pulmonary artery pressure, ventilation pressure	[63]
DUP-697	prevented LPS-induced pulmonary artery pressure, ventilation pressure	[63]
SC-236	prevented LPS-induced pulmonary artery pressure, ventilation pressure	[63]
NS-398	attenuated bradykinin-induced vasoconstriction caused by LPS	[64]
NS-398	decreased LPS-induced pyrexia and mortality at 48, but not 72 h	[65]
NS-398	COX-2 KO mice had augmented LPS-induced airway responsiveness	]66]
NS-398	inhibited bronchoconstriction caused by combination of TNF-α, IL-1β, and IFN-γ	[67]
NS-398	abrogated LPS-induced vascular hyperreactivity, only a partial effect on airway hyperreactivity	[68]
SC-236	maintained perfusion distribution after oleic acid and LPS administration	[69]

monary arteries. Wollin and colleagues investigated the effect of GM-CSF in this model to mimic the administration of GM-CSF that is used to treat septic patients when they are neutropenic after chemotherapy for cancer therapy [62]. In this model, the combination of GM-CSF and concomitant LPS administration induced release of thromboxane and TNF into the venous perfusate and also caused airway bronchoconstriction. Both COX-2 inhibition by NS-398 and the administration of a thromboxane receptor antagonist prevented bronchoconstriction but had no effect on TNF release. Anti-TNF antibodies had no effect on bronchoconstriction suggesting that the bronchoconstriction induced by the combined administration of LPS with GM-CSF is likely TNF-independent. In an *in vivo* model, these same investigators found that rats made neutropenic by the intravenous injection of vinblastine were protected against the lung injury caused by the co-administration of GM-CSF and LPS [62].

Isolated rat lungs have been used in other models to study the effects of LPS on lung physiology. For instance, isolated rat lungs were perfused with either buffered-saline or plasma and concentrations of LPS ranging from 1,000 to 10,000 ng/ml during a 2-h priming period. This was associated with LPS-induced production of thromboxane, PGI_2, and TNFα and with increased pulmonary artery pressure, ventilation pressure, and lung weight [63]. These changes were inhibited by the administration of the COX-2 inhibitors NS-398, DUP-697, and SC-236 [63]. With LPS priming, anti-TNFα antibodies had no effect on either the augmented pulmonary artery pressure, ventilation pressure, or lung weight when buffered-saline perfusate was administered; however, anti-TNFα antibodies greatly reduced the heightened ventilation pressure but had no effect on the pulmonary artery pressure response and lung edema formation when plasma perfusate was used [63]. These results suggest that COX-2, probably through thromboxane generation, mediates LPS-induced changes in vascular and smooth muscle physiology. In this system, COX-1 mRNA expression in the lung was unchanged by LPS administration; however, COX-2 lung mRNA expression was increased [63]. Immunohistochemical staining revealed that COX-2 expression was present prior to LPS infusion and subsequently increased in bronchial epithelial cells, bronchial and vascular smooth muscle cells, cells within bronchial associated lymphoid tissue and myocytes of the large hilar veins [63]. LPS administration led to *de novo* expression of COX-2 in endothelial cells and most alveolar macrophages. In contrast, LPS administration led to a decrease of COX-2 expression in perivascular and peribronchial macrophages [63].

In another model examining the effect of LPS on isolated rat lungs, COX-2 blocked sepsis-induced endothelial dysfunction. In this experiment, rats were treated with indomethacin, a nonselective COX inhibitor, or NS-398 by intraperitoneal injection 1 h prior to administration of either LPS or saline [64]. Three hours after either LPS or saline was given, lung isolation was perfomed and pulmonary arterial vasoconstriction was measured in response to either bradykinin or hypoxemia. Both indomethacin and NS-398 attenuated bradykinin-induced vasoconstriction in a dose-dependent manner. However, only indomethacin blocked hypoxic pulmonary vasoconstriction that occurred with LPS administration, while NS-398 had no effect [64].

In a murine model of endotoxemia, LPS administered intraperitoneally induced PGE_2 production in both the blood and lung [65]. This augmented PGE_2 production was decreased by NS-398 treatment administered by intraperitoneal injection [65]. NS-398 significantly decreased pyrexia 1 and 2 h after LPS administration, but had no effect on lung myeloperoxidase activity. Twenty-four hours after LPS administration, NS-398 treatment significantly improved mortality rate from endotoxemia, but there was no effect on survival at 72 h [65]. In a model of bacterial peritonitis induced by cecal ligation and puncture, this same group found PGE_2 was increased in BAL fluid, but not in plasma [65]. Since PGE_2 was increased in BAL fluid, they hypothesized that NS-398 might also have a protective effect; however, the COX-2 inhibitor had no effect on mortality in this model of bacterial peritonitis [65].

A model of local pulmonary endotoxemia has also been developed in which LPS was administered via the aerosol route to determine the role of COX products in airway physiology [66]. In this model, COX-1 and COX-2 deficient mice had augmented airway responsiveness to methacholine compared to wild type mice, although the number of BAL fluid total cells and neutrophils were not different between the groups. In addition, neither the lung histopathology, BAL fluid total protein, nor BAL leukotriene levels were different between the COX-1 deficient, COX-2 deficient, and wild type mice [66].

These results regarding LPS-induced bronchoconstriction are in contrast to a model in which administration of cytokines that are produced as a result of endotoxemia induce bronchoconstriction [67]. In a model utilizing viable lung slices with intact microanatomy harvested from rats, the combination of TNF-α, IL-1β, and IFN-γ induced COX-2 expression, generation of thromboxane, and airway constriction [67]. This bronchoconstriction was hypothesized to be caused by the induction of thromboxane by these cytokines, as administration of NS-398 inhibited bronchoconstriction as well as thromboxane production. The hypothesis that thromboxane was the critical element in the bronchoconstriction was confirmed when the thromboxane receptor antagonist SQ29.548 fully inhibited the bronchoconstriction caused by combined TNF-α, IL-1β, and IFN-γ administration [67].

Isolated, blood-free perfused lungs from mice have also been used to determine the effect of LPS on lung function [68]. LPS administration through the pulmonary vasculature caused airway hyperresponsiveness to methacholine as well as pulmonary vascular hyperreactivity to platelet activating factor, while it had no effect on either airway resistance or pulmonary artery pressure. Thromboxane is a key element in these LPS-induced physiologic effects as witnessed by the complete inhibition of the airway hyperresponsiveness and pulmonary vascular hyperreactivity by the thromboxane receptor antagonist SQ29.548. COX-2 inhibition with NS-398 completed abrogated LPS-induced vascular hyperreactivity but had only a partial effect on airway hyperreactivity caused by LPS administration. This suggests that the thromboxane generated by a COX-2-independent pathway mediates LPS-induced airway hyperresponsiveness [68].

One of the other pulmonary consequences of sepsis is acute lung injury. Administration of oleic acid has been used as a model of acute lung injury because it shares some pathophysiologic similarities to the adult respiratory distress syndrome (ARDS), one of which is the development of lung edema [69]. One mechanism to protect against hypoxemia that can result from lung edema is the redistribution of pulmonary perfusion away from injured edematous lung regions to maintain a relatively normal ventilation/perfusion pattern. In the model of oleic acid injury in which LPS is administered prior to oleic acid, this redistribution of pulmonary perfusion does not occur, resulting in decreased oxygenation. The prevention of redistribution of pulmonary perfusion that occurs with LPS administration is associated with increases in PGI_2 in both plasma and tissue. In a canine model of LPS administration prior to oleic acid-induced lung injury, the COX-2 inhibitor SC-236 maintained perfusion distribution after the lung injury occurred and inhibited the increased levels of plasma and tissue PGI_2 that occurred with placebo administration [69].

As summarized above, COX-2 inhibition has profound effects on LPS-mediated physiology in these animal models. These include protection against LPS-induced bronchoconstriction, lung inflammation, pulmonary artery vasoconstriction, pulmonary vascular hyperreactivity, and maintenance of protective redistribution of pulmonary perfusion in the setting of lung edema. Whether these protective effects of COX-2 inhibition are translatable into human sepsis is unknown until clinical trials using COX-2 inhibitors are performed in this condition.

Ischemia/reperfusion injury

Hemorrhagic shock results in an inflammatory response that can lead to organ injury and potentially multi-organ failure. One of the consequences of hemorrhagic shock is COX-2 induction, possibly through NF-κB activation and a subsequent increase in NF-κB binding to the COX-2 promoter, thus upregulating COX-2 gene transcription [70–72]. For example, a seven-fold increase in COX-2 expression has been reported in a rat model in which lungs were subjected to hemorrhagic shock [73]. In this model, COX-2 mRNA upregulation in the lungs was independent of reperfusion and occurred during ischemic phase of hemorrhagic shock [73]. To test the contribution of COX-2 products in ischemia-reperfusion injury, a selective COX-2 inhibitor, FK-3311 was used in a canine model [74]. In these experiments, FK-3311 was administered 15 min prior to ischemia and 15 min before reperfusion and this resulted in improved left pulmonary vascular resistance, cardiac output, arterial oxygen pressure, and wet-to-dry ratio measurements compared to placebo-treated animals [74]. Treatment with FK-3311 inhibited neutrophil infiltration of the lungs and improved survival. FK-3311 treatment reduced serum thromboxane, while levels of 6-keto-$PGF_{1\alpha}$, a prostacyclin (PGI_2) metabolite, were unchanged. Although the beneficial effects of FK-3311 in this model are not fully elucidated, inhibition of thromboxane is a potential mechanism [74]. In a canine model of ischemia-reperfusion using orthotopic lung transplantation, FK-3311improved pulmonary gas exchange, reduced neutrophilic infiltration, and protected against damage as assessed by histologic analysis as compared to vehicle treated animals. In this model, thromboxane was significantly decreased while the PGI_2 metabolite was not changed [75]. Therefore, in two separate models of ischemia-reperfusion injury in dog models, selective COX-2 inhibition is protective.

Idiopathic pulmonary fibrosis

Idiopathic pulmonary fibrosis (IPF) is a disease characterized by the proliferation of lung fibroblast and subsequent collagen deposition by these cells. A possible role for COX-2 products to regulate IPF was hypothesized after discovery that fibroblasts from patients with pulmonary fibrosis are unable to increase COX-2 expression in response to non-specific stimuli such as PMA, or IL-1 compared to fibroblasts from cancer patients [76]. Several different prostanoid products have been implicated in potentially modulating IPF. For instance, PGE_2 prevents fibroblast proliferation and collagen production. A possible role for PGE_2 in pulmonary fibrosis was speculated when it was discovered that lung fibroblasts from patients with pulmonary fibrosis have a two-fold decrease in basal PGE_2 production compared to fibroblasts from patients having surgery for lung cancer resection [76]. Therefore, this deficiency in the ability to produce PGE_2 as a result from the decreased induction of COX-2 may contribute to the inflammatory and fibrogenic response in pulmonary fibrosis. However, another group has found that PGE_2 production is not different from IL-1β-stimulated fibroblasts from IPF patients or control subjects [77]. In this setting fibroblasts from IPF patients have greater IL-1β stimulated production of thromboxane and a lower PGI_2 to thromboxane ratio compared to fibroblasts from control subjects [77]. Several animal models have been developed to dissect the role of COX-2 products in pulmonary fibrosis. Wild type mice administered bleomycin intratracheally have a mild fibrotic reaction 14 days after bleomycin administration which is patchy and with moderate interstitium thickening [78]. However, bleomycin-treated mice deficient in COX-2 experience an aggressive fibroproliferative response with increased neutrophilic and lymphocytic inflammation and complete loss of alveolar architecture [78]. The role of COX-2 in pulmonary fibrosis has also been investigated in another model in which vanadium pentoxide was used as the fibrotic stimulus [79]. The lungs from COX-2 deficient mice were severely inflamed three days after vanadium pentoxide intratracheal administration with pulmonary fibrosis developing two weeks after vanadium pentoxide administration, while these changes were not present in either COX-1 deficient or wild type mice. After vanadium pentoxide treatment, bronchoalveolar PGE_2 levels were increased in the wild type and COX-1 deficient mice, but not in the COX-2 deficient mice [79]. An *in vitro* model using cells from either COX-1 deficient, COX-2 deficient, or wild type mice also suggest that COX-2 regulates fibrogenesis [80]. Alveolar epithelial cells from wild type and COX-1 deficient mice, but not from COX-2 deficient mice, suppress serum-induced fibroblast proliferation. Alveolar epithelial cells from wild type and COX-1 deficient mice also have increased PGE_2 synthesis after stimulation with either LPS or IL-1, while these same cells from COX-2 deficient mice do not [80]. To our knowledge, there have been no reports examining COX-2 specific inhibitors in models of pulmonary fibrosis; however, these human and animal studies strongly suggest that COX-2 is a critical regulator of this disease.

Pleurisy

Carrageenan administered into the pleural space results in acute tissue inflammation and has been used to model pleurisy [81]. In this model, the pleural cavity inflammatory response is characterized by pleural effusion containing large numbers of neutrophils, in addition to lung tissue infiltration of neutrophils, resultant lipid peroxidation, and increased production of PGE_2, TNF-α, and IL-1β. A role for COX-2 mediation of this inflammatory response was shown when celecoxib inhibited the consequences of carrageenan administration [81]. Celecoxib treatment decreased ICAM-1 and P-selectin expression on bronchial epithelium and resulted in a dose-dependent decrease in pleural exudate volume, accumulation of neutrophils in the pleural space, lung myeloperoxidase activity, and lung malondialdehyde activity. In addition, celecoxib decreased pleural fluid concentrations of TNF-α, IL-1β, PGE_2, and the metabolite of PGI_2. These results suggest that COX-2-derived prostanoids mediate the pro-inflammatory effects in the carrageenan model of pleurisy [81].

Pancreatitis-induced lung injury

The lung is the extra-abdominal organ most frequently damaged in patients with pancreatitis [82]. Approximately one-third of patients with pancreatitis will develop acute lung injury and the acute respiratory distress syndrome. These complications of pancreatitis are responsible for 60% of pancreatitis-related deaths within the first week of onset of illness [82]. A mouse model of pancreatitis induced by the administration of the cholecystokinin analog cerulein was used to determine the role of COX products in pancreatitis and pancreatitits-associated lung injury [83]. Cerulein administration induced pancreatitic COX-2 expression, but had no effect on COX-1 expression [83]. In addition, cerulein caused marked necrosis, vacuolization, and inflammatory cell sequestration within the pancreas and marked edema and neutrophilic inflammation in the lung with augmented myeloperoxidase activity. In this model, COX-2 deficient mice were significantly protected against cerulein-induced pancreatitis with markedly reduced severity of lung injury, while this was not true for either COX-1 deficient mice in comparison to wild type mice. Treatment with NS-398 significantly protected against cerulein-induced pancreatitis but there was no mention of its effect in the associated lung disease [83].

Conclusion

The role of COX-2 products in the genesis and pathophysiology of lung disease is becoming clearer with the ability to selectively inhibit COX-2 pharmacologically in both human and animal studies, and through the use of animal models utilizing COX-2 deficient mice. However, we are just in the infancy of recognizing the diseases for which COX-2 inhibitor therapy might be useful. To date, probably the greatest surprise in this field of investigation is the animal data detailing the role of COX-2 in neoplastic disease and the hope that COX-2 inhibitors will provide therapeutic benefit in the treatment of human lung cancer. Most likely, COX-2 inhibition in other models will yield pleasant surprises and add to the motivation to continued study in this area of research.

Acknowledgments
This work was supported by The American Academy of Allergy, Asthma and Immunology ERT Award, American Lung Association of Tennessee, R01-AI-45512, and R01-HL-069449.

References

1 Lassus P, Wolff H, Andersson S (2000) Cyclooxygenase-2 in human perinatal lung. *Pediatr Res* 47: 602–605
2 Soslow RA, Dannenberg AJ, Rush D, Woerner BM, Khan KN, Masferrer J, Koki AT (2000) COX-2 is expressed in human pulmonary, colonic, and mammary tumors. *Cancer* 89: 2637–2645
3 Hasturk S, Kemp B, Kalapurakal SK, Kurie JM, Hong WK, Lee JS (2002) Expression of cyclooxygenase-1 and cyclooxygenase-2 in bronchial epithelium and nonsmall cell lung carcinoma. *Cancer* 94: 1023–1031
4 Funahashi A, Harland RW, LeFever A (1994) Association of increased prostaglandin E2 content in bronchoalveolar lavage fluid and intrathoracic malignancy. *Chest* 106: 166–172
5 Yoshimatsu K, Altorki NK, Golijanin D, Zhang F, Jakobsson PJ, Dannenberg AJ, Subbaramaiah K (2001) Inducible prostaglandin E synthase is overexpressed in non-small cell lung cancer. *Clin Cancer Res* 7: 2669–2674
6 Rozic JG, Chakraborty C, Lala PK (2001) Cyclooxygenase inhibitors retard murine mammary tumor progression by reducing tumor cell migration, invasiveness and angiogenesis. *Int J Cancer* 93: 497–506
7 Plastaras JP, Guengerich FP, Nebert DW, Marnett LJ (2000) Xenobiotic-metabolizing cytochromes P450 convert prostaglandin endoperoxide to hydroxyheptadecatrienoic acid and the mutagen, malondialdehyde. *J Biol Chem* 275: 11784–11790
8 Koki AT, Masferrer JL (2002) Celecoxib: a specific COX-2 inhibitor with anticancer properties. *Cancer Control* 9(2 Suppl): 28–35
9 Watkins DN, Lenzo JC, Segal A, Garlepp MJ, Thompson PJ (1999) Expression and localization of cyclo-oxygenase isoforms in non-small cell lung cancer. *Eur Respir J* 14: 412–418
10 Hida T, Yatabe Y, Achiwa H, Muramatsu H, Kozaki K, Nakamura S, Ogawa M, Mitsudomi T, Sugiura T, Takahashi T (1998) Increased expression of cyclooxygenase 2 occurs frequently in human lung cancers, specifically in adenocarcinomas. *Cancer Res* 58: 3761–3764
11 Wolff H, Saukkonen K, Anttila S, Karjalainen A, Vainio H, Ristimaki A (1998) Expression of cyclooxygenase-2 in human lung carcinoma. *Cancer Res* 58: 4997–5001
12 Achiwa H, Yatabe Y, Hida T, Kuroishi T, Kozaki K, Nakamura S, Ogawa M, Sugiura T, Mitsudomi T, Takahashi T (1999) Prognostic significance of elevated cyclooxygenase 2 expression in primary, resected lung adenocarcinomas. *Clin Cancer Res* 5: 1001–1005
13 Khuri FR, Wu H, Lee JJ, Kemp BL, Lotan R, Lippman SM, Feng L, Hong WK, Xu XC (2001) Cyclooxygenase-2 overexpression is a marker of poor prognosis in stage I non-small cell lung cancer. *Clin Cancer Res* 7: 861–867
14 Brabender J, Park J, Metzger R, Schneider PM, Lord RV, Holscher AH, Danenberg KD, Danenberg PV (2002) Prognostic significance of cyclooxygenase 2 mRNA expression in non-small cell lung cancer. *Ann Surg* 253: 440–443
15 Hosomi Y, Yokose T, Hirose Y, Nakajima R, Nagai K, Nishiwaki Y, Ochiai A (2000) Increased cyclooxygenase 2 (COX-2) expression occurs frequently in precursor lesions of human adenocarcinoma of the lung. *Lung Cancer* 30: 73–81
16 Hida T, Kozaki K, Ito H, Miyaishi O, Tatematsu Y, Suzuki T, Matsuo K, Sugiura T, Ogawa M, Takahashi T et al (2002) Significant growth inhibition of human lung cancer cells both *in vitro* and *in vivo* by the combined use of a selective cyclooxygenase 2 inhibitor, JTE-522, and conventional anticancer agents. *Clin Cancer Res* 8: 2443–2447
17 Rioux N, Castonguay A (1998) Prevention of NNK-induced lung tumorigenesis in A/J mice by acetylsalicylic acid and NS-398. *Cancer Res* 58: 5354–5360
18 Yao R, Rioux N, Castonguay A, You M (2000) Inhibition of COX-2 and induction of apoptosis: two determinants of nonsteroidal anti-inflammatory drugs' chemopreventive efficacies in mouse lung tumorigenesis. *Exp Lung Res* 26: 731–742
19 Huang M, Stolina M, Sharma S, Mao JT, Zhu L, Miller PW, Wollman J, Herschman H, Dubinett SM (1998) Non-small cell lung cancer cyclooxygenase-2-dependent regulation of cytokine balance in lymphocytes and macrophages: up-regulation of interleukin 10 and down-regulation of interleukin 12 production. *Cancer Res* 58: 1208–1216
20 Hida T, Kozaki K, Muramatsu H, Masuda A, Shimizu S, Mitsudomi T, Sugiura T, Ogawa M, Takahashi T (2000) Cyclooxygenase-2 inhibitor induces apoptosis and enhances cytotoxicity of various anticancer agents in non-small cell lung cancer cell lines. *Clin Cancer Res* 6: 2006–2011
21 Koki AT, Khan NK, WoernerBM, Seibert K, Harmon, JL, Dannenberg AJ, Soslow RA, Masferrer

JL (2002) Characterization of cyclooxygenase-2 (COX-2) during tumorigenesis in human epithelial cancers: evidence for potential clinical utility of COX-2 inhibitors in epithelial cancers. *Prostag Leukotr Ess* 66: 13–18
22 Williams CS, Tsujii M, Reese J, Dey SK, DuBois RN (2000) Host cyclooxygenase-2 modulates carcinoma growth. *J Clin Invest* 105: 1589–1594
23 Kisley LR, Barrett BS, Dwyer-Nield LD, Bauer AK, Thompson DC, Malkinson AM (2002) Celecoxib reduces pulmonary inflammation but not lung tumorigenesis in mice. *Carcinogenesis* 23: 1653–1660
24 Schreinemachers DM, Everson RB (1994) Aspirin use and lung, colon, and breast cancer incidence in a prospective study. *Epidemiology* 5: 138–146
25 Hanson WR, Thomas C (1983) 16, 16-dimethyl prostaglandin E_2 increases survival of murine intestinal stem cells when given before photon radiation. *Radiat Res* 96: 393–398
26 Pillsbury HC, III, Webster WP, Rosenman J (1986) Prostaglandin inhibitor and radiotherapy in advanced head and neck cancers. *Arch Otolaryngol Head Neck Surg* 112: 552–553
27 Milas L, Hanson WR (1995) Eicosanoids and radiation. *Eur J Cancer* 31A: 1580–1585
28 Furuta Y, Hunter N, Barkley T, Jr, Hall E, Milas L (1988) Increase in radioresponse of murine tumors by treatment with indomethacin. *Cancer Res* 48: 3008–3013
29 Milas L, Furuta Y, Hunter N, Nishiguchi I, Runkel S (1990) Dependence of indomethacin-induced potentiation of murine tumor radioresponse on tumor host immunocompetence. *Cancer Res* 50: 4473–4477
30 Pyo H, Choy H, Amorino GP, Kim JS, Cao Q, Hercules SK, DuBois RN (2001) A selective cyclooxygenase-2 inhibitor, NS-398, enhances the effect of radiation *in vitro* and *in vivo* preferentially on the cells that express cyclooxygenase-2. *Clin Cancer Res* 7: 2998–3005
31 Sousa A, Pfister R, Christie PE, Lane SJ, Nasser SM, Schmitz-Schumann M, Lee TH (1997) Enhanced expression of cyclo-oxygenase isoenzyme 2 (COX-2) in asthmatic airways and its cellular distribution in aspirin-sensitive asthma. *Thorax* 52: 940–945
32 Demoly P, Jaffuel D, Lequeux N, Weksler B, Creminon C, Michel FB, Godard P, Bousquet J (1997) Prostaglandin H synthase 1 and 2 immunoreactivities in the bronchial mucosa of asthmatics. *Am J Respir Crit Care Med* 155: 670–675
33 Redington AE, Meng QH, Springall DR, Evans TJ, Creminon C, Maclouf J, Holgate ST, Howarth PH, Polak JM (2001) Increased expression of inducible nitric oxide synthase and cyclo- oxygenase-2 in the airway epithelium of asthmatic subjects and regulation by corticosteroid treatment. *Thorax* 56: 351–357
34 Holtzman MJ (1991) Arachidonic acid metabolism. Implications of biological chemistry for lung function and disease. *Am Rev Respir Dis* 143: 188–203
35 Wenzel SE (1997) Arachidonic acid metabolites: mediators of inflammation in asthma. *Pharmacotherapy* 17(Pt 2): 3S–12S
36 Murray JJ, Tonnel AB, Brash AR, Roberts LJ 2nd, Gosset P, Workman R, Capron A, Oates JA (1986) Release of prostaglandin D2 into human airways during acute antigen challenge. *N Engl J Med* 315: 800–804
37 Gardiner PJ, Collier HO (1980) Specific receptors for prostaglandins in airways. *Prostaglandins* 19: 819–841
38 Armour CL, Johnson PR, Alfredson ML, Black JL (1989) Characterization of contractile prostanoid receptors on human airway smooth muscle. *Eur J Pharmacol* 165: 215–222
39 Hardy C, Robinson C, Lewis RA, Tattersfield AE, Holgate S (1985) Airway and cardiovascular responses to inhaled prostacyclin in normal and asthmatic subjects. *Am Rev Respir Dis* 131: 18–21
40 Gauvreau GM, Watson RM, O'Byrne PM (1999) Protective effects of inhaled PGE2 on allergen-induced airway responses and airway inflammation. *Am J Respir Crit Care Med* 159: 31–36
41 Pavord ID, Wong CS, Williams J Tattersfield AE (1993) Effect of inhaled prostaglandin E2 on allergen-induced asthma. *Am Rev Respir Dis* 148: 87–90
42 Varner AE, Busse WW, Lemanske RF, Jr, (1998) Hypothesis: decreased use of pediatric aspirin has contributed to the increasing prevalence of childhood asthma. *Ann Allergy Asthma Immunol* 81: 347–351
43 Varner AE (1999) The cyclooxygenase-2 theory of atopy and asthma. *Pediatr Asthma Allergy Immunol* 13: 43–50
44 Hartert TV, Dworski RT, Mellen BG, Oates JA, Murray JJ, Sheller JR (2000) Prostaglandin E(2) decreases allergen-stimulated release of prostaglandin D(2) in airways of subjects with asthma. *Am J Respir Crit Care Med* 162: 637–640

45 Dicpinigaitis PV (2001) Effect of the cyclooxygenase-2 inhibitor celecoxib on bronchial responsiveness and cough reflex sensitivity in asthmatics. *Pulm Pharmacol Ther* 14: 93–97
46 Fish JE, Ankin MG, Adkinson NF, Jr, Peterman VI (1981) Indomethacin modification of immediate-type immunologic airway responses in allergic asthmatic and non-asthmatic subjects: evidence for altered arachidonic acid metabolism in asthma. *Am Rev Respir Dis* 123: 609–614
47 Kirby JG, Hargreave FE, Cockcroft DW, O'Byrne PM (1989) Effect of indomethacin on allergen-induced asthmatic responses. *J Appl Physiol* 66: 578–583
48 Sladek K, Dworski R, Fitzgerald GA, Buitkus KL, Block FJ, Marney SR, Jr, Sheller JR (1990) Allergen-stimulated release of thromboxane A2 and leukotriene E4 in humans. Effect of indomethacin. *Am Rev Respir Dis* 141: 1441–1445
49 Gavett SH, Madison SL, Chulada PC, Scarborough PE, Qu W, Boyle JE, Tiano HF, Lee CA, Langenbach R, Roggli VL et al (1999) Allergic lung responses are increased in prostaglandin H synthase-deficient mice. *J Clin Invest* 104: 721–732
50 Rocca B, Spain LM, Pure E, Langenbach R, Patrono C, Fitzgerald GA (1999) Distinct roles of prostaglandin H synthases 1 and 2 in T-cell development. *J Clin Invest* 103: 1469–1477
51 Peebles RS, Jr, Hashimoto K, Morrow JD, Dworski R, Collins RD, Hashimoto Y, Christman JW, Kang KH, Jarzecka K, Furlong J et al (2002) Selective cyclooxygenase-1 and -2 inhibitors each increase allergic inflammation and airway hyperresponsiveness in mice. *Am J Respir Crit Care Med* 165: 1154–1160
52 Oguma T, Asano K, Shiomi T, Fukunaga K, Suzuki Y, Nakamura M, Matsubara H, Sheldon HK, Haley KJ, Lilly CM et al (2002) Cyclooxygenase-2 expression during allergic inflammation in guinea-pig lungs. *Am J Respir Crit Care Med* 165: 382–386
53 Szczeklik A, Nizankowska E, Bochenek G, Nagraba K, Mejza F, Swierczynska M (2001) Safety of a specific COX-2 inhibitor in aspirin-induced asthma. *Clin Exp Allergy* 31: 219–225
54 Cowburn AS, Sladek K, Soja J, Adamek L, Nizankowska E, Szczeklik A, Lam BK, Penrose JF, Austen FK, Holgate ST et al (1998) Overexpression of leukotriene C4 synthase in bronchial biopsies from patients with aspirin-intolerant asthma. *J Clin Invest* 101: 834–846
55 Sampson AP, Cowburn AS, Sladek K, Adamek L, Nizankowska E, Szczeklik A, Lam BK, Penrose JF, Austen KF, Holgate ST (1997) Profound overexpression of leukotriene C4 synthase in bronchial biopsies from aspirin-intolerant asthmatic patients. *Int Arch Allergy Immunol* 113: 355–357
56 Picado C, Valero A (2001) COX-1 sparing drugs in aspirin-sensitive asthma. *Clin Exp Allergy* 31: 179–181
57 Sestini P, Armetti L, Gambaro G, Pieroni MG, Refini RM, Sala A, Vaghi A, Folco GC, Bianco S, Robuschi M (1996) Inhaled PGE_2 prevents aspirin-induced bronchoconstriction and urinary LTE4 excretion in aspirin-sensitive asthma. *Am J Respir Crit Care Med* 153: 572–575
58 Celik G, Bavbek S, Misirligil Z, Melli M (2001) Release of cysteinyl leukotrienes with aspirin stimulation and the effect of prostaglandin E_2 on this release from peripheral blood leucocytes in aspirin-induced asthmatic patients. *Clin Exp Allergy* 31: 1615–1622
59 Woessner KM, Simon RA, Stevenson DD (2002) The safety of celecoxib in patients with aspirin-sensitive asthma. *Arthritis Rheum* 46: 2201–2206
60 Martin-Garcia C, Hinojosa M, Berges P, Camacho E, Garcia-Rodriguez R, Alfaya T, Iscar A (2002) Safety of a cyclooxygenase-2 inhibitor in patients with aspirin- sensitive asthma. *Chest* 121: 1812–1817
61 Bernard GR, Wheeler AP, Russell JA, Schein R, Summer WR, Steinberg KP, Fulkerson WJ, Wright PE, Christman BW, Dupont WD et al (1997) The effects of ibuprofen on the physiology and survival of patients with sepsis. The Ibuprofen in Sepsis Study Group. *N Engl J Med* 336: 912–918
62 Wollin L, Uhlig S, Nusing R, Wendel A (2001) Granulocyte-macrophage colony-stimulating factor amplifies lipopolysaccharide-induced bronchoconstriction by a neutrophil- and cyclooxygenase 2-dependent mechanism. *Am J Respir Crit Care Med* 163: 443–450
63 Ermert M, Merkle M, Mootz R, Grimminger F, Seeger W, Ermert L (2000) Endotoxin priming of the cyclooxygenase-2-thromboxane axis in isolated rat lungs. *Am J Physiol Lung Cell Mol Physiol* 278: L1195-L1203
64 Fischer LG, Hollmann MW, Horstman DJ, Rich GF (2000) Cyclooxygenase inhibitors attenuate bradykinin-induced vasoconstriction in septic isolated rat lungs. *Anesth Analg* 90: 625–631
65 Reddy RC, Chen GH, Tateda K, Tsai WC, Phare SM, Mancuso P, Peters-Golden M, Standiford TJ (2001) Selective inhibition of COX-2 improves early survival in murine endotoxemia but not in

bacterial peritonitis. *Am J Physiol Lung Cell Mol Physiol* 281: L537–L543
66 Zeldin DC, Wohlford-Lenane C, Chulada P, Bradbury JA, Scarborough PE, Roggli V, Langenbach R, Schwartz DA (2001) Airway inflammation and responsiveness in prostaglandin H synthase-deficient mice exposed to bacterial lipopolysaccharide. *Am J Respir Cell Mol Biol* 25: 457–465
67 Martin C, Uhlig S, Ullrich V (2001) Cytokine-induced bronchoconstriction in precision-cut lung slices is dependent upon cyclooxygenase-2 and thromboxane receptor activation. *Am J Respir Cell Mol Biol* 24: 139–145
68 Held HD, Uhlig S (2000) Mechanisms of endotoxin-induced airway and pulmonary vascular hyperreactivity in mice. *Am J Respir Crit Care Med* 162: 1547–1552
69 Gust R, Kozlowski JK, Stephenson AH, Schuster DP (1999) Role of cyclooxygenase-2 in oleic acid-induced acute lung injury. *Am J Respir Crit Care Med* 160: 1165–1170
70 Hierholzer C, Harbrecht B, Menezes JM, Kane J, MacMicking J, Nathan CF, Peitzman AB, Billiar TR, Tweardy DJ (1998) Essential role of induced nitric oxide in the initiation of the inflammatory response after hemorrhagic shock. *J Exp Med* 187: 917–928
71 Meldrum DR, Shenkar R, Sheridan BC, Cain BS, Abraham E, Harken AH (1997) Hemorrhage activates myocardial NFkappaB and increases TNF-alpha in the heart. *J Mol Cell Cardiol* 29: 2849–2854
72 Schmedtje JF, Jr, Ji YS, Liu WL, DuBois RN, Runge MS (1997) Hypoxia induces cyclooxygenase-2 via the NF-kappaB p65 transcription factor in human vascular endothelial cells. *J Biol Chem* 272: 601–608
73 Hierholzer C, Harbrecht BG, Billiar TR, Tweardy DJ (2001) Hypoxia-inducible factor-1 activation and cyclo-oxygenase-2 induction are early reperfusion-independent inflammatory events in hemorrhagic shock. *Arch Orthop Trauma Surg* 121: 219–222
74 Sunose Y, Takeyoshi I, Tsutsumi H, Kawata K, Tokumine M, Iwazaki S, Tomizawa N, Ohwada S, Matsumoto K, Morishita Y (2001) Effects of FK3311 on pulmonary ischemia-reperfusion injury in a canine model. *J Surg Res* 95: 167–173
75 Sunose Y, Takeyoshi I, Tsutsumi H, Ohwada S, Oriuchi N, Matsumoto K, Morishita Y (2001) Effect of a cyclooxygenase-2 inhibitor, FK3311, in a canine lung transplantation model. *Ann Thorac Surg* 72: 1165–1171
76 Wilborn J, Crofford LJ, Burdick MD, Kunkel SL, Strieter RM, Peters-Golden M (1995) Cultured lung fibroblasts isolated from patients with idiopathic pulmonary fibrosis have a diminished capacity to synthesize prostaglandin E2 and to express cyclooxygenase-2. *J Clin Invest* 95: 1861–1868
77 Cruz-Gervis R, Stecenko AA, Dworski R, Lane KB, Loyd JE, Pierson R, King G, Brigham KL (2002) Altered prostanoid production by fibroblasts cultured from the lungs of human subjects with idiopathic pulmonary fibrosis. *Respir Res* 3: 17
78 Keerthisingam CB, Jenkins RG, Harrison NK, Hernandez-Rodriguez NA, Booth H, Laurent GJ, Hart SL, Foster ML, McAnulty RJ (2001) Cyclooxygenase-2 deficiency results in a loss of the anti-proliferative response to transforming growth factor-beta in human fibrotic lung fibroblasts and promotes bleomycin-induced pulmonary fibrosis in mice. *Am J Pathol* 158: 1411–1422
79 Bonner JC, Rice AB, Ingram JL, Moomaw CR, Nyska A, Bradbury A, Sessoms AR, Chulada PC, Morgan DL, Zeldin DC et al (2002) Susceptibility of cyclooxygenase-2-deficient mice to pulmonary fibrogenesis. *Am J Pathol* 161: 459–470
80 Lama V, Moore BB, Christensen P, Toews GB, Peters-Golden M (2002) Prostaglandin E_2 synthesis and suppression of fibroblast proliferation by alveolar epithelial cells is cyclooxygenase-2-dependent. *Am J Respir Cell Mol Biol* 27: 752–758
81 Cuzzocrea S, Mazzon E, Sautebin L, Dugo L, Serraino I, De Sarro A, Caputi AP (2002) Protective effects of Celecoxib on lung injury and red blood cells modification induced by carrageenan in the rat. *Biochem Pharmacol* 63: 785–795
82 Shields CJ, Winter DC, Redmond HP (2002) Lung injury in acute pancreatitis: mechanisms, prevention, and therapy. *Curr Opin Crit Care* 8: 158–163
83 Ethridge RT, Chung DH, Slogoff M, Ehlers RA, Hellmich MR, Rajaraman S, Saito H, Uchida T, Evers BM (2002) Cyclooxygenase-2 gene disruption attenuates the severity of acute pancreatitis and pancreatitis-associated lung injury. *Gastroenterology* 123: 1311–1322

COX-2 Inhibitors
Edited by M. Pairet and J. van Ryn

Reproduction: role of COX-2 and its inhibition

Aarthi R. Mohan and Phillip R. Bennett

Imperial College Parturition Research Group, Institute of Reproductive and Developmental Biology, Hammersmith Hospital Campus, Du Cane Road, London W12 ONN, UK

Introduction

Prostaglandins play a key role in human reproduction. There is increasing evidence to support their involvement, not only in the processes of term and preterm labor, but also in menstruation, ovulation and blastocyst implantation. Our aim is to review the physiological significance of COX-2 and prostaglandins in reproduction, and to discuss the implications of COX-2 inhibition with regard to tocolysis.

COX-2 and male reproduction

Prostaglandins were first discovered as a constituent of human seminal fluid in 1934. They were found to be acidic lipoid factors that modify uterine smooth muscle contractility [1]. Despite a great deal of information on the role of prostaglandins in virtually all mammalian organ systems, very little information is available on their role in male reproduction. McKanna et al. [2] found that COX-2 is the predominant isoform in the adult rat male reproductive system and is heavily localized to the epithelium of the distal vas deferens, where constitutive expression is much greater than in any other organ of the body. COX-2 was not detected in the proximal half of the vas deferens or in the testes, epididymis, seminal vesicles or prostate. It was shown that COX-2 levels were not affected by the elimination of luminal sperm by vasectomy, whereas the levels were severely depleted by castration. However, replacement with androgens post-castration restored the levels, indicating that COX-2 expression in the vas deferens is androgen dependent. A role in penile erection has also been suggested in view of the fact that the distal vas encompasses a large submucosal venous plexus, in the rodent, connected to the penile corpora cavernosa. A similar venous plexus has not been described in primates, however, but the injection of prostaglandins into the corpora cavernosa has been used as an effective remedy for human erectile impotence [3].

It has been widely emphasised that COX-2 expression in most tissues is low in normal adult animals and can be increased in response to injury and various inflammatory stimuli. However, Harris et al. [4] found sustained COX-2 expression in the kidney cortex of normal adult rats. The findings previously described in normal rat vas deferens are, therefore, another example of localized COX-2 expression in the absence of inflammation or dysplasia.

COX-2 and menstruation

Van Voorhis et al. [5] have studied the localization of COX in the female reproductive tract. In the uterus, COX was primarily localized to endometrial epithelium with surface epithelium containing more immunoreactive COX than glandular epithelium. Fallopian tube secretory epithelial cells also contained COX while ciliated epithelial cells did not. It was shown that a higher percentage of glands in endometriosis and adenomyosis, as compared with ectopic endometrium, contained COX suggesting that these tissues are also capable of prostaglandin synthesis.

COX was first thought to be hormone dependent when $PGF_{2\alpha}$ and PGE_2 levels were measured at different stages of the menstrual cycle [6]. There was found to be a marked increase in the ability of the endometrium to synthesise $PGF_{2\alpha}$ in the luteal phase of the cycle. The concentration of prostaglandins is determined not only by their rate of synthesis but also by their rate of metabolism. Progesterone stimulates the metabolism of $PGF_{2\alpha}$ and PGE_2 to the inactive 13-14-dihydro-15-keto form. Therefore, during the luteal phase there is an increased ability of the endometrium to metabolise these prostaglandins due to increased prostaglandin dehydrogenase activity.

Administration of an antiprogestogen in the early luteal phase of the menstrual cycle inhibits the appearance of prostaglandin dehydrogenase and this may be one factor leading to the increase in uterine activity and menstrual bleeding when an antiprogestogen is given in the mid luteal phase of the cycle. However, as stated above, the withdrawal of progesterone alone would be expected to cause release of prostaglandins. In fact, when stromal cells are maintained *in vitro* the addition of the progesterone antagonist mifepristone does induce a dose-related release of $PGF_{2\alpha}$, suggesting that under normal circumstances progesterone is inhibiting the release of prostaglandins.

COX-2 and ovulation

Ovulation is accompanied by induction of prostaglandin synthesis as a consequence of the gonadotrophin LH surge [7, 8]. COX-2 is induced in the ovary and has been found to be necessary for the successful rupture of the ovarian follicle, perhaps by mediating the activation of proteolytic enzymes necessary for the release of the ovum [9]. The processes of ovulation and implantation are considered analogous to 'pro-inflammatory' mechanisms; hence the past speculation of prostaglandin involvement in these processes [10, 11].

Prostaglandins are formed predominantly in the granulosa cells of the pre-ovulatory follicle, and highest concentrations are found at the time of ovulation. Sirois et al. [12] showed that in bovine granulosa cells of preovulatory follicles, COX-2 mRNA and protein is transiently induced by gonadotrophins just prior to ovulation. The number of ovulations is reduced in COX-2 knockout mice, however, follicular development is normal in these mice and administration of exogenous gonadotrophins did not restore the ovulation number [13]. The selective COX-2 inhibitor, meloxicam, was shown to inhibit ovulation in rabbits in a dose- and time-dependent manner [14]. COX-2 inhibition has also been linked to a cause of reversible female infertility. The luteinised unruptured follicle (LUF) syndrome, an anovulatory condition characterized by clinical signs of ovulation but the absence of follicular rupture and ovum release, has been described by Stone et al. [15]. They have identified a possible link between regular NSAID use and reversible female infertility using both animal and clinical studies.

COX-2, implantation and placentation

Prostaglandins play a major role in blastocyst implantation. In the rodent, increased uterine vascular permeability at sites of blastocyst apposition is one of the earliest events in the implantation process. This is preceded by generalized uterine edema and luminal closure, and coincides with the initial attachment reaction between the trophectoderm and luminal epithelium [16]. It is thought that vasoactive prostaglandins are involved in these processes. Chakraborty et al. [17] found that the COX genes are differentially regulated in the peri-implantation mouse uterus. During the pre-implantation period, COX-1 is expressed in the uterine epithelium until the initiation of the attachment reaction, after which expression is down regulated. This COX-1 expression coincides with the generalized uterine oedema required for luminal closure. COX-2, however, is expressed in the luminal epithelium and subepithelial stromal cells at the anti-mesenteric pole exclusively surrounding the blastocyst at the time of the attachment reaction. These results suggest that ovarian steroids influence uterine COX-1 expression, and prostaglandins produced by COX-1 are involved in decidualization and/or continued localized endometrial vascular permeability. However, the COX-2 gene is regulated by the implanting blastocyst during early pregnancy, and COX-2 generated prostaglandins are involved in angiogenesis for the establishment of placentation. There may be a possible role for COX-2 inhibitors as contraceptives in view of their ability to inhibit ovulation, however, their long-term side-effects on other body systems may prove to be a problem.

COX knock-out experiments and fetal development

There have been a number of reports of transgenic mice with knock-outs for COX-1, COX-2 and the prostaglandin E (EP4) and F (FP) receptors [18–21]. Langenbach et al. [18] have shown, surprisingly, that COX-1 knock-out mice show less indomethacin-induced gastrointestinal lesions than wild-type mice, despite the apparent importance of COX-1 mediated prostaglandin synthesis in the stomach. The reason for this could be that the pharmacological inhibition of COX leaves peroxidase activity intact. However, when homozygous COX-1 knock-out mice were crossbred, there was a high mortality of their offspring. Heterozygous COX-1 knock-out mice when mated with homozygotes produced litters of normal size and survival, indicating that COX-1 expression in only 50% of these pups or their placentas is sufficient for survival. These studies showed that COX-1 is an essential requirement for mouse parturition but not for ovulation or spermatogenesis.

Morham et al. [19] first reported a COX-2 knock-out mouse and showed that homozygous COX-2 knock-out mice begin to show signs of nephropathy by six weeks of age, which increases in severity until death at around 8 weeks. All the other tissues in these mice were normal. Examination of COX-2 deleted mice embryos showed that kidney maturation had prematurely ceased after only a small percentage of nephrons had developed and the majority of glomeruli and tubules remained small and immature. During postnatal life, the small number of nephrons was seen to atrophy, with the development of glomerulosclerosis, interstitial inflammation and fibrosis leading, ultimately, to kidney failure. In those mice that did survive to adulthood, kidney lesions ranged from mild to severe, with the more severe lesions being more common in female than male homozygotes. This report clearly illustrates an important role for COX-2 in the embryological development of the kidney.

The dilatation of the ductus arteriosus (DA) *in utero* is an active process maintained primarily by PGE_2. The prostaglandin receptors that may have a role in dilatation of the DA include the PGE_2 receptors, EP2 [22] and EP4 [23], and a receptor for prostacyclin [24]. Nguyen et al. [20] showed abnormal function of the DA in EP4 receptor-deficient mice, with greater than 95% of these mice dying within 48 h of birth. Loftin et al. [25] reported that the absence of either or both COX isoforms in mice did not result in premature closure of the DA *in utero*. However, 35% of COX-2 (–/–) mice died with a patent DA within 48 h of birth. In contrast, the absence of only the COX-1 isoform did not affect ductal closure. The mortality (35%) and patent DA incidence due to absence of COX-2 was, however, significantly increased (79%) when only one copy of the gene encoding COX-1 was also inactivated. All of the mice deficient in both isoforms died with a patent DA within 12 h of birth, indicating that in COX-2 deficient mice, the contribution of COX-1 to DA closure is gene dosage-dependent. Loftin et al. found that indomethacin, a non-steroidal anti-inflammatory drug, induces ductal closure in both wild type and heterozygous COX-2-deficient mice, irrespective of their COX-1 genotype. However, indomethacin did not induce ductal closure in fetuses that lacked COX-2 expression. Therefore, indomethacin-induced closure of the DA requires fetal COX-2 expression, and neither maternal nor fetal COX-1 expression is required for the DA to be sensitive to indomethacin. It has been suggested that COX-2 in the DA may produce constrictor prostaglandins important for DA contraction. Therefore, contraction of the DA following maternal indomethacin exposure may result from the inhibition of dilatory PGE_2 synthesis in the placenta without sufficient inhibition of ductal COX-2 to attenuate DA contraction.

Sugimoto et al. [21] found that mice with deleted FP receptors had normal estrous cycles, normal ovulation and normal implantation. The onset of labor is stimulated by regression of the corpus luteum and withdrawal of ovarian progesterone in mice, allowing upregulation of the uterine oxytocin receptor. However, pregnant homozygous FP knock-out mice failed to initiate labor because of failure of regression of the corpus luteum. Ovariectomy in these mice ceased the progesterone inhibition permitting the onset of labor.

There have been numerous reports of the roles of COX-1 and COX-2 in rodents, with respect to their menstrual cycle, ovulation and implantation, which may be similar to other species and the primate. However, there are fundamental differences between the two groups in the general mechanism of the onset of parturition and in the roles of COX-1 and 2 in labor.

COX-2 in human labor

Prostaglandins play a central role in human parturition. A number of studies have shown that the fetal membranes are a major source of intrauterine prostaglandins in labor [26, 27]. Prostaglandins mediate cervical ripening and directly stimulate uterine contraction [28]. They indirectly increase fundally dominant myometrial contractility by upregulation of gap junctions and oxytocin receptors and synchronization of contractions [29, 30].

As discussed above, prostaglandins are formed from the precursor arachidonic acid, which is itself a substrate for at least three enzyme groups; COX, lipoxygenase and epoxygenase. Prior to labor, endogenous arachidonic acid metabolism in the amnion is principally via the lipoxygenase enzyme pathways. Bennett et al. [31] have shown that with labor there is an increase in arachidonic acid metabolism and a change in the ratio of COX to lipoxygenase metabolism to favor synthesis of PGE_2. It has been suggested that the lipoxygenase metabolites of arachidonic acid may play a role in pre-labor (Braxton-Hicks) contractions [32, 33].

It was previously thought that phospholipase activity was the limiting factor in prostaglandin production in the fetal membranes. However, any stimulus to prostaglandin production must also increase COX activity since it has a short half-life and undergoes destruction after a limited number of reactions [34]. The change in the ratio of COX: lipoxygenase metabolism with labor suggests that the activity of the two enzymes, COX and phospholipase must be independently controlled [31, 35].

In sheep, the placental cotyledon is the principal source of prostaglandins in late gestation [36] and upregulation of COX-2 expression takes place gradually as pregnancy advances and precedes the labor-associated peak of maternal estrogen concentrations. COX-1 expression within the sheep uterus does not change with gestational age [37, 38].

The principle sources of uterine prostaglandins in humans are the fetal membranes and the myometrium. Both COX-1 and COX-2 expression has been demonstrated in fetal membranes using *in situ* hybridization [39]. COX-2 expression is mainly within the amnion epithelial layer, the reticular layer of the chorion and the decidua. COX-2 is not expressed in the trophoblast layer of the chorion. COX-1 expression is more diffuse and is seen in the amnionic epithelium, amnionic mesoderm and also in the chorionic mesoderm and decidua. The levels of expression of the two COX isoforms are very different. In amnion at term, expression of COX-2 mRNA is 100-fold greater than that of COX-1 mRNA [40]. COX-2 expression also doubles in association with labor [39–41]. Slater et al. [40] showed that the mRNA levels of COX-1 do not change significantly through gestation and with labor onset, whereas COX-2 mRNA increases steadily throughout gestation with a significant increase at term. COX-2 protein concentrations in amnion also increase from the first trimester to term with a further increase with the onset of labor. Although COX-1 mRNA concentrations remain low and do not change with labor, COX-1 protein concentrations are similar to those of COX-2, although there are no labor or gestation-associated changes [42]. This apparent discrepancy between mRNA and protein concentrations for COX-1 and COX-2 may be explained in terms of the known functions of the enzymes. COX is thought to be a 'suicide enzyme', undergoing destruction during catalysis [43]. Increased prostaglandin synthesis via COX-2 would therefore mean increased turnover of the enzyme itself. This would mean a large increase in mRNA concentrations to drive protein synthesis, but because the protein itself is also being actively destroyed, the COX-2 protein concentration measured at any one time would be increased to a much lesser extent. As COX-1 is not active at this time in the fetal membranes, the protein may be present at concentrations similar to that of COX-2 but COX-1 mRNA concentrations would be very low. Enzyme kinetic studies suggest that the increase in COX activity with the onset of labor is due to an increase in the synthesis of the COX-2 enzyme.

COX-2 mRNA and protein concentrations within the human fetal membranes increase in the third trimester of pregnancy many weeks before the onset of labor. Slater et al. [44] reported a similar increase in COX activity, suggesting that the fetal membranes, in particular the amnion, have the capacity to synthesize prostaglandins several weeks prior to the onset of labor at term. It is thought that this is opposed by prostaglandin dehydrogenase (PGDH) activity. Sangha et al. [45] have shown a reduction in PGDH expression with the onset of labor, and have also demonstrated that a subgroup of patients presenting in 'idiopathic' pre-term labor have reduced expression of chorionic PGDH both at the mRNA and protein levels. Van Meir et al. [46, 47] have shown a labor-associated reduction in PGDH activity in chorion near to the cervix and chorioamnionitis, leading to pre-term delivery, is associated with reduced chorionic PGDH expression. They have suggested that the opposing effects of progesterone and cortisol may control PGDH activity in the fetal membranes.

The role of progesterone and IL-8 in labor

Progesterone represses myometrial contractility by changing the electrical properties of the plasma membrane and restricting the propagation of the action potential. It increases calcium binding to intracellular sites [48], prevents the formation of gap junctions [49] and oxytocin receptors [50] and inhibits arachidonic acid release [51]. The corpus luteum regresses in the first trimester of human pregnancy and progesterone synthesis is taken over by the placenta and chorion. Unlike the rodent, a corpus luteum-dependent species, there is no detectable progesterone withdrawal in the human to signal the onset of labor. However, there is upregulation of a range of pro-labor genes, including gap junction proteins, oxytocin receptor, prostaglandin receptors and interleukin-8 (IL-8), which are normally repressed by the presence of progesterone. IL-8 is a cytokine whose expression increases and acts as a chemoattractant for neutrophils that invade the cervical tissue, fetal membranes and myometrium. Elliott et al. [52] have shown that IL-8 is produced by the term human placenta, with increased production during labor and by treatment with the antigestagen onapristone. It was also shown that biopsies of first-trimester cervices produced IL-8 [53] and that this production is increased if the woman is pre-treated with the progesterone antagonist mifepristone prior to collection of the biopsy. In both guinea pigs and rabbits, application of IL-8 to the cervix induces cervical ripening [54].

The promoter of the COX-2 gene has been cloned and sequenced and has been shown to have some similarities with the IL-8 promoter. Several putative transcription factor binding sites have been identified, including a cyclic AMP response element (CRE), a nuclear factor IL-6 (NF-IL6) site and two nuclear factor-κB (NF-κB) sites. Each of these has been demonstrated to regulate COX-2 transcription in various cells. It has been shown that the promoter region of the IL-8 gene is regulated in a cell-line specific manner by NF-κB and either an AP-1 or CCAAT enhancer-binding protein site (C/EBP). NF-κB is the transcription factor family classically associated with inflammation. The presence of NF-κB sites in both COX-2 and IL-8 promoters, and the view that the process of labor may be similar to an inflammatory reaction, has led to the suggestion that NF-κB may play a role in labor.

We have used primary cultures of amnion cells to study the roles of the NF-κB sites in COX-2 regulation. Mutation of the downstream NF-κB DNA-binding site showed significantly less reporter expression than the wild type COX-2 promoter only in post-labor amnion cells. However, mutation of the upstream NF-κB DNA-binding site significantly reduced reporter expression in both pre- and post-labor amnion cells. We have also performed similar studies using reporter constructs of the IL-8 gene promoter with site mutations in the NF-κB, C/EBP and AP-1 sites. Mutation of the NF-κB site reduced both basal and IL-1β-stimulated promoter activity in primary amnion cells whilst mutation of the C/EBP or the AP-1 sites did not appear to have a significant effect. NF-κB, therefore, appears to be essential for the transcriptional regulation of both COX-2 and IL-8 in primary amnion cells. In myometrium, NF-κB plays a role in IL-1β stimulated COX-2 expression but is not activated with labor. NF-κB is, therefore, an important regulator of the labor-associated inflammatory mediators and plays a central role in the onset of human parturition.

Preterm labor and COX-2 inhibitors

Preterm birth occurs in 5–10% of all pregnancies but is associated with 70% of all neonatal deaths and up to 75% of neonatal morbidity. Premature infants are at increased risk of cerebral palsy, deafness, blindness and respiratory problems. The social and financial costs are enormous, but most distressing is the impact of long-term handicap on the individuals and their caretakers. Unfortunately, despite research directed to the development of drugs to inhibit myometrial contractions, there has been no reduction in the incidence of preterm birth. Present therapies cannot prevent preterm delivery, but at best provide sufficient delay in order to attempt treatments, which will ameliorate the consequences of prematurity. It is probable that improved therapies for preterm labor will be targeted less directly at myometrial contractility, but at the earlier biochemical events of labor, including the increase in synthesis of prostaglandins and inflammatory cytokines.

Preterm labor is a multi-factorial syndrome. Most theories for the biochemical etiology of preterm labor have, as an important endpoint, increased prostaglandin synthesis mediated via increased COX-2 expression. Upregulation of inflammatory cytokines and prostaglandins that occur over a period of several weeks leads to cervical ripening and membrane rupture and, ultimately, to myometrial contractility and labor [55]. The most common cause of early preterm labor is generally thought to be genital tract infection. Bennett et al. [56] reported that prostaglandins are not synthesized by bacteria themselves, but increase prostaglandin synthesis both by release of phospholipase A [32] and via the endotoxin lipopolysaccharide (LPS) induced upregulation of COX-2 expression. It has been shown that LPS increases prostaglandin synthesis in fetal membranes and increases COX-2 expression.

Preterm labor due to multiple pregnancy may be due to increased placental mass, so that the production of endocrine factors, such as CRH, which upregulate COX-2 synthesis is reached at an earlier gestational age. COX-2 and other contraction-associated proteins are stretch-dependent which may also contribute to early delivery in multiple pregnancy and hydramnios. Where there is fetal growth retardation or placental dysfunction, increased fetal cortisol may upregulate placental CRH synthesis [57] leading to a premature increase in COX-2 expression and the onset of preterm labor.

Studies in the sheep, rat and wallaby, have shown that stretch increases myometrial expression of the oxytocin receptor [58, 59], COX-2 [58] and the gap junction protein connexin-43 [60]. We have shown that, in human myometrium, stretch increases the expression of the oxytocin receptor and COX-2, and that these are associated with activation of AP-1 and C/EBP but not NF-κB. We have also shown that stretch leads to the activation of p38 MAPK and ERK-1 and –2 and that inhibition of ERK blocks stretch induced COX-2 expression (unpublished data).

Beta-sympathomimetics, the most commonly used tocolytics, exhibit tachyphylaxis and cause potentially life-threatening maternal side-effects [61, 62]. Although they are effective in inhibition of preterm contractions and delay of delivery, the use of non-selective COX inhibitors like indomethacin is limited by fetal side-effects, including constriction of the DA, renal tubular dysfunction and oligohydramnios, and neonatal complications including intracranial hemorrhage and necrotising enterocolitis [63]. Since prostaglandin synthesis in the fetus is constant, it is probably mainly mediated by COX-1. Therefore, a COX-2 selective inhibitor may have the efficacy of indomethacin without the fetal side-effects.

Nimesulide is a non-steroidal anti-inflammatory drug, which has been shown not to affect prostaglandin synthesis in tissues in which COX-1 is the predominant isoform [64] but does inhibit COX-2 mediated prostaglandin synthesis [65, 66]. *In vitro*, nimesulide displays 30–100 fold selectivity for COX-2 compared with COX-1 [66, 67]. Studies in animals and in explanted human myometrial cells have demonstrated its efficacy at inhibiting myometrial contractility. Sawdy et al. [68] have described the successful clinical use of nimesulide as a tocolytic agent from 16 to 34 weeks with no effect on amniotic fluid volume or patency of the DA. Locatelli et al. [69] evaluated the efficacy and safety of oral nimesulide for the treatment of preterm labor in a pilot study of five women. They showed that tocolysis with nimesulide was associated with fetal side-effects similar to those described with indomethacin therapy. Oligohydramnios occurred within 9 days of therapy in all treated patients. This could be explained by the fact that nimesulide could have a lower COX-2 selectivity than previously reported. Also, nimesulide could lose its COX-2 selectivity in a dose-time-dependent manner. Studies in sheep have shown that at high doses, nimesulide administration results in inactivation of both COX-1 and –2 [70]. The doses used in clinical practice may result in concentrations where COX-2 selectivity is lost.

Stika et al. [71] have carried out a trial comparing the safety of celecoxib, a selective cyclo-oxygenase-2 inhibitor, with indomethacin administered for the treatment of preterm labor. However, exposure to either drug was only for 48 h, there were only 12 women in each arm of the trial, and it is doubtful whether all of the 24 women in the trial were actually in preterm labor. The trial suggested that the two drugs were equally effective in the maintenance of tocolysis, but the trial was not adequately powered to show this. Both drugs were associated with a transient decrease in amniotic fluid volume, however, celecoxib produced fewer fetal side-effects when compared with indomethacin.

Further randomized, controlled trials using larger numbers of patients focussing on long term safety and drug efficacy are warranted before the widespread introduction of selective COX-2 inhibitors for the treatment of preterm labor.

References

1 Euler US von (1936) On the specific vaso-dilating and plain muscle stimulating substances from accessory genital glands in man and certain animals (prostaglandin and vesiglandin). *J Physiol (Lond)* 88: 213–234

2 McKanna JA, Zhang M-Z, Wang J-L, Cheng H-F, Harris RC (1998) Constitutive expression of cyclooxygenase-2 in rat vas deferens. *Am J Physiol* 275 (*Regulatory Integrative Comp Physiol* 44): R227–R233

3 Wagner G, Kaplan HS (1993) *The New Injection Treatment for Impotence: Medical and Psychological Aspects.* Brunner/Mazel, New York

4 Harris RC, McKanna JA, Akai Y, Jacobson HR, DuBois RN, Breyer MD (1994) Cyclooxygenase-2 is associated with the macula densa of rat kidney and increases with salt restriction. *J Clin Invest* 94: 2504–2510

5 Van Voorhis BJ, Huettner PC, Clark MR, Hill JA (1990) Immunohistological localization of prostaglandin H synthase in the female reproductive tract and endometriosis. *Am J Obstet Gynaecol* 163 (1 Pt 1): 57–62

6 Downie J, Poyser NL, Wunderlich M (1974) Levels of prostaglandins in human endometrium during the normal menstrual cycle. *J Physiol* 236: 465–472

7 LeMaire WJ, Marsh JM (1975) Interrelationships between prostaglandins, cyclic AMP and steroids ovulation. *J Reprod Fertil* 22 (Suppl): 53–74

8 Armstrong DT (1981) Prostaglandins and follicular functions. *J Reprod Fertil* 62: 283–291

9 Tsafriri A (1995) Ovulation as a tissue remodelling process. Proteolysis and cumulus expansion. *Adv Exp Med Biol* 377: 121–140

10 McMaster MT, Dey SK, Andrews GK (1993) Association of monocytes and neutrophils with early events of blastocyst implantation in mice. *J Reprod Fertil* 99: 561–569

11 Espey LL (1994) Current status of the hypothesis that mammalian ovulation is comparable to an inflammatory reaction. *Biol Reprod* 50: 233–238

12 Sirois J (1994) Induction of prostaglandin endoperoxide synthase-2 by human chorionic gonadotropin in bovine preovulatory follicles *in vivo*. *Endocrinology* 135: 841–848

13 Lim H, Paria BC, Das SK, Dinchuk JE, Langenbach R, Trzaskos JM, Dey SK (1997) Multiple Female Reproductive Failures in Cyclooxygenase 2-Deficient Mice. *Cell* 91: 197–208

14 Salhab AS, Gharaibeh MN, Shomaf MS, Amro BI (2001) Meloxicam inhibits rabbit ovulation. *Contraception* 63: 329–333

15 Stone S (2002) Nonsteroidal anti-inflammatory drugs and reversible female infertility: is there a link? *Drug Saf* 25: 545–551

16 Kennedy TG (1983) Embryonic signals and the initiation of blastocyst implantation. *Aust J Biol Sci* 36: 531–543

17 Chakraborty I, Das SK, Wang J, Dey SK (1996) Developmental expression of the cyclooxygenase-1 and cyclooxygenase-2 genes in the peri-implantation mouse uterus and their differential

regulation by the blastocyst and ovarian steroids. *J Mol Endocrinol* 16: 107–122
18 Langenbach R, Morham SG, Tiano HF, Loftin CD, Ghanayem BI, Chulada PC, Mahler JF, Lee CA, Goulding EH, Kluckman KD et al (1995) Prostaglandin synthase 1 gene disruption in mice reduces arachidonic acid-induced inflammation and indomethacin-induced gastric ulceration. *Cell* 83: 483–492
19 Morham SG, Langenbach R, Loftin CD, Tiano HF, Vouloumanos N, Jennette JC, Mahler JF, Kluckman KD, Ledford A, Lee CA et al (1995) Prostaglandin synthase 2 gene disruption causes severe renal pathology in the mouse. *Cell* 83: 473–482
20 Nguyen MT, Camenisch T, Snouwaert JN, Hicks E, Coffman TM, Anderson PAW, Malouf NN, Koller BH (1997) The prostaglandin receptor EP_4 triggers remodelling of the cardiovascular system at birth. *Nature* 390: 78–81
21 Sugimoto Y, Yamasaki A, Segi E, Tsuboi K, Aze Y, Nishimura T, Oida H, Yoshida N, Tanaka T, Katasuyama M et al (1997) Failure of parturition in mice lacking the prostaglandin F receptor. *Science* 277: 681–683
22 Bhattacharya M, Asselin P, Hardy P, Guerguerian AM, Shichi H, Hou X, Varma DR, Bouayad A, Fouron JC, Clyman R et al (1999) Developmental changes in prostaglandin E(2) receptor subtypes in porcine ductus arteriosus. Possible contribution in altered responsiveness to prostaglandin E(2). *Circulation* 100: 1751–1756
23 Smith GC, Coleman RA, McGrath JC (1994) Characterization of dilator prostanoid receptors in the fetal rabbit ductus arteriosus. *Pharmacol Exp Ther* 271: 390–396
24 Smith GC, McGrath JC (1993) Characterisation of the effect of oxygen tension on response of fetal rabbit ductus arteriosus to vasodilators. *Cardiovasc Res* 27: 2205–2211
25 Loftin CD, Trivedi DB, Tiano HF, Clark JA, Lee CA, Epstein JA, Morham SG, Breyer MD, Nguyen MT, Hawkins BM et al (2001) Failure of ductus arteriosus closure and remodelling in neonatal mice deficient in cyclooxygenase-1 and cyclooxygenase-2. *Proc Natl Acad Sci* 98: 1059–1064
26 Kierse MJ, Mitchell MD, Turnbull AC (1977) Changes in prostaglandin F and 13,14-dihydro-15-keto-prostaglandin F concentrations in amniotic fluid at the onset of and during labour. *British Journal of Obstetrics and Gynaecology* 84: 743–746
27 Skinner KA, Challis JRG (1985) Changes in the synthesis and metabolism of prostaglandins by human fetal membranes and decidua at labour. *American Journal of Obstetrics and Gynaecology* 151: 519–523
28 Crankshaw DJ, Dyal R (1994) Effects of some naturally occurring prostanoids and some cyclooxygenase inhibitors on the contractility of the human lower uterine segment *in vitro*. *Can J Physiol Pharmacol* 72: 870–874
29 Liggins G (1989) Initiation of labour. *Biol Neonate* 55: 366–375
30 Garfield RE, Hertzberg EL (1990) Cell-to-cell coupling in the myometrium: Emil Bozler's prediction. *Prog Clin Biol Res* 327: 673–681
31 Bennett PR, Slater D, Sullivan M, Elder MG, Moore GE (1993) Changes in amniotic arachidonic acid metabolism associated with increased cyclooxygenase gene expression. *Br J Obstet Gynaecol* 100: 1037–1042
32 Bennett PR, Elder MG, Myatt L (1987) The effects of lipoxygenase metabolites of arachidonic acid on human myometrial contractility. *Prostaglandins* 33: 837–844
33 Walsh SW (1989) 5-Hydroxyeicosatetraenoic acid, leukotriene C4 and prostaglandin F2α in amniotic fluid before and during term and preterm labor. *Am J Obstet Gynaecol* 161: 1352
34 Marshall P, Kulmacz R, Lands W (1979) Constraints on prostaglandin synthesis in tissues. *J Biol Chem* 262: 3510–3517
35 Bennett PR, Rose MP, Myatt L, Elder MG (1987) Preterm labor: Stimulation of arachidonic acid metabolism in human amnion cells by bacterial products. *Am J Obstet Gynecol* 156: 649–655
36 Rice GE, Payne MJ, Wong MH, Thorburn GD (1992) Immunoreactive prostaglandin G/H synthase content increases in ovine cotyledons during late gestation. *Placenta* 13: 429–437
37 Wimsatt J, Nathanielsz PW, Sirois J (1993) Induction of prostaglandin endoperoxide synthase isoform-2 in ovine cotyledonary tissues during late gestation. *Endocrinology* 133: 1068–1073
38 Rice GE, Freed KA, Aitken MA, Jacobs RA (1995) Gestational- and labour-associated changes in the relative abundance of prostaglandin G/H synthase-1 and –2 mRNA in ovine placenta. *J Mol Endocrinol* 14: 237–245
39 Slater DM, Berger LC, Newton R, Moore GE, Bennett PR (1995) Expression of cyclooxygenase types 1 and 2 in human fetal membranes at term. *Am J Obstet Gynecol* 172: 77–82

40 Slater D, Berger L, Newton R, Moore G, Bennett P (1994) The relative abundance of type 1 to type 2 cyclo-oxygenase mRNA in human amnion at term. *Biochem Biophys Res Commun* 198: 304–308
41 Hirst JJ, Teixeira FJ, Zakar T, Olson DM (1995) Prostaglandin endoperoxide-H synthase-2 expression increases in human gestational tissues with spontaneous labour onset. *Reproduction, Fertility and Development* 7: 633–637
42 Sawdy RJ, Slater DM, Dennes WJB, Sullivan MHF, Bennett PR (2000) The roles of the cyclo-oxygenases types one and two in human fetal membranes at term. *Placenta* 21: 54–57
43 Marshall P, Kulmacz R and Lands W (1979) Constrants on prostaglandin synthesis in tissues. *J Biol Chem* 262: 3510–3517
44 Slater D, Dennes W, Sawdy R, Allport V, Bennett P (1999) Expression of cyclo-oxygenase types-1 and –2 in human fetal membranes throughout pregnancy. *J Mol Endocrinol* 22: 125–130
45 Sangha RK, Walton JC, Ensor CM, Tai HH, Challis JR (1994) Immunohistochemical localization, messenger ribonucleic acid abundance, and activity of 15-hydroxyprostaglandin dehydrogenase in placenta and fetal membranes during term and preterm labor. *J Clin Endocrinol MeTab* 78: 982–989
46 Van Meir CA, Ramirez MM, Matthews SG, Calder AA, Keirse MJ, Challis JR (1997) Chorionic prostaglandin catabolism is decreased in the lower uterine segment with term labour. *Placenta* 18: 109–114
47 Van Meir CA, Sangha RK, Walton JC, Matthews SG, Keirse MJ, Challis JR (1996) Immunoreactive 15-hydroxyprostaglandin dehydrogenase (PGDH) is reduced in fetal membranes from patients at preterm delivery in the presence of infection. *Placenta* 17: 291–297
48 Carsten ME (1979) Calcium accumulation by human uterine microsomal preparations: effects of progesterone and oxytocin. *Am J Obstet Gynecol* 133: 598–601
49 Garfield RE, Merrett D, Grover AK (1980) Gap junction formation and regulation in myometrium. *Am J Physiol* 239: C217–C228
50 Fuchs AR, Periyasamy S, Soloff MS (1983) Systemic and local regulation of oxytocin receptors in the rat uterus, and their functional significance. *Can J Biochem Cell Biol* 61: 615–624
51 Wilson T, Liggins GC, Aimer GP, Watkins EJ (1986) The effect of progesterone on the release of arachidonic acid from human endometrial cells stimulated by histamine. *Prostaglandins* 31: 343–360
52 Elliott CL, Kelly RW, Critchley HO, Riley SC, Calder AA (1998) Regulation of interleukin 8 production in the term human placenta during labor and by antigestagens. *Am J Obstet Gynecol* 179: 215–220
53 Elliott CL, Brennard JE, Calder AA (1998) The effects of mifepristone on cervical ripening and labor induction in primigravidae. *Obstet Gynecol* 92: 804–809
54 Chwalisz K, Garfield RE (1994) Antiprogestins in the induction of labor. *Ann NY Acad Sci* 734: 387–413
55 Bennett PR, Elder MG, Myatt L (1990) Secretion of phospholipases by bacterial pathogens may initiate preterm labor. *Am J Obstet Gynecol* 163 (1 Pt 1): 241–242
56 Bennett PR, Allport V, Loudon J, Elliott C (2001) Prostaglandins, the fetal membranes and the cervix: the endocrinology of parturition: basic science and clinical application. *Front Horm Res* 27: 147–164
57 Challis JR, Matthews SG, Van Meir C, Ramirez MM (1995) Current topic: the placental corticotrophin-releasing hormone-adrenocorticotrophin axis. *Placenta* 16: 481–502
58 Wu WX, Ma XH, Nathanielsz PW (1999) Tissue-specific ontogenic expression of prostaglandin H synthase 2 in the ovine myometrium, endometrium, and placenta during late gestation and at spontaneous term labor. *Am J Obstet Gynecol* 181: 1512–1519
59 Parry LJ, Bathgate RA (2000) The role of oxytocin and regulation of uterine oxytocin receptors in pregnant marsupials. *Exp Physiol* 85: 91S–99S
60 Ou CW, Orsino A, Lye SJ (1997) Expression of connexin-43 and connexin-26 in the rat myometrium during pregnancy and labor is differentially regulated by mechanical and hormonal signals. *Endocrinology* 138: 5398–5407
61 Morrison JJ (1996) Physiology and pharmacology of uterine contraction. In: J Studd (ed.): *The Yearbook of the Royal College of Obstetricians and Gynaecologists*. RCOG Press, London, 45–61
62 Carroll SG, Seibre NJ, Nicolaides KH (1996) *Preterm Prelabour Amniorrhexis*. Parthenon Publishing Group, New York
63 Norton ME, Merrill J Cooper BA (1993) Neonatal complications after administration of

indomethacin for preterm labour. *N Engl J Med* 329: 1602–1607
64 Bianco S, Robuschi M, Gambaro G, Spagnotto S, Petrigni G (1991) Bronchial inflammation and NSAIDs. *Drug Invest* 3 (Suppl 2): 33–38
65 Taniguchi Y, Ikesue A, Yokoyama K, Noda K, Deguchi H, Nakamura T, Toda A, Shimeno H (1995) Selective inhibition by nimesulide, a novel non-steroidal anti-inflammatory drug, with prostaglandin endoperoxide synthase-2 activity *in vitro*. *Pharmacol Sci* 1: 173–175
66 Miralpeix M, Camacho M, Lopez-Belmonte J, Canalias F, Beleta J, Palacios JM, Vila L (1997) Selective induction of cyclo-oxygenase-2 activity in the permanent human endothelial cell line HUV-EC-C: biochemical and pharmacological characterization. *Br J Pharmacol* 121: 171–180
67 Yamada M, Niki H, Yamashita M, Mue S, Ohuchi K (1997) Prostaglandin E_2 production dependent upon cyclo-oxygenase-1 and cyclo-oxygenase-2 and its contradictory modulation by auranofin in rat peritoneal macrophages. *J Pharmacol Exp Ther* 281: 1005–1012
68 Sawdy R, Slater D, Fisk N, Edmonds DK, Bennett P (1997) Use of a cyclo-oxygenase type-2-selective non-steroidal anti-inflammatory agent to prevent preterm delivery. *Lancet* 350: 265–266
69 Locatelli A, Vergani P, Bellini P, Strobelt N, Ghidini A (2001) Can a cyclo-oxygenase type-2 selective tocolytic agent avoid the fetal side-effects of indomethacin? *Br J Obstet Gynecol* 108: 325–326
70 Poore KR, Young IR, Hirst JJ (1999) Efficacy of the selective prostaglandin synthase type 2 inhibitor nimesulide in blocking basal prostaglandin production and delaying glucocorticoid-induced premature labor in sheep. *Am J Obstet Gynecol* 180: 1244–1253
71 Stika CS, Gross GA, Laguizamon G, Gerber S, Levy R, Mathur A, Bernhard LM, Nelson DM, Sadovsky Y (2002) A prospective randomised safety trial of celecoxib for treatment of preterm labor. *Am J Obstet Gynecol* 187: 653–660

COX-2 Inhibitors
Edited by M. Pairet and J. van Ryn

COX-2 in cancer

Kirsi Saukkonen[1, 4], Christianne J. Buskens[5], Anna Sivula[1, 4], Bastiaan P. van Rees[6], Tiina-Liisa Erkinheimo[2, 4], Johanna Rintahaka[1, 4], Leena Pohjola[1, 4], J. Jan B. van Lanschot[5], G. Johan A. Offerhaus[6], Caj Haglund[3] and Ari Ristimäki[1, 4]

[1] *Departments of Pathology,* [2]*Obstetrics and Gynecology and* [3]*Surgery, Helsinki University Central Hospital and* [4]*Molecular and Cancer Biology Research Program, Biomedicum Helsinki, University of Helsinki, P.O. Box 63, FIN-00014 Helsinki, Finland*
[5] *Departments of Surgery and* [6]*Pathology, Academic Medical Center, University of Amsterdam, NL-1105 AZ Amsterdam, The Netherlands*

Role of COX-2 in colorectal carcinogenesis

Multiple epidemiological studies indicate that the use of aspirin and other NSAIDs are associated with reduced risk of malignancies especially in the digestive tract. In addition, another NSAID sulindac causes regression of colorectal adenomatous polyps in patients with familial adenomatous polyposis (FAP) [1, 2]. Recent studies suggest that COX-2 is a rational target of NSAIDs in prevention of colorectal cancer. First, elevated levels of COX-2 mRNA and protein, but not those of COX-1, were found in colorectal adenocarcinomas and in their adenomatous precursors [3–5]. Second, selective COX-2 inhibitors suppress neoplasia formation in rodent models of colorectal cancer [6–10]. Importantly, genetic disruption of COX-2 suppresses the polyp formation in $Apc^{\Delta716}$-knockout and in Min mice, which are models for FAP [6, 11]. However, it should be noted that disruption of COX-1 gene also reduced the polyp burden in the Min mouse model [11]. Finally, elevated COX-2 expression was shown to associate with poor prognosis in colorectal carcinoma [12].

Several ongoing clinical trials are investigating the chemopreventive or -therapeutic (adjuvant) effects of COX-2 selective inhibitors [1, 2]. The first one of these studies has been published, which indicates that a selective COX-2 inhibitor celecoxib reduced polyp burden in patients who suffer from FAP based on a 6-month, randomized controlled clinical trial [13]. The US Food and Drug Administration granted accelerated approval in December 1999 for celecoxib for adjunctive therapy in patients with FAP. However, it should be kept in mind that the absolute decrease in polyp number in this trial in the celecoxib treatment group was quite modest (28% using 400 mg twice a day) and that no data exist that this treatment would reduce the incidence of colorectal cancer in FAP patients. Thus, at the moment this drug treatment cannot replace standard follow-up and treatment practices.

We have recently characterized COX-2 expression in several different types of human adenocarcinomas. Our data show that COX-2 is primarily expressed in neoplastic epithelial cells in human adenocarcinomas, it is present in both noninvasive precursor lesions and invasive carcinomas, and elevated level of COX-2 can be a marker for poor prognosis in certain types of adenocarcinomas. Our aim is to review data on COX-2 expression in several adenocarcinomas and evaluate the role of COX-2 in human carcinogenesis.

Esophageal adenocarcinoma

Adenocarcinoma of the esophagus is a highly lethal disease and its incidence has markedly increased during the last few decades [14–16]. Surgical resection is currently the preferred curative option, but postoperative mortality and morbidity are significant even after "curative" resection and thus the 5-year survival rates rarely exceed 25%. It is now generally accepted that esophageal adenocarcinoma develops from a premalignant lesion of the esophagus, also referred to as Barrett's esophagus. Barrett's esophagus is a metaplastic change of the normal squamous epithelium of the esophagus to a columnar type epithelium due to longstanding gastro-esophageal reflux [17].

Epidemiological studies suggest that the use of NSAIDs is associated with reduced incidence of esophageal cancer. Of the five published observational studies, four demonstrated a protective effect of NSAIDs (Tab. 1) [18–22]. The role of COX-2 in esophageal carcinogenesis has been emphasized by studies that show that COX-2 expression is elevated in the Barrett's metaplasia-dysplasia-adenocarcinoma sequence. In these studies, the normal squamous epithelium of the esophagus was consistently negative or only weakly positive, while elevated expression of COX-2 was found in 78–100% of the adenocar-

Table 1. Epidemiological studies on the protective effect of NSAIDs for esophageal cancer

Design	Sample size	Endpoint	RR (95% CI)	Reference
Cohort				
	635,031	Mortality	0.59 (0.34–1.03)	Thun et al. 1993 [18]
	14,407	Incidence	0.10 (0.01–0.76)	Funkhouser et al. 1995 [19]
Case control	Cases: 650 Controls: 695	Incidence	0.37 (0.24–0.58)	Farrow et al., 1998 [20]
	Cases: 12,174 Controls: 34,934	Incidence	0.64 (0.41–0.98)	Langman et al. 2000 [21]
	Cases: 1,149 Controls: 5,952	Incidence	0.80 (0.5–1.4)	Coogan et al. 2000 [22]

RR = Risk ratio; CI = Confidence interval.

Table 2. COX-2 expression in normal squamous epithelium of the esophagus and in the metaplasia-dysplasia-adenocarcinoma sequence

Normal	Metaplasia	Dysplasia	Adenocarcinoma	Reference
0%	81%	ND	100%	Wilson et al., 1998 [23]
0%	0%	ND	78%	Zimmerman et al. 1999 [24]
Significant increase in expression of COX-2				Shirvani et al. 2000 [25]
0%	75%	83% (LG) 100% (HG)	100%	Morris et al. 2001 [26]
0%	50%	ND	99%	Buskens et al. 2002 [27]

ND = Not determined; LG = Low-grade; HG = High-grade.

cinomas (Tab. 2) [23–27]. Furthermore, Wilson et al. reported an elevated COX-2 mRNA expression in 81% of the Barrett's metaplasias, and in two studies expression of COX-2 was shown to increase progressively in each stage of esophageal carcinogenesis from Barrett's metaplasia via dysplasia to adenocarcinoma on both mRNA and protein level [25, 26]. Importantly, we demonstrated recently that elevated expression of the COX-2 protein is associated with reduced survival in patients undergoing potentially curative esophageal resection due to Barrett carcinoma (Fig. 1) [27]. Our data show that patients with tumors with high COX-2 expression had a more aggressive course of disease because they were more likely to develop locoregional recurrences and distant metastases. Multivariate analysis indicated that elevated COX-2 expression was an independent prognostic variable for patient survival [27].

Currently several clinical trials are being performed on chemoprevention of esophageal cancer by using COX-2 selective inhibitors [1, 2]. In a phase II multicenter study, Forastiere et al. (Baltimore, Maryland, USA) randomize patients with low- or high-grade columnar dysplasia between the daily selective COX-2 inhibitor celecoxib and no additional treatment, with dysplasia regression as primary endpoint. In addition, randomized trials have been planned and/or recently started using COX-2 selective inhibitors after potentially curative esophageal resection as an adjuvant treatment modality. In the Amsterdam Medical Center, we are currently testing the value of celecoxib as a neo-adjuvant therapy in patients with a Barrett carcinoma by analyzing its effect on various biomarkers.

Gastric adenocarcinoma

Gastric cancer is one of the most frequent and lethal malignancies worldwide and the 5-year survival rate is only about 20% [28]. Etiological background of

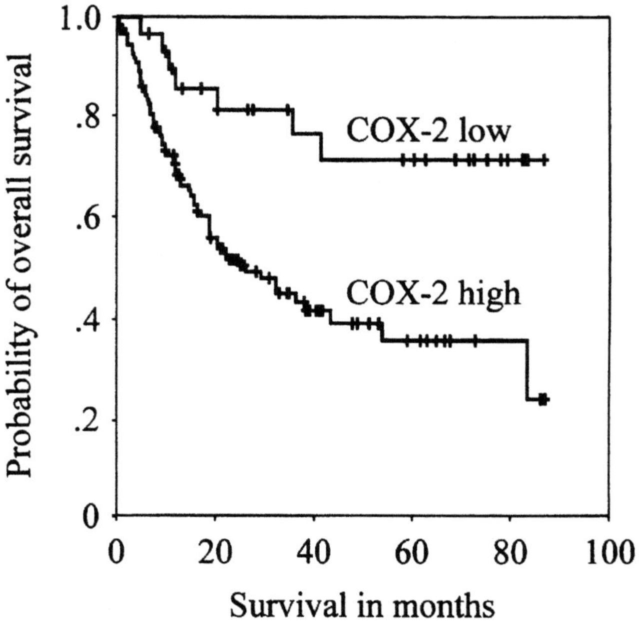

Figure 1. Survival of 145 patients with esophageal adenocarcinoma according to COX-2 protein expression as detected by immunohistochemistry. There were 30 patients with low (scores 0–1) and 115 with high (scores 2–3) COX-2 expression. A statistically significant difference was observed between the 2 groups ($p = 0.002$; log-rank test). Reproduced from Buskens et al. (2002) [27] with permission from the copyright holder.

stomach cancer is complex and combinations of environmental (diet), host factors (including *Helicobacter pylori* infection, partial gastrectomy and gastric adenomas) and genetic factors (hereditary nonpolyposis colorectal cancer syndrome and mutations of E-cadherin) play a role in gastric carcinogenesis. Gastric adenocarcinoma can be divided into histological subtypes, denoted as intestinal and diffuse types by Lauren's classification [29]. The intestinal variant has certain precursor lesions that lead to invasive carcinoma, which include chronic atrophic gastritis, intestinal metaplasia and dysplasia.

Four observational epidemiological studies have been published in which the association between NSAID use and gastric cancer was studied [18, 21, 30, 31]. Three of these studies showed a protective effect of NSAIDs against gastric cancer (Tab. 3). COX-1 was expressed in normal gastric mucosa and COX-2 expression was low or below the detection limit [32]. Expression of COX-2 was elevated in gastric adenocarcinomas when compared to the adjacent non-neoplastic mucosa, suggesting that COX-2 may play a role in gastric carcinogenesis (Tab. 4) [33–44]. Our group has shown that COX-2 protein was expressed in 58% (25/43) of intestinal-type carcinomas, but only in one of 18 diffuse type-tumors (6%) [42]. In agreement with these data, colorectal tumors with a histological pattern of signet ring cells, a typical feature of diffuse type

Table 3. Epidemiological studies on the protective effect of NSAIDs for gastric cancer

Design	Sample size	Endpoint	RR (95% CI)	Reference
Cohort				
	635,031	Mortality	0.53 (0.34–0.81)	Thun et al. 1993 [18]
	12,668	Incidence	0.93 (0.49–1.74)	Schreinemachers et al. 1994 [30]
Case control	Cases: 448 Controls: 610	Incidence	0.60 (0.41–0.90)	Zaridze et al. 1999 [31]
	Cases: 12,174 Controls: 34,934	Incidence	0.51 (0.33–0.79)	Langman et al. 2000 [21]

RR = Risk ratio; CI = Confidence interval.

Table 4. COX-2 expression in gastric adenocarcinoma

Method	COX-2 positive tumors (n)	COX-2 expression	Reference
NB	73% (11)	In contrast to COX-2, expression of COX-1 was not elevated in the tumors.	Ristimäki et al. 1997 [33]
RT-PCR	51% (37)	Correlated with depth of invasion and diameter of the tumor.	Uefuji et al. 2001 [34]
WB	73% (6)	In contrast to COX-2, expression of $cPLA_2$ and COX-1 were not elevated in the tumors.	Soydan et al., 1997 [35]
	83% (23)	Did not correlate with histology of the tumors.	Uefuji et al., 1998 [36]
	70% (100)	Correlated with LN metastasis and advanced tumor stage.	Yamamoto et al. 1999 [37]
	67% (15)	Correlated with LN metastasis.	Murata et al. 1999 [38]
	74% (42)	Correlated with LN metastasis and microvessel density.	Uefuji et al. 2000 [39]
IHC	47% (50)	More frequent in corpus cancer than in cardia cancer.	Ratnasinghe et al. 1999 [40]
	100% (104)	Was also detected in (dysplastic) gastric adenomas.	Lim et al. 2000 [41]
	43% (61)	More frequent in intestinal type than in diffuse type cancers, and was also detected in gastric dysplasias.	Saukkonen et al. 2001 [42]
	61% (33)	More frequent in intestinal type than in diffuse type cancers, and was also detected in intestinal metaplasias.	Yamagata et al. 2001 [43]
	62% (53)	Increased during progression from nonneoplastic epithelium to dysplasia.	van Rees et al. 2002 [44]

NB = Northern blot; RT-PCR = Reverse transcriptase-polymerase chain reaction; WB = Western blot; IHC = Immunohistochemistry; $cPLA_2$ = Cytosolic phospholipase A_2; LN = Lymph node.

gastric cancer, contain low levels of COX-2 immunoreactivity [45]. Recently, Yamagata et al. confirmed our results and reported that COX-2 expression is associated with intestinal type gastric cancer [43]. Since a different set of genetic alterations may take place in the sequence leading to the two histologically distinct types of gastric carcinoma [28], it is possible that a cellular event(s) typical for intestinal type carcinogenesis is responsible for induction of COX-2 expression. Furthermore, overexpression of COX-2 protein is less frequent in gastric and colorectal cancers with microsatellite instability (MSI) than in sporadic ones [37, 45, 46], which suggests that genetic changes typical for sporadic tumors more readily induce COX-2 overexpression than those associated with MSI. Interestingly, COX-2 expression is not restricted to invasive gastric adenocarcinomas, since dysplastic precursor lesions of intestinal type gastric cancer express COX-2, suggesting that COX-2 may contribute to an early event in gastric carcinogenesis [41, 42, 44].

Recent studies have shown that elevated expression of COX-2 protein is associated with invasion into gastric wall lymphatic vessels as well as with metastasis to lymph nodes [37–39]. Thus, similar to esophageal adenocarcinoma (see above) overexpression of COX-2 by gastric tumors may induce invasive and metastatic processes that can lead to a more aggressive behavior of the disease. However, in contrast to adenocarcinomas of the esophagus and the colorectum, expression of COX-2 was not found to correlate with prognosis in gastric cancer patients [38, 46]. This may indicate that the prognostic value of COX-2 is restricted to a certain subgroup of gastric cancer patients (for example early cancers), or that the role of COX-2 is different in the stomach when compared to adenocarcinomas arising in other parts of the gastrointestinal tract.

Lung adenocarcinoma

Lung cancer is one of the most frequent and fatal malignancies, the 5-year survival rate being approximately 14% [47]. The role of external risk factors such as cigarette smoke, exposure to radiation and asbestos are particularly important in the pathogenesis of lung cancer. The great majority of primary lung cancers (90–95%) are carcinomas, which are divided into two groups: small cell lung carcinoma (SCLC) and non-small cell lung carcinoma (NSCLC); adenocarcinoma and squamous cell lung cancer belong to the NSCLC group. The histological distinction between NSCLC and SCLC is important, since there are substantial differences between the two groups in both treatment and prognosis. In general, SCLC is a more aggressive disease and has already metastasized at the time of diagnosis. However, SCLC responds initially well to the treatment with chemotherapy and/or radiation therapy.

The protective role of NSAIDs is less clear for lung cancer than in gastrointestinal cancers [18, 21, 30]. However, a recent case control study showed that daily intake of NSAIDs for at least 2 years prior to interview among 489 lung cancer patients and 978 control subjects was associated with a 68% reduction in the relative risk of lung cancer (RR, 0.32; 95% CI, 0.23–0.44) [48].

Non-neoplastic human lung epithelium expresses only low or nondetectable levels of COX-2 mRNA and protein [49–51]. Elevated levels of COX-2 mRNA and protein expression have been detected in human lung NSCLC by Northern blot [51], RT-PCR [52, 53], *in situ* hybridization [54], immunoblotting [55] and immunohistochemistry [49–51, 56–61] (Tab. 5). Among

Table 5. COX-2 expression in lung adenocarcinoma

Method	COX-2 positive tumors (n)	COX-2 expression	Reference
RT-PCR	100% (19)	AdCa > SCC > SCLC	Ochiai et al. 1999 [52]
(Real-time)	100% (33)	Was an independent predictor of worse survival in patients with NSCLC	Brabender et al. 2002 [53]
ISH	60% (62)	Correlated with shortened survival of patients with stage I disease	Khuri et al. 2001 [54]
WB	82% (17)	No correlation between COX-2 and Bcl-2 expression	Sweeney et al. 2002 [55]
IHC	70% (23)	Was also detected in atypical adenomatous hyperplasia, and was higher in LN metastasis than in primary tumors. AdCa > SCC > SCLC	Hida et al. 1998 [49]
	100% (8)	Lung cancer cell derived PGE_2 mediate immunosuppression *in vitro*	Huang et al. 1998 [50]
(IHC and NB)	90% (21)	Was also detected in atypical alveolar epithelium, and was higher in well- than in poorly differentiated tumors. AdCa > SCC	Wolff et al. 1998 [51]
	72% (130)	Correlated with shortened survival of patients with stage I disease	Achiwa et al. 1999 [56]
(IHC and ISH)	80% (10)	AdCa > SCC	Watkins et al. 1999 [57]
	90% (88)	Was also detected in atypical adenomatous hyperplasia	Hosomi et al. 2000 [58]
	92% (13)	AdCa > SCC	Soslow et al. 2000 [59]
	41% (51)	AdCa > SCC	Hastürk et al. 2002 [60]
	95% (102)	Co-localized with laminin-5 at the invasive front of early-stage tumors	Niki et al. 2002 [61]

RT-PCR = Reverse transcriptase-polymerase chain reaction; ISH = *In situ* hybridization; WB = Western blot; IHC = Immunohistochemistry; NB = Northern blot; AdCa = Adenocarcinoma; SCC = Squamous cell carcinoma; SCLC = Small cell lung cancer; NSCLC = Non-small cell lung cancer; LN = Lymph node.

NSCLCs the level of COX-2 expression has been consistently found to be higher in adenocarcinomas than in squamous cell carcinomas, while SCLCs express only low or undetectable levels of COX-2 [49, 51, 52, 57, 59, 60]. Interestingly, COX-2 overexpression appears to be a marker for poor survival among patients with early stage adenocarcinomas of the lung, but not in more advanced lung tumors [53, 54, 56]. To this end it is interesting to note that we found highest level of COX-2 mRNA and protein in the well-differentiated adenocarcinomas of the lung [51], and that COX-2 is already expressed in precursor lesions of this disease [49, 51, 58]. Thus, COX-2 expression is an early event in the carcinogenic sequence leading to adenocarcinoma of the lung and may play a more significant role at an early stage than in advanced carcinomas. Mechanistically COX-2 may be related to invasion and metastasis, since co-localization of COX-2 and laminin-5 was found at the invasive front of early-stage lung adenocarcinomas [61] and since higher COX-2 levels were found in lymph node metastases than in primary tumors [49]. Currently phase II trials are performed using COX-2 selective inhibitors as a neoadjuvant treatment in conjunction with preoperative chemotherapy or as an adjuvant modality with postoperative chemotherapy of NSCLC.

Breast cancer

Breast cancer is the most frequently diagnosed cancer in women [47]. The protective role of NSAIDs is not clear for breast cancer [18, 21, 30], but a recent meta-analysis suggested that the incidence of this disease might be reduced among NSAID users [62]. Furthermore, experimental work suggests that COX-2 may be linked to mammary gland carcinogenesis, since selective COX-2 inhibitors suppress tumorigenesis in rodent models of breast cancer [63–68]. Moreover, expression of COX-2 as such was sufficient for formation of invasive breast tumors in a transgenic mouse model [69].

COX-2 mRNA and protein are expressed in breast adenocarcinoma specimens, but not in benign epithelium of the breast [59, 70, 71]. Our results suggest that COX-2 is mainly expressed by the neoplastic cells of the tumors when compared to the stromal compartment, and that elevated COX-2 expression is more frequent in ductal type than in lobular type carcinomas (40% *versus* 30% positive, respectively) [72]. Localization of COX-2 to the cytoplasm of the cancer cells is consistent with observations in other studies [59, 70, 73]. In addition to invasive carcinomas, *in situ* carcinomas of ductal type expressed COX-2 [73]. A recent study also showed that COX-2 levels are elevated in breast cancer samples that overexpress HER-2 [74].

Our data show that expression of COX-2 protein is associated with an unfavorable outcome as detected by immunohistochemistry in tissue array of 1576 invasive breast cancer specimens (Fig. 2) [72]. In this material, elevated expression of COX-2 correlated with poor prognostic markers, such as tumor size over 2 cm, axillary lymph node metastasis, high histological grade, high

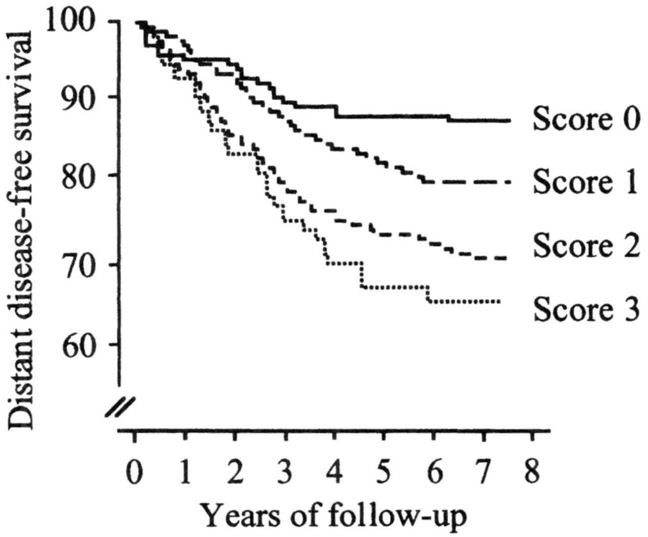

Figure 2. Distant disease-free survival (DDFS) of 1576 breast cancer patients according to COX-2 protein expression as detected by immunohistochemistry. DDFS of patients with no COX-2 expression (score 0; $n = 133$) or with weak (score 1; $n = 854$), moderate (score 2; $n = 511$) or strong (score 3; $n = 78$) COX-2 expression. Elevated expression of COX-2 protein correlated with reduced survival ($p < 0.0001$; log-rank test for trend). Reproduced from Ristimäki et al. (2002) [72] with permission from the copyright holder.

proliferation rate, negative estrogen receptor immunoreactivity, high p53 immunopositivity and amplification of HER-2 oncogene. Interestingly, the prognostic value of COX-2 expression tends to be more marked in certain subgroups of patients, e.g., in cancers with estrogen receptor positivity, low level of p53 expression and no amplification of the HER-2 oncogene (Tab. 6).

Thus, our data suggest that the tumor promoting effect of COX-2 may not be evenly distributed throughout the material and that certain subgroups of patients might benefit more from COX-2 targeted treatment. The fact that elevated expression of COX-2 is associated with poor survival in estrogen receptor positive tumors is of particular interest, since COX-2-derived prostanoids have been implicated in the enhancement of stromal cell aromatase expression [75, 76]. It is thus possible that elevated COX-2 expression in estrogen receptor positive cancers could enhance a growth-promoting microenvironment for the tumor cells by inducing estrogen production via the aromatase pathway. It should be pointed out that induction of aromatase expression is most likely not the only tumor promoting effect of COX-2 (see below), and that our results do not rule out the possibility that COX-2 inhibitors would work as adjuvant drugs in combination with for example HER-2 inhibitors. To this end, several ongoing and/or proposed trials study chemopreventive, neoadjuvant (biomarker studies) and adjuvant (combined with anti-estrogen and/or anti-HER-2) effects of selective COX-2 inhibitors.

Table 6. Five-year distant disease-free survival (DDFS) according to COX-2 expression level

Parameter	COX-2 score	N-value (%)	DDFS (CI 95%)	P-value*
ER positive	0–1	658 (66.6%)	86 (84–89)	<0.0001
	2–3	330 (33.4%)	76 (72–81)	
ER negative	0–1	213 (47.7%)	70 (64–77)	NS; 0.396
	2–3	234 (52.3%)	68 (62–74)	
p53 negative/low	0–1	679 (63.5%)	85 (82–88)	<0.0001
	2–3	391 (36.5%)	74 (70–79)	
p53 positive	0–1	99 (40.9%)	64 (54–74)	NS; 0.842
	2–3	143 (59.1%)	68 (60–76)	
HER-2 negative	0–1	732 (63.3%)	86 (84–89)	<0.0001
	2–3	425 (36.7%)	77 (73–81)	
HER-2 positive	0–1	130 (49.6%)	63 (54–71)	NS; 0.729
	2–3	132 (50.4%)	59 (50–68)	

*Log-rank test
CI = Confidence interval
NS = Not significant
ER = Estrogen receptor
Modified from Ristimäki et al. (2002) [72] with the permission from the copyright holder.

Mechanism of COX-2 -induced carcinogenesis

It is not known whether COX-2 contributes to human carcinogenesis by removing excess arachidonic acid, by producing prostanoids or by metabolizing other compounds. It has, however, been demonstrated that in experimental conditions many effects of COX-2 can be facilitated by PGE_2, which include stimulation of proliferation (in many cases by inhibition of apoptosis), enhanced invasion and metastasis, promotion of angiogenesis and induction of immunosuppression (Fig. 3) [77, 78]. The central role of PGE_2 in tumorogenesis has been further emphasized using homozygous deletion of one of the PGE_2 receptors EP2, which lead to reduced number and size of intestinal polyps in the $Apc^{\Delta 716}$ model [79]. Our clinical data support the role of COX-2 in promoting formation of metastases in both esophageal and breast adenocarcinomas [27, 72]. In case of gastric cancer other groups have found COX-2 expression to correlate with lymph node metastases and angiogenesis [37–39].

The mechanism by which COX-2 is upregulated in human cancers is largely unknown. One possibility is that cancer cells become intrinsically more active in expressing COX-2, since COX-2 gene expression is induced by activation of oncogenes, such as Ras in case of colorectal cancer cells. This leads to subsequent activation of signal transducers (MAPKs and pAkt) that can stimulate COX-2 transcription and increase the stability of the transcript (Fig. 4) [77, 78]. Furthermore, since COX-2 gene has been shown to be

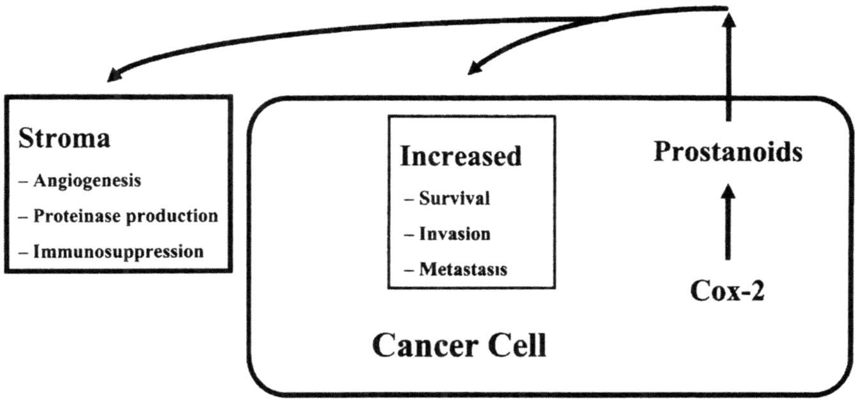

Figure 3. COX-2-derived prostanoids (in most cases PGE_2) inhibit tumor cell apoptosis, enhance production of angiogenic factors, such as vascular endothelial growth factor, increase invasion and metastasis, possibly via activating matrix metalloproteinases and act as immunosuppressive agents. Target cells can be either the cancer cell itself (autocrine) or stromal cells (paracrine).

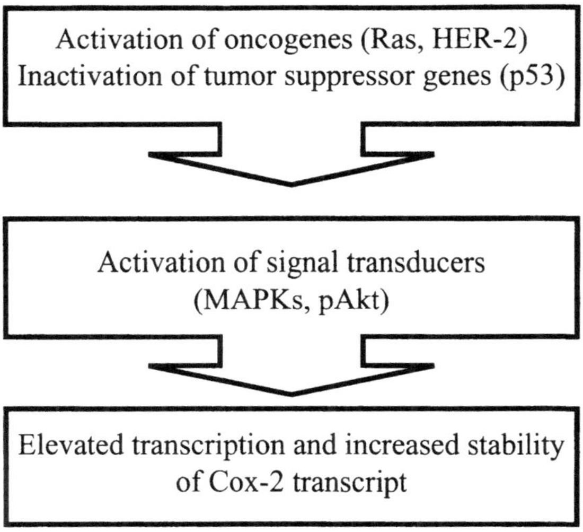

Figure 4. Regulation of COX-2 expression in cancer cells is primarily regulated by intracrine phenomenons such as dysregulation of oncogenes (activating mutation or gene amplification) or tumor suppressor genes (inactivating mutation, gene silencing or deletion). This leads to subsequent activation of signal transducers (MAPKs and pAkt) that can stimulate COX-2 transcription and increase stability of the transcript.

induced in p53 defective cells and downregulated by wild type p53 [80, 81], a direct link between a defective p53 pathway and elevated levels of COX-2 expression in cancer cells may exist. Thus, inactivation of tumor suppressor genes and activation of oncogenes may induce COX-2 expression in human carcinomas. Our results support this hypothesis, because elevated COX-2 expression was significantly more common in invasive breast cancers with high expression of p53 (a marker for inactivating mutations of p53) or with amplification of the HER-2 oncogene [72]. Other researchers have also been able to correlate COX-2 expression to overexpression of HER-2 in case of breast cancer [74] and to mutations of p53 in the case gastric cancer [82]. However, because elevated COX-2 expression is not restricted to p53- and HER-2-positive tumors, other factors and/or mechanism are probably linked to elevated COX-2 expression. Since the stroma of human adenocarcinomas often harbors inflammatory cells, it is possible that for example macrophages release factors that regulate tumor cell COX-2 expression in a paracrine manner.

Conclusions

COX-2 is widely expressed in different types of human adenocarcinomas. However, even within a single organ site the frequency of COX-2 expression may differ considerably depending on histology (intestinal > diffuse gastric cancer; ductal > lobular breast cancer) [42, 43, 72] and genotype (sporadic > MSI positive gastric and colorectal cancers) of the tumor [37, 45, 46]. Even more importantly, the prognostic significance of COX-2 may not be evenly distributed even within one subtype of adenocarcinoma in a single organ site. The prognostic value of COX-2 seems to depend on tumor stage in the case of lung adenocarcinoma [54, 56] and on molecular make up of the tumor in the case of breast adenocarcinoma [72]. Interestingly, it was recently suggested that a COX-2 variant (Val511Ala) might be associated with a reduced risk for colorectal neoplasia [83]. Thus, certain polymorphisms in COX-2 gene may mimic the effect of NSAIDs in reduction of colorectal cancer. While prognostic significance of COX-2 has been most clearly shown for certain types of adenocarcinomas, it was recently published that elevated bone marrow COX-2 levels are associated with reduced survival in chronic myeloid leukemia [84]. Thus, COX-2 selective inhibitors might also have a role in the treatment of non-epithelial malignancies.

Clinically it will be important to determine whether selective COX-2 inhibitors will deliver safer and at least as effective anti-cancer properties as nonselective NSAIDs. In addition, subtypes of premalignant lesions and invasive cancers (based on histology, stage and/or genotype) that are especially sensitive to these drugs should be recognized. Finally, it will be important to perform biomarker studies using selective COX-2 inhibitors in a neoadjuvant setting, integrating them into already existing radio- and/or chemotherapy protocols as adjuvant treatments and at the same time continue research on alternative and hopefully less toxic combination treatments of cancer. One such novel treatment option could be the combination of NSAIDs with inhibitors of Erb/HER family of growth factor receptors [68, 84–86]. All this should help to target the use of COX-2 selective drugs to those cancer patients who most likely will benefit from this treatment. Furthermore, understanding the mechanisms of COX-2 overexpression and the discovery of its role in tumor formation can eventually lead to development of new treatment modalities for cancer.

Acknowledgements
Original work was supported by the Academy of Finland, the Finnish Cancer Foundation and the Helsinki University Central Hospital Research Funds.

References

1 Anderson WF, Umar A, Viner JL, Hawk ET (2002) The role of cyclooxygenase inhibitors in cancer prevention. *Curr Pharm Des* 8: 1035–1062
2 Thun MJ, Henley SJ, Patrono C (2002) Nonsteroidal anti-inflammatory drugs as anticancer agents: mechanistic, pharmacologic, and clinical issues. *J Natl Cancer Inst* 94: 252–266
3 Eberhart CE, Coffey RJ, Radhika A, Giardiello FM, Ferrenbach S, DuBois RN (1994) Up-regulation of cyclooxygenase 2 gene expression in human colorectal adenomas and adenocarcinomas. *Gastroenterology* 107: 1183–1188
4 Kargman SL, O'Neill GP, Vickers PJ, Evans JF, Mancini JA, Jothy S (1995) Expression of prostaglandin G/H synthase-1 and -2 protein in human colon cancer. *Cancer Res* 55: 2556–2559
5 Sano H, Kawahito Y, Wilder RL, Hashiramoto A, Mukai S, Asai K, Kimura S, Kato H, Kondo M, Hla T (1995) Expression of cyclooxygenase-1 and -2 in human colorectal cancer. *Cancer Res* 55: 3785–3789
6 Oshima M, Dinchuk JE, Kargman SL, Oshima H, Hancock B, Kwong E, Trzaskos JM, Evans JF, Taketo MM (1996) Suppression of intestinal polyposis in $Apc^{\Delta 716}$ knockout mice by inhibition of cyclooxygenase 2 (COX-2). *Cell* 87: 803–809
7 Reddy BS, Rao CV, Seibert K (1996) Evaluation of cyclooxygenase-2 inhibitor for potential chemopreventive properties in colon carcinogenesis. *Cancer Res* 56: 4566–4569
8 Kawamori T, Rao CV, Seibert K, Reddy BS (1998) Chemopreventive activity of celecoxib, a specific cyclooxygenase-2 inhibitor, against colon carcinogenesis. *Cancer Res* 58: 409–412
9 Jacoby RF, Seibert K, Cole CE, Kelloff G, Lubet RA (2000) The cyclooxygenase-2 inhibitor celecoxib is a potent preventive and therapeutic agent in the min mouse model of adenomatous polyposis. *Cancer Res* 60: 5040–5044
10 Reddy BS, Hirose Y, Lubet R, Steele V, Kelloff S, Paulson S, Seibert K, Rao C (2000) Chemoprevention of colon cancer by spesific cyclooxygenase-2 inhibitor, celecoxib, administered during different stages of carcinogenesis. *Cancer Res* 60: 293–297
11 Chulada PC, Thompson MB, Mahler JF, Doyle CM, Gaul BW, Lee C, Tiano HF, Morham SG, Smithies O, Langenbach R (2000) Genetic disruption of Ptgs-1, as well as Ptgs-2, reduces intestinal tumorigenesis in Min mice. *Cancer Res* 60: 4705–4708
12 Sheehan KM, Sheahan K, O'Donoghue DP, MacSweeney F, Conroy RM, Fitzgerald DJ, Murray FE (1999) The relationship between cyclooxygenase-2 expression and colorectal cancer. *JAMA* 282: 1254–1257
13 Steinbach G, Lynch PM, Phillips RK, Wallace MH, Hawk E, Gordon GB, Wakabayashi N, Saunders B, Shen Y, Fujimura T et al (2000) The effect of celecoxib, a cyclooxygenase-2 inhibitor, in familial adenomatous polyposis. *N Engl J Med* 342: 1946–1952
14 Blot WJ, Devesa SS, Kneller RW, Fraumeni JF Jr, (1991) Rising incidence of adenocarcinoma of the esophagus and gastric cardia. *JAMA* 265: 1287–1289
15 Pera M, Cameron AJ, Trastek VF, Carpenter HA, Zinsmeister AR (1993) Increasing incidence of adenocarcinoma of the esophagus and esophagogastric junction. *Gastroenterology* 104: 510–513
16 Dolan K, Sutton R, Walker SJ, Morris AI, Campbell F, Williams EM (1999) New classification of esophageal and gastric carcinomas derived from changing patterns in epidemiology. *Br J Cancer* 80: 834–842
17 Spechler SJ, Goyal RK (1986) Barrett's esophagus. *N Engl J Med* 315: 362–371
18 Thun MJ, Namboodiri MM, Calle EE, Flanders WD, Heath CW Jr, (1993) Aspirin use and risk of fatal cancer. *Cancer Res* 53: 1322–1327

19 Funkhouser EM, Sharp GB (1995) Aspirin and reduced risk of esophageal carcinoma. *Cancer* 76: 1116–1119
20 Farrow DC, Vaughan TL, Hansten PD, Stanford JL, Risch HA, Gammon MD, Chow WH, Dubrow R, Ahsan H, Mayne ST et al (1998) Use of aspirin and other nonsteroidal anti-inflammatory drugs and risk of esophageal and gastric cancer. *Cancer Epidemiol Biomarkers Prev* 2: 97–102
21 Langman MJS, Cheng KK, Gilman EA, Lancashire RJ (2000) Effect of anti-inflammatory drugs on overall risk of common cancer: Case control study in general practice research database. *BMJ* 320: 1642–1646
22 Coogan PF, Rosenberg L, Palmer JR, Strom BL, Zauber AG, Stolley PD, Shapiro S (2000) Nonsteroidal anti-inflammatory drugs and risk of digestive cancers at sites other than the large bowel. *Cancer Epidemiol Biomarkers Prev* 1: 119–123
23 Wilson KT, Fu S, Ramanujam KS, Meltzer SJ (1998) Increased expression of inducible nitric oxide synthase and cyclooxygenase-2 in Barrett's esophagus and associated adenocarcinoma. *Cancer Res* 58: 2929–2934
24 Zimmermann KC, Sarbia M, Weber AA, Borchard F, Gabbert HE, Schror K (1999) Cyclooxygenase-2 expression in human esophageal carcinoma. Cancer Res 59: 198–204
25 Shirvani VN, Ouatu-Lascar R, Kaur BS, Omary MB, Triadafilopoulos G (2000) Cyclooxygenase-2 expression in Barrett's esophagus and adenocarcinoma: *Ex vivo* induction by bile salts and acid exposure. *Gastroenterology* 118: 487–496
26 Morris CD, Armstrong GR, Bigley G, Green H, Attwood SEA (2001) Cyclooxygenase-2 expression in the Barrett's metaplasia-dysplasia-adenocarcinoma sequence. *Am J Gastroenterol* 96: 990–996
27 Buskens CJ, van Rees BP, Sivula A, Reitsma JB, Haglund C, Bosma PJ, Offerhaus GJ, van Lanschot JJ, Ristimäki A (2002) Prognostic significance of elevated cyclooxygenase-2 expression in patients with adenocarcinoma of the esophagus. *Gastroenterology* 122: 1800–1807
28 Stadtländer CT, Waterbor JW (1999) Molecular epidemiology, pathogenesis and prevention of gastric cancer. *Carcinogenesis* 20: 2195–2208
29 Laurèn P (1965) The two histological main types of gastric carcinoma: diffuse and so-called intestinal-type carcinoma. *Acta Path Microbiol Scand* 64: 31–49
30 Schreinemachers DM, Everson RB (1994) Aspirin use and lung, colon, and breast cancer incidence in a prospective study. *Epidemiology* 5: 138–146
31 Zaridze D, Borisova E, Maximovitch D, Chkhikvadze V (1999) Aspirin protects against gastric cancer: results of a case-control study from Moscow, Russia. *Int J Cancer* 82: 473–476
32 Kargman SL, O'Neill GP, Vickers PJ, Evans JF, Mancini JA, Jothy S (1995) Expression of prostaglandin G/H synthase-1 and -2 protein in human colon cancer. *Cancer Res* 55: 2556–2559
33 Ristimäki A, Honkanen N, Jänkälä H, Sipponen P, Härkonen M (1997) Expression of cyclooxygenase-2 in human gastric carcinoma. *Cancer Res* 57: 1276–1280
34 Uefuji K, Ichikura T, Mochizuki H (2001) Expression of cyclooxygenase-2 in human gastric adenomas and adenocarcinomas. *J Surg Oncol* 76: 26–30
35 Soydan AS, Gaffen JD, Weech PK, Tremblay NM, Kargman S, O'Neill G, Bennett A, Tavares IA (1997) Cytosolic phospholipase A_2, cyclo-oxygenases and arachidonate in human stomach tumours. *Eur J Cancer* 33: 1508–1512
36 Uefuji K, Ichikura T, Mochizuki H, Shinomiya N (1998) Expression of cyclooxygenase-2 protein in gastric adenocarcinoma. *J Surg Oncol* 69: 168–172
37 Yamamoto H, Itoh F, Fukushima H, Hinoda Y, Imai K (1999) Overexpression of cyclooxygenase-2 protein is less frequent in gastric cancers with microsatellite instability. *Int J Cancer* 84: 400–403
38 Murata H, Kawano S, Tsuji S, Tsuji M, Sawaoka H, Kimura Y, Shiozaki H, Hori M (1999) Cyclooxygenase-2 overexpression enhances lymphatic invasion and metastasis in human gastric carcinoma. *Am J Gastroenterol* 94: 451–455
39 Uefuji K, Ichikura T, Mochizuki H (2000) Cyclooxygenase-2 expression is related to prostaglandin biosynthesis and angiogenesis in human gastric cancer. *Clin Cancer Res* 6: 135–138
40 Ratnasinghe D, Tangrea JA, Roth MJ, Dawsey SM, Anver M, Kasprzak BA, Hu N, Wang QH, Taylor PR (1999) Expression of cyclooxygenase-2 in human adenocarcinomas of the gastric cardia and corpus. *Oncol Rep* 6: 965–968
41 Lim HY, Joo HJ, Choi JH, Yi JW, Yang MS, Cho DY, Kim HS, Nam DK, Lee KB, Kim HC (2000) Increased expression of cyclooxygenase-2 protein in human gastric carcinoma. *Clin Cancer Res* 6: 519–525

42 Saukkonen K, Nieminen O, van Rees B, Vilkki S, Härkönen M, Juhola M, Mecklin JP, Sipponen P, Ristimäki A (2001) Expression of cyclooxygenase-2 in dysplasia of the stomach and in intestinal-type gastric adenocarcinoma. *Clin Cancer Res* 7: 1923–1931
43 Yamagata R, Shimoyama T, Fukuda S, Yoshimura T, Tanaka M, Munakata A (2001) Cyclooxygenase-2 expression is increased in early intestinal-type gastric cancer and gastric mucosa with intestinal metaplasia. *Eur J Gastroenterol Hepatol* 14: 359–363
44 van Rees BP, Saukkonen K, Ristimäki A, Polkowski W, Tytgat GN, Drillenburg P, Offerhaus GJ (2002) Cyclooxygenase-2 expression during carcinogenesis in the human stomach. *J Pathol* 196: 171–179
45 Karnes WE, Jr, Shattuck-Brandt R, Burgart, LJ, DuBois RN, Tester DJ, Cunningham JM, Kim CY, McDonnell SK, Schaid DJ, Thibodeau SN (1998) Reduced COX-2 protein in colorectal cancer with defective mismatch repair. *Cancer Res* 58: 5473–5477
46 Lee TL, Leung WK, Lau JY, Tong JH, Ng EK, Chan FK, Chung SC, Sung JJ, To KF (2001) Inverse association between cyclooxygenase-2 overexpression and microsatellite instability in gastric cancer. *Cancer Lett* 168: 133–140
47 Jemal A, Thomas A, Murray T, Thun M (2002) Cancer statistics. *CA Cancer J Clin* 52: 23–47
48 Harris RE, Beebe-Donk J, Schuller HM (2002) Chemoprevention of lung cancer by non-steroidal anti-inflammatory drugs among cigarette smokers. *Oncol Rep* 9: 693–695
49 Hida T, Yatabe Y, Achiwa H, Muramatsu H, Kozaki K, Nakamura S, Ogawa M, Mitsudomi T, Sugiura T, Takahashi T (1998) Increased expression of cyclooxygenase 2 occurs frequently in human lung cancers, specifically in adenocarcinomas. *Cancer Res* 58: 3761–3764
50 Huang M, Stolina M, Sharma S, Mao JT, Zhu L, Miller PW, Wollman J, Herschman H, Dubinett SM (1998) Non-small cell lung cancer cyclooxygenase-2-dependent regulation of cytokine balance in lymphocytes and macrophages: up-regulation of interleukin 10 and down-regulation of interleukin 12 production. *Cancer Res* 58: 1208–1216
51 Wolff H, Saukkonen K, Anttila S, Karjalainen A, Vainio H, Ristimäki A (1998) Expression of cyclooxygenase-2 in human lung carcinoma. *Cancer Res* 58: 4997–5001
52 Ochiai M, Oguri T, Isobe T, Ishioka S, Yamakido M (1999) Cyclooxygenase-2 (COX-2) mRNA expression levels in normal lung tissues and non-small cell lung cancers. *Jpn J Cancer Res* 90: 1338–1343
53 Brabender J, Park J, Metzger R, Schneider PM, Lord RV, Holscher AH, Danenberg KD, Danenberg PV (2002) Prognostic significance of cyclooxygenase 2 mRNA expression in non-small cell lung cancer. *Ann Surg* 235: 440–443
54 Khuri FR, Wu H, Lee JJ, Kemp BL, Lotan R, Lippman SM, Feng L, Hong WK, Xu XC (2001) Cyclooxygenase-2 overexpression is a marker of poor prognosis in stage I non-small cell lung cancer. *Clin Cancer Res* 7: 861–867
55 Sweeney CJ, Marshall MS, Barnard DS, Heilman DK, Billings SD, Cheng L, Marshall SJ, Yip-Schneider MT (2002) Cyclo-oxygenase-2 expression in primary cancers of the lung and bladder compared to normal adjacent tissue. *Cancer Detect Prev* 26: 238–244
56 Achiwa H, Yatabe Y, Hida T, Kuroishi T, Kozaki K, Nakamura S, Ogawa M, Sugiura T, Mitsudomi T, Takahashi T (1999) Prognostic significance of elevated cyclooxygenase 2 expression in primary, resected lung adenocarcinomas. *Clin Cancer Res* 5: 1001–1005
57 Watkins DN, Lenzo JC, Segal A, Garlepp MJ, Thompson PJ (1999) Expression and localization of cyclo-oxygenase isoforms in non-small cell lung cancer. *Eur Respir J* 14: 412–418
58 Hosomi Y, Yokose T, Hirose Y, Nakajima R, Nagai K, Nishiwaki Y, Ochiai A (2000) Increased cyclooxygenase 2 (COX-2) expression occurs frequently in precursor lesions of human adenocarcinoma of the lung. *Lung Cancer* 30: 73–81
59 Soslow RA, Dannenberg AJ, Rush D, Woerner BM, Khan KN, Masferrer J, Koki AT (2000) COX-2 is expressed in human pulmonary, colonic, and mammary tumors. *Cancer* 89: 2637–2645
60 Hastürk S, Kemp B, Kalapurakal SK, Kurie JM, Hong WK, Lee JS (2002) Expression of cyclooxygenase-1 and cyclooxygenase-2 in bronchial epithelium and nonsmall cell lung carcinoma. *Cancer* 94: 1023–1031
61 Niki T, Kohno T, Iba S, Moriya Y, Takahashi Y, Saito M, Maeshima A, Yamada T, Matsuno Y, Fukayama M et al (2002) Frequent co-localization of Cox-2 and laminin-5 gamma2 chain at the invasive front of early-stage lung adenocarcinomas. *Am J Pathol* 160: 1129–1141
62 Khuder SA, Mutgi AB (2001) Breast cancer and NSAID use: a meta-analysis. *Br J Cancer* 84: 1188–1192
63 Alshafie GA, Abou-Issa HM, Seibert K, Harris RE (2000) Chemotherapeutic evaluation of cele-

coxib, a cyclooxygenase-2 inhibitor, in rat mammary tumor model. *Oncology Reports* 7: 1377–1381
64 Harris RE, Alshafie GA, Abou-Issa H, Seibert K (2000) Chemoprevention of breast cancer in rats by celecoxib, a cyclooxygenase 2 inhibitor. *Cancer Res* 60: 2101–2103
65 Nakatsugi S, Ohta T, Kawamori T, Mutoh M, Tanigawa T, Watanabe K, Sugie S, Sugimura T, Wakabayashi K (2000) Chemoprevention by nimesulide, a selective cyclooxygenase-2 inhibitor, of 2-amino-1-methyl-6-phenylimidazo[4,5-b]pyridine (PhIP)-induced mammary gland carcinogenesis in rats. *Jpn J Cancer Res* 91: 886–892
66 Kundu N, Fulton AM (2002) Selective cyclooxygenase (COX)-1 or COX-2 inhibitors control metastatic disease in a murine model of breast cancer. *Cancer Res* 62: 2343–2346
67 Connolly EM, Harmey JH, O'Grady T, Foley D, Roche-Nagle G, Kay E, Bouchier-Hayes DJ (2002) Cyclo-oxygenase inhibition reduces tumour growth and metastasis in an orthotopic model of breast cancer. *Br J Cancer* 87: 231–237
68 Howe LR, Subbaramaiah K, Patel J, Masferrer JL, Deora A, Hudis C, Thaler HT, Muller WJ, Du B, Brown AM et al (2002) Celecoxib, a selective cyclooxygenase 2 inhibitor, protects against human epidermal growth factor receptor 2 (HER-2)/neu-induced breast cancer. *Cancer Res* 62: 5405–5407
69 Liu CH, Chang SH, Narko K, Trifan OC, Wu MT, Smith E, Haudenschild C, Lane TF, Hla T (2001) Overexpression of cyclooxygenase-2 is sufficient to induce tumorigenesis in transgenic mice. *J Biol Chem* 267: 18563–18569
70 Parrett ML, Harris RE, Joarder FS, Ross MS, Clausen KP, Robertson FM. Cyclooxygenase-2 expression in human breast cancer (1997) *Int J Oncol* 10: 503–507
71 Hwang D, Scollard D, Byrne J, Levine E (1998) Expression of cyclooxygenase-1 and cyclooxygenase-2 in human breast cancer. *J Natl Cancer Inst* 90: 455–460
72 Ristimäki A, Sivula A, Lundin J, Lundin M, Salminen T, Haglund C, Joensuu H, Isola J (2002) Prognostic significance of elevated cyclooxygenase-2 expression in breast cancer. *Cancer Res* 62: 632–635
73 Half E, Tang XM, Gwyn K, Sahin A, Wathen K, Sinicrope FA (2002) Cyclooxygenase-2 expression in human breast cancers and adjacent ductal carcinoma *in situ*. *Cancer Res* 62: 1676–1681
74 Subbaramaiah K, Norton L, Gerald W, Dannenberg AJ (2002) Cyclooxygenase-2 is overexpressed in HER-2/neu-positive breast cancer: evidence for involvement of AP-1 and PEA3. *J Biol Chem* 277: 18649–18657
75 Brueggemeier RW, Quinn AL, Parrett ML, Joarder FS, Harris RE, Robertson FM (1999) Correlation of aromatase and cyclooxygenase gene expression in human breast cancer specimens. *Cancer Lett* 140: 27–35
76 Brodie AM, Lu Q, Long BJ, Fulton A, Chen T, Macpherson N, DeJong PC, Blankenstein MA, Nortier JW, Slee PH et al (2001) Aromatase and COX-2 expression in human breast cancers. *J Steroid Biochem Mol Biol* 79: 41–47
77 Dannenberg AJ, Altorki NK, Boyle JO, Dang C, Howe LR, Weksler BB, Subbaramaiah K (2001) Cyclo-oxygenase 2: a pharmacological target for the prevention of cancer. *Lancet Oncol* 2: 544–551
78 Gupta RA, Dubois RN (2001) Colorectal cancer prevention and treatment by inhibition of cyclooxygenase-2. *Nature Rev Cancer* 1: 11–21
79 Sonoshita M, Takaku K, Sasaki N, Sugimoto Y, Ushikubi F, Narumiya S, Oshima M, Taketo MM (2001) Acceleration of intestinal polyposis through prostaglandin receptor EP2 in Apc(Delta 716) knockout mice. *Nat Med* 7: 1048–1051
80 Subbaramaiah K, Altorki N, Chung WJ, Mestre JR, Sampat A, Dannenberg AJ (1999) Inhibition of cyclooxygenase-2 gene expression by p53. *J Biol Chem* 274: 10911–10915
81 Han JA, Kim JI, Ongusaha PP, Hwang DH, Ballou LR, Mahale A, Aaronson SA, Lee SW (2002) P53-mediated induction of Cox-2 counteracts p53- or genotoxic stress-induced apoptosis. *EMBO J* 21: 5635–5644
82 Leung WK, To KF, Ng YP, Lee TL, Lau JY, Chan FK, Ng EK, Chung SC, Sung JJ (2001) Association between cyclo-oxygenase-2 overexpression and missense p53 mutations in gastric cancer. *Br J Cancer* 84: 335–339
83 Lin HJ, Lakkides KM, Keku TO, Reddy ST, Louie AD, Kau IH, Zhou H, Gim JS, Ma HL, Matthies CF et al (2002) Prostaglandin H synthase 2 variant (Val511Ala) in African Americans may reduce the risk for colorectal neoplasia. *Cancer Epidemiol Biomarkers Prev* 11: 1305–1315
84 Giles FJ, Kantarjian HM, Bekele BN, Cortes JE, Faderl S, Thomas DA, Manshouri T, Rogers A, Keating MJ, Talpaz M et al (2002) Bone marrow cyclooxygenase-2 levels are elevated in chron-

ic-phase chronic myeloid leukaemia and are associated with reduced survival. *Br J Haematol* 119: 38–45

85 Torrance CJ, Jackson PE, Montgomery E, Kinzler KW, Vogelstein B, Wissner A, Nunes M, Frost P, Discafani CM (2000) Combinatorial chemoprevention of intestinal neoplasia. *Nat Med* 6: 1024–1028

86 Mann M, Sheng H, Shao J, Williams CS, Pisacane PI, Sliwkowski MX, DuBois RN (2001) Targeting cyclooxygenase 2 and HER-2/neu pathways inhibits colorectal carcinoma growth. *Gastroenterology* 120: 1713–1719

Index

GPSR Compliance

The European Union's (EU) General Product Safety Regulation (GPSR) is a set of rules that requires consumer products to be safe and our obligations to ensure this.

If you have any concerns about our products, you can contact us on ProductSafety@springernature.com

In case Publisher is established outside the EU, the EU authorized representative is:

Springer Nature Customer Service Center GmbH
Europaplatz 3
69115 Heidelberg, Germany

Batch number: 10732220

Printed by Printforce, the Netherlands